Biochemical Calculations *and* Biostatistics

Dr. E. Padmini

Since
BOOKS AND ALLIED
1960

CELEBRATING 60 YEARS 1960-2020

[For the U.G. & P.G. students of Biochemistry, Biotechnology, Genetics, Microbiology, Zoology, Botany, Environmental Science, Biomedical Engineering & Medical Sciences.]

Biochemical Calculations *and* Biostatistics

Dr. E. Padmini

Reader in Biochemistry
P.G. Department of Biochemistry
Bharathi Women's College, Chennai.
(Affiliated to Madras University)

BOOKS & ALLIED (P) LTD.

No.1E(1), 'Shubham Plaza'
83/1, Dr. S. C. Banerjee Road, Beliaghata, Kolkata 700010
Tel : (033)23740062 ● ***e-mail*** : booksandallied1960@gmail.com
website : www.booksandallied.com

Biochemical Calculations
***and* Biostatistics**

First Published : June, 2007
Reprinted : 2013, 2015
Reprinted : 2022

Publisher : Ruba Sen
BOOKS & ALLIED (P) LTD.
NO. 1E(1), 'SHUBHAM PLAZA'
83/1, Dr. S. C. Banerjee Road, Beliaghata, Kolkata 700010

Typesetter : BOOKS & ALLIED (P) LTD.
NO. 1E(1), 'SHUBHAM PLAZA'
83/1, Dr. S. C. Banerjee Road, Beliaghata, Kolkata 700010

Printer : EUREKA
91/A, Baithakkhana Road, Kolkata 700009

ISBN 978-81-87134-06-0

***Price :* ₹ 265.00**

Dedicated

to

Sri Mahaswami

of

Kanchi

UNIVERSITY OF MADRAS
Department of Biochemistry
Guindy Campus, Chennai – 600 025. India

Dr. S. Niranjali Devaraj
Professor & Head

FOREWORD

"BIOCHEMICAL CALCULATIONS AND BIOSTATISTICS" is an excellent book conceived and authored by exceptionally experienced expert in Biochemistry. This book provides a good collection of some of the important and fundamental aspects of biochemical calculations. It also traces through careful and well balanced selection of chapters in biostatistics. The selection of chapters important to the curriculum of Life Science students highlights the understanding and learning of the basic calculations in biochemistry and biostatistics.

Each chapter has a detailed introduction to the fundamentals along with problems with author's explanatory steps prefacing each of them. The sample calculations are illustrated clearly in an easy-to-follow, step-by-step format.

The book is written for UG and PG students of Life Sciences and is also useful reference for the researchers and teachers. It is a unique and commendable work of the recent times in biosciences and deserves serious study and appreciation. In my opinion, this book authored by Dr. E. Padmini is a must-have for all students of Life Sciences.

S. Niranjali Devaraj

PREFACE

Contemporary biochemical inquiry requires a basic appreciation and knowledge of biochemical calculations and statistical analysis. Over many years, my teachers, students and colleagues have aided directly or indirectly in guiding me to the material that is prescribed in this book. Having taught Biochemistry for UG and PG students for the past 18 years and with 10 years of research experience I have considered the basic topics in biochemical calculations and statistics for an introductory treatment in this book. With no authoritarian intent, I suggest these book sections for biochemistry, biotechnology, microbiology, zoology, botany, environmental, medical, and biomedical engineering students.

In order to be useful as a reference as well as to allow for difference in content among courses for which it might be used, this book contains enough material that would be included in a term course. The material in this book requires no previous mathematical competence beyond very elementary algebra. The data in the examples are intended to demonstrate the biochemical and statistical procedures that will make the students understand and appreciate the subject. Reflecting the magnificent diversity of scientific endeavours that can be found in biological sciences, this book presents a short collection of biochemical calculations and a broad collection of statistical techniques. This book is written for two purposes, one is to emphasize on the importance of learning biological sciences along with calculations and the other is to understand this science that requires statistical analysis (actual calculations) being consulted long after formal instruction has ended.

A book of this nature requires and benefits from the assistance of many people. I acknowledge every one of them particularly my family, publisher and his representatives and close friends who spent long hours during preparation and proof correction.

March 2007

Dr. E. Padmini

(e-mail : epi622001@yahoo.com)

CONTENTS

PART–I

BIOCHEMICAL CALCULATIONS

PART–II

BIOSTATISTICS

Part I

Biochemical Calculations

1

Introduction to Units

The systems of basic units employed in Science are SI System, English system and cgs system. The first and most important of these is the SI (Systems International "d" Units) which has as its three basic units the meter (m), the kilogram (kg) and the second (s). The English System has the foot (ft), pound (lb), seconds (s) and the cgs system has the centimeter (cm), gram (gm) and seconds (s).

At present the SI System is most prevalent in science but the older English and cgs systems will be used for sometime. Much of the physical and chemical data are given in the latter two systems. Hence the science students not only should be proficient with the SI system but must also be able to understand and use the other two systems to limited extent.

1.1 SI System of Units

The basic quantities used in the SI System are the unit of length in meter (m), the unit of time in second (s), the unit of mass in the kilogram (kg); the unit of temperature is the Kelvin (K) and the unit of an element is the kilogram mole (kg mol). The other standard units are derived from the basic quantities.

A few of the standard prefixes for multiples of the basic units are as follows.

$$\text{Giga (G)} = 10^9$$
$$\text{Mega (M)} = 10^6$$
$$\text{kilo (k)} = 10^3$$
$$\text{centi (c)} = 10^{-2}$$
$$\text{milli (m)} = 10^{-3}$$
$$\text{micro } (\mu) = 10^{-6}$$
$$\text{nano (n)} = 10^{-9}$$

Temperature is defined in Kelvin (K) as the preferred unit in the SI System. In practice wide use is made of the degree Celsius (°C) scale which is defined by

$$t\ °\text{C} = T\ (\text{K}) - 273.15$$

The difference between the boiling point of water and the melting point of ice at 1 atm is 100 °C or 180 °F. Thus a 1.8 °F change is equal to a 1 °C change. The following equation can be used to convert from one scale to another.

$$°\text{F} = 32 + 1.8(°\text{C})$$

$$°\text{C} = \frac{1}{1.8}(°\text{F} - 32)$$

The standard unit of time is the second (s) but time can be in nondecimal units of minutes (min) hours (h) or days (d).

There are many methods used to express composition in gases, liquids and solids. One of the most useful is molar units, since chemical reactions are simpler to express in terms of molar units.

A mole (mol) of a pure substance is defined as the amount of that substance whose mass is numerically equal to its molecular weight.

A mole fraction of a particular substance is simply the moles of this substance divided by the total number of moles. The weight or mass fraction is the mass of the substance divided by the total mass. These two compositions, which hold for gases, liquids and solids, can be expressed as follows for component A in a mixture

$$X_A = \frac{\text{moles of A}}{\text{total moles}}$$

$$W_A = \frac{\text{mass A}}{\text{total mass}}$$

Example

A bottle holds 50 g of water and 50 g of sodium hydroxide. Calculate the weight fraction and mole fraction of sodium hydroxide.

Solution: 50 g water + 50 g NaOH = 100 g of solution

$$\text{Weight fraction} = \frac{50}{100} = 0.5$$

$$\text{Gram moles} = \frac{50}{40} = 1.25 \text{ (Molecular weight of NaOH = 40)}$$

$$\text{Mole fraction} = \frac{1.25}{4.03} = 0.310 \text{ (Gram Mole of Water 2.78 + 1.25 = 4.03)}$$

In order to make the students of biochemistry become more familiar with the basic concepts, the definitions and most common conventions are given below.

Molarity (M) = the number of moles of solute per litre of solution.

> **Note:** Mole concentrations are expressed with square brackets e.g. $[CO_2]$ = molarity of CO_2 ions. M can be calculated by knowing the weight of the dissolved solute and its molecular weight mol. wt.

Thus

$$\text{moles (M)} = \frac{\text{wt. in g}}{\text{mol. wt}} \quad \text{where wt. in g represents weight in grams.}$$

Avogadro's number = No. of molecules/g. mole
= No. of atoms/g. atom
= No. of ions/g-ion
= 6.023×10^{23}

In actual practice one Avogadros number of particles is called a mole, irrespective of the substance being ionic, monoatomic or molecular in nature

For e.g. 1 g-ion/litre = 6.023×10^{23} ions/litre

Activity (a) = the effective or apparent molarity of a solute. The relation is

$$a = \gamma[\text{M}]$$

where γ is called the activity coefficient. Due to the interaction between solute molecules that prevent their full expression, γ is generally less than one.

Normality (N) = the number of equivalents of solute per litre of solution.

Again N can be calculated only if we know the weight of the dissolved solute and its equivalent weight EW, and the relationship is

$$\frac{\text{wt. in g}}{\text{EW}} = \text{equivalents} = \text{N}$$

In general $\text{EW} = \dfrac{\text{mol. wt}}{n}$

where n = the number of replaceable H^+ or OH^- per molecule (for acids and bases) i.e. n = the number of electrons lost or gained per molecule (for oxidizing and reducing agents).

$$\text{Thus, N} = \frac{\text{wt. in g}}{\text{EW}} = \frac{\text{wt. in g}}{(\text{mol. wt}/n)}$$

$$= n\left(\frac{\text{wt. in g}}{\text{mol. wt}}\right)$$

$$= nM \qquad \text{i.e.} \quad N = nM$$

Weight/Volume per cent (W/V) = the weight in gram of a solute per 100 ml of solution

For routine laboratory solutions for which exact concentrations are not important, weight/volume per cent is often used.

Milligram per cent (mg%) = the weight in milligram of a solute per 100 ml of solution.

Milligram per cent is used in clinical laboratories.

Osmolarity = the molarity of particles in the solution.

Osmolarity is widely considered in physiological studies where tissues and cells must be immersed in a solution of the same osmolarity as the cytoplasm in order to prevent the uptake or release of water.

Weight/Weight per cent (% W/W) = the weight in gram of a solute per 100 g of solution

wt. in g = $\text{vol}_{\text{ml}} \times \rho_{\text{g/ml}} \times$ % as decimal

where, wt. in g = weight of pure substance required in gram

vol_{ml} = volume of stock solution required in ml

= fraction of total weight that is pure substance.

where, ρ = density = weight per unit volume

sp. gr. = specific gravity = density relative to water

> **Note:** In order to calculate the volume of the stock solution required for a given preparation, we must know its density or specific gravity.

Molarity (m) = the number of moles of solute per 1000 g of solvent

Molarity is used in some physical, chemical calculations.

In order to interrelate m and M we need to know % (w/w)

Molefraction (MF) = the fraction of the total number of moles present represented by the relevant compound.

Per cent saturation = the concentration of salt in a solution as a per cent of the maximum concentration possible at a given temperature;

Under specific conditions we also need to know the specific volume $\bar{v}$.

$\bar{v}$ = volume occupied by one gm of salt (ml/g)

= the reciprocal of density

2

Acid Base Chemistry

The properties of biological molecules can be well explained only with understanding of acid-base chemistry. Many of the low molecular weight metabolites and macromolecular components of the living cells are in water and thus they have the protential to ionize. Water can be considered an acid because of its ability to ionize to a proton and a negatively charged hydroxide ion:

$$\text{H–O–H} \longrightarrow \text{H}^+ + \text{OH}^-$$

The frequency with which this occurs in pure water is very low. The molar concentrations of H^+ and OH^- are approximately 10^{-7} each in pure water. Hydrogen-ion concentrations of aqueous solutions range from greater than 1 M to less than 10^{-14} M. For convenience, H^+ concentrations are expressed on a logarithmic scale as pH. The pH of a solution is the negative log of its H^+ concentration, or

$$\text{pH} = -\log\,[\text{H}^+]$$

Therefore, a H^+ concentration of 10^{-7} M has a pH of 7.0. In pure water, there are always equal numbers of H^+ and OH^-, and the pH is defined to be neutral. Acids contribute another source of H^+ causing the pH to decrease from neutrality. Conversely, bases absorb H^+ from solution and thus make the pH higher than neutrality. The first precise definition of an an acid and base was given by Svante Arrhenius, and is referred to as Arrhenius Theory.

Here an acid is defined as a substance that ionizes in water to give hydrogen ions, and a base as a substance that ionizes in water to give hydroxide ions.

Hydrochloric acid, HCl, is a strong acid, and is very soluble in water. It dissociates into its component ions in the following manner:

$$\text{HCl (g)} \longrightarrow \text{H}^+\text{ (aq)} + \text{Cl}^-\text{ (aq)}$$

The hydrogen ion interacts strongly with a lone pair of electrons on the oxygen of a water molecule. The resulting ion, H_3O^+ is called the hydronium ion.

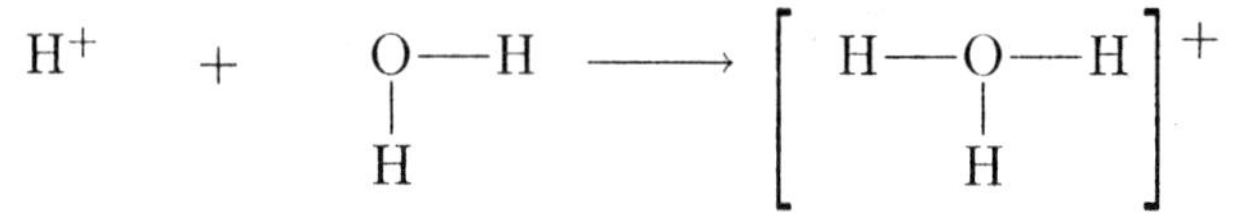

Acid solutions are formed when an acid transfers a proton to water. The reaction of HCl with water can be written in either of the following ways:

$$HCl\ (aq) \quad + \quad H_2O\ (l) \quad \rightleftharpoons \quad H_3O^+\ (aq) \quad + \quad Cl^-\ (aq)$$

$$HCl\ (aq) \quad \rightleftharpoons \quad H^+\ (aq) \quad + \quad Cl^-\ (aq)$$

According to Bronsted-Lowry concept acids are substances that are capable of donating a proton, and bases are substances capable of accepting a proton. So, in the example above, HCl acts as a Bronsted acid by donating a proton in water, and water in turn acts as a Bronsted base by accepting a proton from HCl. Water can act as an acid or a base. Here is another example:

$$NH_3(aq) \quad + \quad H_2O(l) \quad \rightleftharpoons \quad NH_4^+(aq) \quad + \quad OH^-(aq)$$

Here, H_2O acts as a Bronsted acid by donating a proton to NH_3 which acts as a Bronsted base. Using the Arrhenius definition, we say that the resulting solution is basic because it contains OH^- ions. Thus we say that the NH_3 molecule is basic (a proton acceptor).

Let's look at the reaction of NH_3 and H_2O again: to understand conjugate Acid-Base pair

$$NH_3 \quad + \quad H_2O \quad \rightleftharpoons \quad NH_4^+ \quad + \quad OH^-$$

The reverse of this reaction is:

$$NH_4^+ \quad + \quad OH^- \quad \rightleftharpoons \quad NH_3 \quad + \quad H_2O$$

In this case, NH_4^+ acts as an acid which donates a proton to OH^-. OH^- acts as a base. An acid and a base that are related by the gain and loss of a proton are called a conjugate acid-base pair. For example, NH_4^+ is the conjugate acid of NH_3, and $NH_3$is the conjugate base of NH_4^+. Every acid has associated with it a conjugate base. Likewise, every base has associated with it a conjugate acid.

$$NH_3(aq) \quad + \quad H_2O\ (l) \quad \rightleftharpoons \quad NH_4^+(aq) \quad + \quad OH^-\ (aq)$$

base | acid | conj. acid | conj. base

$$NH_4^+(aq) \quad + \quad OH^-(aq) \quad \rightleftharpoons \quad NH_3(aq) \quad + \quad H_2O(l)$$

acid | base | conj. base | conj. acid

Acids and bases can be either neutral or charged species: H_2O (acid or base, neutral), O_2^- (base, charged), $C_2H_3O_2^-$ (acid, charged). For any reaction:

$$HA \quad + \quad H_2O \quad \rightleftharpoons \quad H_3O^+ \quad + \quad A^-$$

If HA is a strong acid because it gives up its proton readily, then A^- is a weak base because it has little affinity for the proton.

If HA is a weak acid because it donates very few protons to the water, then A^- has a high affinity for a proton, and A^- is a stronger base than water.

In terms of Bronsted conjugate acid-conjugate base pure water is amphoteric as it yields both H^+ and OH^- ions, it can both donate and accept protons. The ionization of H_2O can be given by a 'dissociation constant' (K_d) ionization constant K_i and specific constant for water K_w

Simple dissociation: $HOH \quad \rightleftharpoons \quad H^+ \quad + \quad OH^-$.

$$K_d = \frac{[H^+][OH^-]}{[HOH]}$$

Water also produces two conjugate acid-conjugate base pairs.

$\frac{HOH}{OH^-}$ and $\frac{H_3O^+}{HOH}$ by the following reaction

$$HOH \quad + HOH \quad = \quad H_3O^+ \quad + \quad OH^-$$

$$K_i = \frac{[H_3O^-][OH^-]}{[HOH]^2}$$

Here for every mole of H^+ or H_3O^+ one mole of OH^- is produced. In pure H_2O $[H^+]$ or $[OH^-] = 10^{-7}$ M.

So molarity of HOH can be worked out.

$$M = \frac{\text{number of moles of } H_2O \left(\frac{\text{wt.}}{\text{MW}}\right)}{\text{litre}}$$

$$\therefore\ M = \frac{\frac{1000}{18}}{1 \text{ litre}} = 55.6 \quad (1 \text{ litre of } H_2O = 1000 \text{ g})$$

Substitute these values in K_d and K_i

$$K_d = \frac{10^{-7} \times 10^{-7}}{55.6} = \frac{10^{-14}}{55.6} = 1.8 \times 10^{-16}$$

$$K_i = \frac{10^{-7} \times 10^{-7}}{(55.6)^2} = \frac{10^{-14}}{3.09 \times 10^3} = 3.24 \times 10^{-18}$$

In almost all biochemical problems the molarity of H_2O is constant in dilute solutions. So the dissociation of water combines the two factors K_d and K_i.

$$K_w = K_d \times [H_2O](1.8 \times 10^{-16})(55.6) = 1 \times 10^{-14}$$

Similarly for ionization of H_2O

$$K_w = K_i \times [H_2O]^2 = (3.24 \times 10^{-18})(55.6)^2 = (3.24 \times 10^{-18}) \times (3.09 \times 10^3)$$

$$K_w = 1 \times 10^{-14}$$

So the specific constant for water K_w remains the same whether it is a simple dissociation or ionization as conjugate acid-base conjugate.

2.1 Composition and Action of Buffered Solutions

Buffered solutions or buffers are solutions which resist a change in pH when small amounts of acid or base are added. Buffers contain an acidic species to neutralize OH^- ions and a basic species to neutralize H^+ ions. However, these two species must be able to coexist in a solution, without completely neutralizing each other. Buffers are therefore made of weak acid-base conjugate pairs, such as $HC_2H_3O_2$ and $C_2H_3O_2^-$.

If we have a weak acid, HX, and its conjugate base, X^-, the following equilibrium occurs:

$$HX\ (aq) \rightleftharpoons H^+\ (aq) + X^-\ (aq)$$

For this reaction,

$$K_a = \left\{ \frac{[H^+][X^-]}{[HX]} \right\}$$

If an amount of H^+ ions is added, the above reaction will shift to the left. This will cause the $[H^+]$ to decrease, to close to what it was before, and thus the pH will stay fairly constant. If OH^- ions are added they will remove H^+ ions to form water, thus increasing the pH. However, the equilibrium reaction will shift to the right as H^+ ions are removed. The $[H^+]$ will therefore remain fairly constant, as will the pH.

The most effective buffering solutions are those which have similar concentrations of HX and X^-, because then the buffer has the capacity to absorb both acid and base, with the same effectiveness in either direction. When choosing an appropriate conjugate acid-base pair to form a buffer at a specific pH, the most effective buffers have the desired pH within 1.0 of the conjugate acid's pKa.

Ways to make a buffer:

- Adding a conjugate base to a weak acid.
- Adding a strong acid to a weak base.
- Adding a strong base to a weak acid.

2.2 Buffering Capacity and pH

The buffering capacity is the amount of acid or base a buffer can accept without the pH changing appreciably. The greater the amounts of the conjugate acid-base pair, the more resistant they are to change in pH. The acid-dissociation-constant expression for $[H^+]$ is:

$$[H^+] = K_a \left\{ \frac{[HX]}{[X^-]} \right\}$$

assuming that the amount of added acid or base is less than 5% of the conjugate acid base molarity. We can use this to determine the pH of a buffer. We first take the negative log of both sides:

$$-\log[H^+] = -\log K_a \left\{ \frac{[HX]}{[X^-]} \right\}$$

$$= -\log K_a - \log \left\{ \frac{[HX]}{[X^-]} \right\}$$

We know $-\log[H^+] = pH$ and $-\log K_a = pK_a$

$$pH = pK_a - \log\left\{\frac{[HX]}{[X^-]}\right\}$$

$$= pK_a + \log\left\{\frac{[X^-]}{[HX]}\right\}$$

In general:

$$pH = pK_a + \log\left\{\frac{[base]}{[acid]}\right\}$$

This is known as the Henderson-Hasselbalch equation.

A titration is a procedure for carrying out a chemical reaction between two solutions by the controlled addition from a burette of one solution (the titrant) to the other, allowing measurements to be made throughout the reaction. For a reaction between an acid and a base, a titration is useful for measuring the pH at various points throughout the reaction. A titration curve is a graph of the pH as a function of the amount of titrant (acid or base) added.

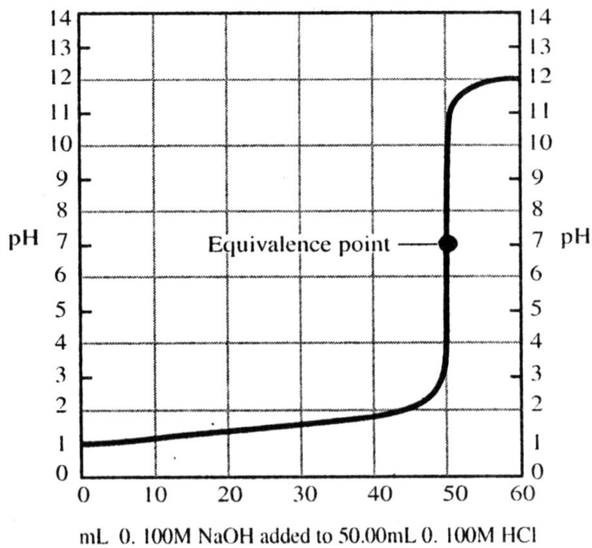

Fig. 2-1

2.3 Strong Acid-Strong Base Titrations

An example of a titration curve, produced when a strong base is added to a strong acid is given. This curve shows how pH varies as 0.2 M NaOH is added to 50.0 ml of 0.2 M HCl.

The equivalence point of the titration is the point at which exactly enough titrant has been added to react with all of the substance being titrated with no titrant left over. In other words, at the equivalence point, the number of moles of titrant added so far corresponds exactly to the number of moles of substance being titrated according to the reaction stoichiometry. (In an acid-base titration, there is a 1 : 1 acid : base stoichiometry, so the equivalence point is the point where the moles of titrant added equals the moles of substance initially in the solution being titrated.) The pH increases slowly at first, then rapidly as it nears the equivalence point.

2.4 Titrations Involving a Weak Acid or Weak Base

Titration curve of a weak acid being titrated by a strong base:

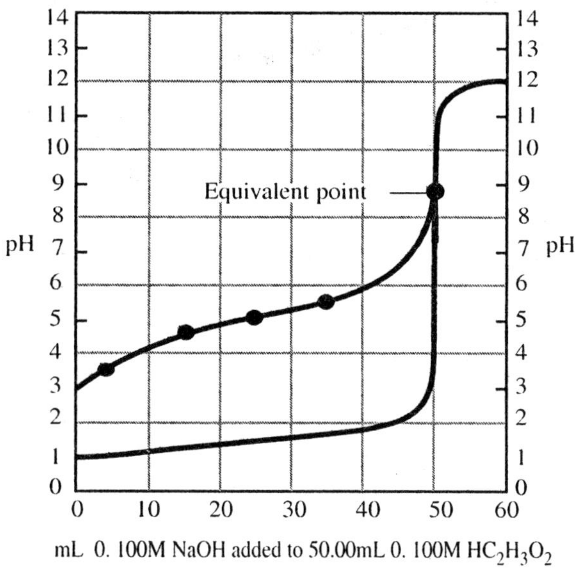

Fig. 2-2

Here, 0.100 M NaOH is being added to 50.0 ml of 0.100 M acetic acid. There are three major differences between this curve (in dots) and the one given earlier:

1. The weak-acid solution has a higher initial pH.
2. The pH rises more rapidly at the start, but less rapidly near the equivalence point.
3. The pH at the equivalence point does not equal 7.00.

Point of Emphasis

The equivalence point for a weak acid-strong base titration has a pH > 7.00. For a strong acid-weak base or weak acid-strong base titration, the pH will change rapidly at the very beginning and then have a gradual slope until near the equivalence point. The gradual slope results from a buffer solution being produced by the addition of the strong acid or base, which resists rapid change in pH until the added acid or base exceeds the buffer's capacity and the rapid pH change occurs near the equivalence point.

Titration curve of a weak base being titrated by a strong acid:

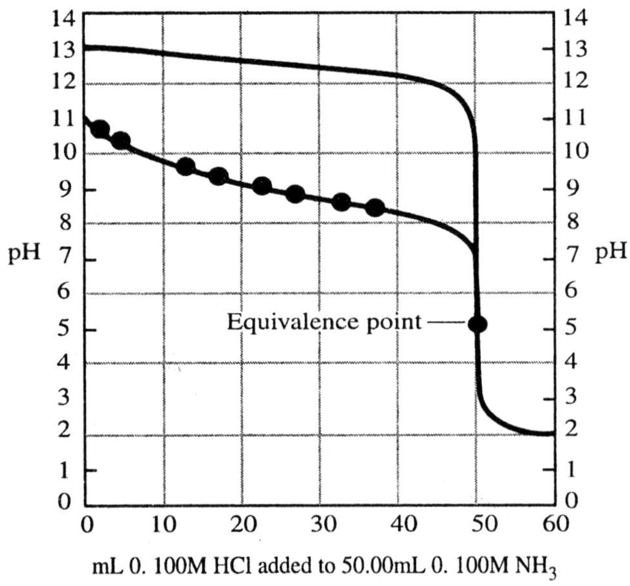

Fig. 2-3

Here, 0.100 M HCl is being added to 50.0 mL of 0.100 M ammonia solution. As in the weak acid-strong base titration, there are three major differences between this curve (in dots) and a strong base-strong acid one:

1. The weak-acid solution has a lower initial pH.
2. The pH drops more rapidly at the start, but less rapidly near the equivalence point.
3. The pH at the equivalence point does not equal 7.00.

Polyprotic acids have the equivalence point for a weak base-strong acid titration at pH < 7.00.

Titrations of Polyprotic acids

An example of a polyprotic acid is H_2CO_3 which neutralizes in two steps:

$$H_2CO_3\ (aq) \quad + \quad OH^-\ (aq) \longrightarrow H_2O\ (l) \quad + \quad HCO_3^-\ (aq)$$

$$HCO_3^-\ (aq) \quad + \quad OH^-\ (aq) \longrightarrow H_2O\ (l) \quad + \quad CO_3^{2-}\ (aq)$$

The titration curve for these reactions will look like this, with two equivalence points.

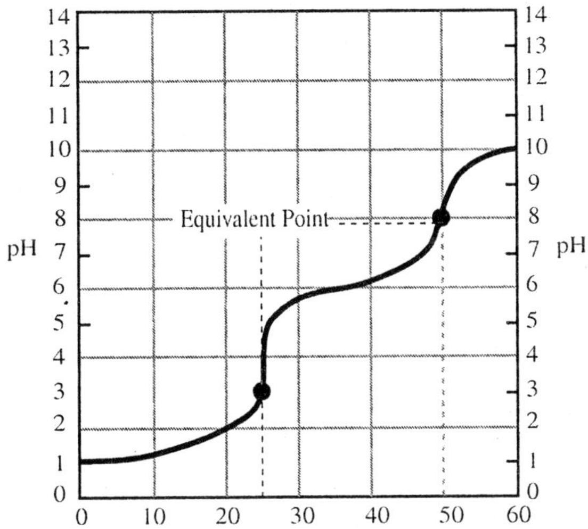

Fig. 2-4

2.5 Uses of Titration

Use titration data or a titration curve to calculate reaction quantities such as the concentration of the substance being titrated. The most common use of titrations is for measuring unknown concentrations. This is done by titrating a known volume of the unknown solution with a solution of known concentration (where the two react in a predictable manner) and finding the volume of titrant needed to reach the equivalence point using some method appropriate to the particular reaction. Then, the volume and concentration of titrant can be used to calculate the moles of titrant added, which, when used with the reaction stoichiometry, gives the number of moles of substance being titrated. Finally, this quantity, along with the volume of substance being titrated, gives the unknown concentration. For acid-base titrations, the equivalence point can be found very easily.

A pH meter is simply placed in the solution being titrated and the pH is measured after various volumes of titrant have been added to produce a titration curve. The equivalence point can then be read off the curve.

Example

If 80.0 ml of 0.200 M NaOH are required to reach the equivalence point in a titration of 50.0 ml of hydrofluoric acid, what is the concentration of the hydrofluoric acid?

Solution: The neutralization reaction goes to completion:

$$HF + OH^- \longrightarrow F^- + H_2O$$

The number of moles of NaOH added was:

$$n_{NaOH} = [NaOH] \times V_{NaOH}$$
$$n_{NaOH} = 0.200\ M \times 80.0\ mL$$
$$n_{NaOH} = 16\ mmol$$

Since each NaOH produces 1 OH^-, $n_{OH} = n_{NaOH} = 16$ mmol. From the 1 : 1 stoichiometry between HF and OH^-, $n_{HF} = n_{OH} = 16$ mmol. So, the concentration of the original hydrofluoric acid solution was:

$$[HF] = n_{HF}\ V_{HF}$$
$$[HF] = 16\ mmol/50\ ml$$
$$[HF] = 0.320\ M$$

In the same way, knowing the equivalence point can also be used to calculate other unknown quantities of interest in acid base reactions, such as concentration of titrant or volume of solution being titrated, provided that enough other information is known to perform the calculations.

2.6 Acid-Base Properties of Amino Acids

Amino acids act as acids and bases and following information about them can be obtained from their titration curves. When in an aqueous solution, amino acids can act as both acids and bases as they are amphoteric.

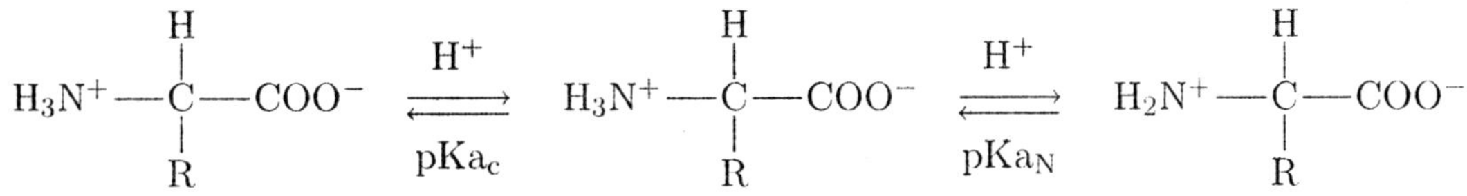

To fully describe the ionization state of an amino acid, both the individual charges and the net charge must be considered: If only positive charges or only negative charges are present, the molecule is described as either a cation or an anion respectively. However, both positive and negative charges can be present at the same time. When this happens, the molecule is called a dipolar ion or zwitterion. All amino acids exist as zwitterions at pH 7.0. Depending on whether the side chain is positively-charged, negatively-charged or uncharged, an amino acid zwitterion may or may not have a net charge. Note that there is no pH at which both groups are electrically neutral.

The net charge on an amino acid is the sum of all the individual charges. The net charge can be described as monopositive, dipositive, mononegative, dinegative or isoelectric. Isoelectric means that the molecule has both positive and negative charges but no net charge, as opposed to neutral which means that no charges are present.

The ionization state of an amino acid can be described completely and unambiguously by describing both the net charge and the individual charges (for example, the mononegative zwitterion form of aspartate). However, for a given amino acid, there is only one form that has a given net charge, so the net charge alone can be used to describe the amino acid. (For example, the mononegative form of aspartate is by definition a zwitterion, whereas the mononegative form of lysine is an anion).

The titration of an amino acid is simply a polyprotic acid titration, with the pH values at the midpoints giving the pKa's of the dissociable groups.

1. $pH < pKa_c$	Almost all monopositive form	Avg. net charge just below +1
2. $pH = pKa_c$	Half monopositive, half isoelectric	Avg. net charge = +0.5
3. $pH = \frac{1}{2}(pKa_c + pKa_N)$	All isoelectric form	Avg. net charge = 0
4. $pH = pKa_N$	Half isoelectric, half mononegative	Avg. net charge = 0.5
5. $pH > pKa_N$	Almost all mononegative	Avg. net charge just above 1

Each amino acid has different pKa_c and pKa_N values, and those with dissociable side chains also have a pKa_R.

2.6.1 Isoelectric point and average molecular charge

The isoelectric point (pI) of a molecule such as an amino acid, peptide or protein is the pH at which it has a net charge of zero. We saw above that an amino acid with a non-dissociable side chain had a net charge of zero when the pH was halfway between the two pKa values. Thus, for an amino acid with a non-dissociable side chain,

$$pI = \frac{1}{2}(pKa_c + pKa_N)$$

For amino acids with acidic and basic side chains, the pI is halfway between the pKa values on either side of the isoelectric form.

Example

Write equations for the dissociation of aspartate and calculate its pI.

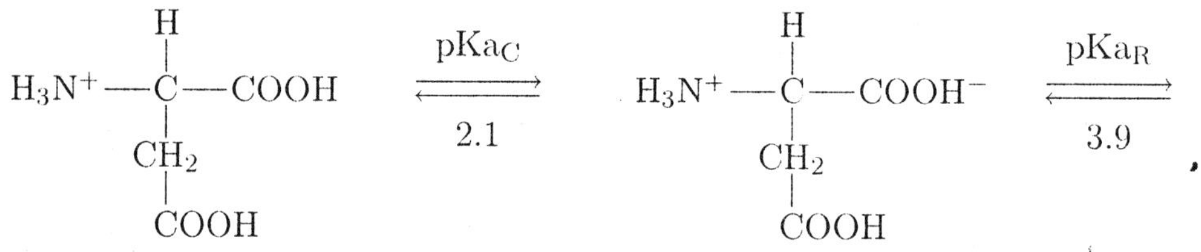

Monopositive (cation) Isoelectric (zwitterion)

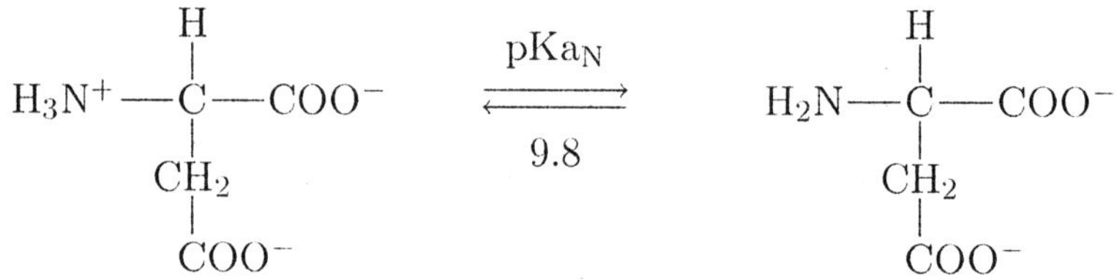

Mononegative (zwitterion) Dinegative (Anion)

The isoelectric form is found after the first dissociation, between pKa_c and pKa_R

$$pI = \frac{1}{2}(pKa_c + pKa_R)$$

$$pI = \frac{1}{2}(2.1 + 3.9)$$

$$pI = 3.0$$

Looking at these two examples, we can see a connection between the net charge on the amino acid and the relative values of the pH and the pI:

- If the pH is less than the pI, the amino acid will have a net positive charge.
- If the pH is greater than the pI, the amino acid will have a net negative charge.
- If the pH equals the pI, the amino acid will have no net charge.

At certain pH's, the amino acid will have an integral charge, but most pH's are between these points and give fractional charges. There is, in fact, a continuous (though not uniform) charge gradient as the pH is changed. However, in discussing the charge on amino acids, we must remember that each individual molecule has an integral charge, not a fractional charge. The fractional charge comes from the proportions of two integrally-charged species in equilibrium, giving an average net fractional charge to the solution.

For example, in a 3:1 mixture of mononegative to isoelectric forms, 3 out of every 4 molecules (75) has a charge of 1, and 1 out of every 4 (25) is neutral. The average net charge is then

$$q_{\text{net}} = (0.75)(1) + (0.25)(0) = 0.75$$

The ratio between a conjugate acid and base can be obtained from the Henderson-Hasselbalch equation:

$$\text{pH} = \text{pKa} + \log \frac{[\text{base}]}{[\text{acid}]}$$

At any given pH, there will be at most two forms of an amino acid present in significant quantities, and they will be a conjugate acid-base pair. This means that their ratio can be calculated with the Henderson-Hasselbalch equation, and the average net charge on the amino acid can be calculated.

The importance of pH in biological systems

All living things are water-based systems, which means that they depend heavily on aqueous equilibria, especially acid-base equilibria. Therefore, all the acid-base and pH concepts we have discussed so far are extremely important to biochemistry for following reasons.

- It gives a qualitative measure for many problems in cell biology and related fields.

- Proton dissociable groups are found in macromolecules (such as proteins) as well as the in the small molecules.

- The cell environment is always buffered at approximately pH 7.

- Experiments such as biological enzymatic assays require a certain pH.

- Just as in other acid-base systems, biological macromolecules act as acids and bases by donating and accepting protons. However, due to the size of these molecules, they often contain several different groups that accept or donate protons instead of just one such group. Thus, macromolecules are referred to as having acidic and basic groups rather than as being acids and bases. These acidic and basic groups act as weak acids and bases, with K_a values which determine the extent of dissociation of the group depending on the pH of the system. Therefore, changes in the pH around the macromolecule will determine which groups are protonated and which are not, which in turn determines properties of the molecule. This is especially important for enzymes, which are proteins that act as catalysts for important biological reactions. Most enzymes only work within a certain pH range.

For example, consider an enzyme with a carboxyl group. The structure of that group will depend on the pH:

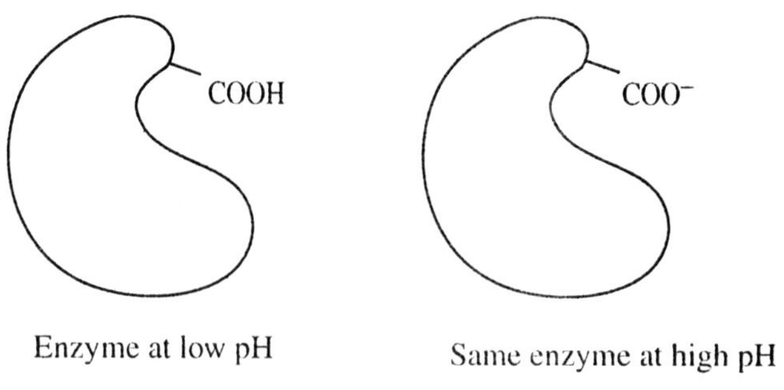

Fig. 2-5

If the enzyme needs to be protonated in order to be active, then the enzyme will only work in the pH range in which the majority of the enzyme molecules have their carboxyl group protonated. In this way, pH determines which enzymes are active and thus which biochemical reactions can occur.

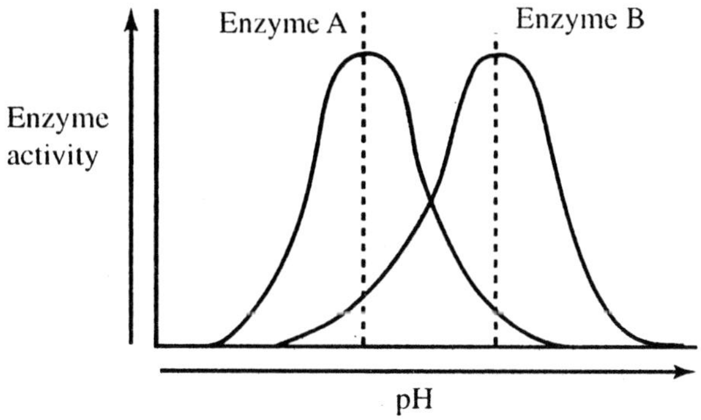

Fig. 2-6

2.7 Buffer Systems in Living Organisms

Because all biological processes are dependent on pH, cells and organisms must maintain a specific and constant pH in order to keep their enzymes in the optimum state of protonation.

1. Cytoplasm controlled by the phosphate buffer system:

$$H_2PO_4 \rightleftharpoons H^+ + HPO_4^{2-}$$

   This system provides the maximum buffering capacity near pH 6.86 (the pKa of H_2PO_4). It provides the buffering effect in intracellular fluid, and is important in urine.

2. Blood must be maintained at pH~7.4 (pH < 6.9 and > 7.6 are life-threatening). The pH of the blood is controlled by the bicarbonate buffer system:

$$CO_{2(g)} \rightleftharpoons CO_{2(aq)} + H_2O_{(l)} \rightleftharpoons H_2CO_{3(aq)} \rightleftharpoons H^+_{(aq)} + HCO_{3(aq)}$$

The tissues release CO_2 into the blood, where it is converted to H_2CO_3. In this form, it is carried to the lungs, where it is converted back to gaseous CO_2 for exhalation. pH depends on $[H_2CO_3]$ and $[HCO_3]$, and $[H_2CO_3]$ depends on the CO_2 dissolved in the blood.

Breathing

Tissues release H^+ into blood:

- $[H^+]$ increases
- $[H_2CO_3]$ increases
- $[CO_2]$ dissolved in blood increases
- pressure of CO_2 in lungs increases
- exhalation occurs, restoring equilibrium

Tissues release OH^- into blood (or remove H^+ from blood):

- $[H^+]$ decreases
- $[H_2CO_3]$ decreases
- $[CO_2]$ dissolved in blood decreases
- pressure of CO_2 in lungs decreases
- inhalation occurs, restoring equilibrium

Major Buffer Systems of the Human Body

Bicarbonate buffer	$CO_2 + H_2O \rightleftharpoons H_2CO_3 \rightleftharpoons H^+ + HCO_3$	In blood plasma
Hemoglobin	$Hb{-}H \rightleftharpoons Hb + H^+$	Interior of red blood cells
Phosphate buffer	$H_2PO_4 \rightleftharpoons H^+ + HPO_4^2$	Most important in urine
Protein	$Pr{-}H \rightleftharpoons Pr + H^+$	Intracellular fluid

Table for Acids and Bases

Table 1: Acids and bases: physical and chemical data

	Formula	Mol. wt.	m. p.(°C)	b.p.(°C)	Specific gravity (20°C)	Molarity of the conc.solution	ml litre^{-1} to give a 1 M solution
Acids							
Acetic	CH_3COOH	60.05	17	118	1.05	17.4	57.5
Formic	$HCOOH$	46.03	8	100	1.222	23.6(90%)	42.4 (90%)
						25.9(98%)	38.5(98%)
Hydrochloric	HCl	36.46	–	–	1.18	11.6	85.9
Nitric	HNO_3	63.01	−42	86	1.42	15.7	63.7
Orthophos phoric*	H_3PO_4	98.00	22	261	1.70	14.7	67.8
Sulfuric	H_2SO_4	98.07	10	330	1.84	18.3	54.5
Base							
Ammonium hydroxide	NH_4OH	17.03**	–	(25%)	0.91	13.3	75.1
				(35%)	0.88	18.1	55.2
Potassium hydroxide	KOH	56.11	360	1320	2.0	–	–
Sodium hydroxide	$NaOH$	40.00	318	1390	2.12	–	–

*Commonly called phosphoric acid.

**This is the molecular weight of NH_3.aq. The formula weight is 35.05.

Table 2: Zwitterionic (Good's) buffers: physical and chemical data

Buffer*	**Mol. wt.**	**Solubility (0°C) (mol/litre H_2O)**	**pK_a**		**Useful pH range**
			at 20°C	**Δ°C**	
ACES	182.2	0.22	6.88	−0.020	6.4—7.4
ADA**	190.2	0.09	6.62	−0.011	6.4—7.4
BES	213.3	3.2	7.17	−0.016	6.6—7.6
Bicine	163.2	1.1	8.35	−0.018	7.8—8.8
CAPS	221.3	0.47	10.4	−0.032	9.7—11.1
CHES	207.1	1.15	9.55	−0.011	9.0—10.1
Glycylglycine	132.1	1.01 (25°C)	8.25	−0.026	7.7—8.8
HEPES	238.3	2.25	7.55	−0.014	7.0—8.0
HEPPS (EPPS)	252.3	1.6	8.0	−0.011	7.6—8.6
MES	213.3	0.65	6.15	−0.011	5.8—6.5
MOPS	209.3	3.1	7.2	−0.011	6.5—7.9
PIPES	302.4	1.4	6.82	−0.009	6.4—7.2
TAPS	243.3	v. sol.	8.4 (25°C)	−0.018	7.7—9.1
TES	229.3	2.6	7.5	−0.020	7.0—8.0
Tricine	179.2	0.8	8.15	−0.021	7.6—8.8
Tris	121.1	4.54 (25°C)	8.08	−0.028	7.0—9.0***

* **Chemical names:** ACES = N-(2-acetamido)-2-aminoethanesulfonic acid

ADA = N-(2-acetamido) iminodiacetic acid

BES = N,N-bis(2-hydroxyethyl)-2-aminoethanesulfonic acid

Bicine = N,N-bis(2-hydroxyethyl)glycine

CAPS = 3-(cyclohexylamino)-1-propanesulfonic acid

CHES = 2-(cyclohexylamino)-ethanesulfonic acid

HEPES 4-(2-hydroxyethyl)piperazine-1-ethanesulfonic acid

HEPPS 4-(2-hydroxyethyl)piperazine-1-propanesulfonic acid

MES = 2-morpholinoethanesulfonic acid monohydrate

MOPS = 3-morpholinopropanesulfonic acid

PIPES = Piperazine-1,4-bis(2-ethanesulfonic acid)

TAPS = N-[tris(hydroxymethyl)methyl]-3-aminopropanesulfonic acid

TES = N-[tris(hydroxymethyl)methyl]-2-aminopropanesulfonic acid

Tricine = N-[tris(hydroxymethyl)methyl]glycine

Tris = Tris(hydroxymethyl)aminomethane.

** = ADA may cause skin irritation. Avoid contact with the dust.

*** = Refers to Tris-HCl

Table 3: Components of non-zwitterionic buffers: physical and chemical data

Compound	Formula	Mol. wt.	Solubility (20°C)(g/100ml H_2O)	pKa (20°C)		
				1	2	3
Boric acid[a]	H_3PO_4	61.8	6.4	9.14	12.74	13.80
Citric acid, anhydrous[c]	$C_6H_8O_7$[c]	192.4	146.0	3.14	4.77	6.39 (18 °C)
Imidazole	$C_3H_4N_2$[c]	68.1	v. sol.	6.95 (25°C)	–	–
Potassium dihydrogen phosphate (potassium phosphate, monobasic)	KH_2PO_4	136.1	136.1	2.12	7.21	12.67(25° C)
di-Potasium hydrogen phosphate (potassium phosphate, dibasic)	K_2HPO_4	174.2	174.2	2.12	7.21	12.67(25°C)
Potassium hydrogen phthalate	$C_8H_5KO_4$[c]	204.2	204.2	2.89	5.51 (25°C)	
Sodium acetate						
anhydrous	CH_3COONa	82.0	119.0 (0°C)	4.75 (25°C)	–	–
trihydrate	$CH_3COONa.3H_2O$	136.1	76.2 (0°C)	–	–	–
Sodium cacodylate	$(CH_3)_2AsO_2Na.3H_2O$	214.0	200	6.21	–	
Sodium cacodylate[b]						
anhydrous	Na_2CO_3	106.0	7.1 (0°C)	6.37	10.25	–
decahydrate	$Na_2CO_3.10H_2O$	286.1	21.5 (0°C)			
tri-Sodium citrate	$Na_3C_6H_5O_7.2H_2O$[c]	294.1	72.0	3.09	4.75	5.41(18°C)
Sodium dihydrogen phosphate (sodium phosphate, monobasic)	$NaH_2PO_4.2H_2O$	156.0	v. sol	2.12	7.21	12.67
Sodium hydrogen carbonate	$NaHCO_3$	84.0	9.6	6.37	10.25	–

Table 4: Components of non-zwitterionic buffers: physical and chemical data (cont.)

Compound	Formula	Mol. wt.	Solubility (20°C) (g/100ml H_2O)	pKa (20°C) 1	2	3
di-Sodium hydrogen phosphate						
anhydrous	Na_2HPO_4	142.0		2.12	7.21	12.67 (25°C)
dihydrate	$Na_2HPO_4.2H_2O$	178.0	60.0			
dodecahydrate	$Na_2HPO_4.12H_2O$	358.1	4.2			
(sodium phosphate, dibasic)						
Sodium tetraborate						
anhydrous	$Na_2B_4O_7$	201.2	1.1 (0°C)	9.14	12.74	13.80
decahydrate	$Na_2B_4O_7.10H_2O$	381.4	2.0 (0°C)			

a: Also called orthoboric acid.

b: Sodium cacodylate is hydroxydimethylarsine oxide, sodium salt, or dimethylarsinic acid, sodium salt.

Table 5: L-amino acids: physical and chemical data

Amino acid	Abbreviations	Mol. wt.	m.p.[a] (°C)	Solubility (25°C) (g/100 ml H_2O)	pKa (25°C) -COOH	pKa (25°C) $-NH_2$	pKa (25°C) R-group	pI (25°C)
Alanine	Ala A	89.10	>200	16.65	2.35	9.87	—	6.11
Arginine	Arg R	174.20	244	15.00[b]	1.82	8.99	12.48	10.76
Aspargine	Asn N	132.12	236	2.99	2.14	8.72	—	5.43
Aspartic acid	Asp D	133.11	270	0.5	1.99	9.90	3.90	2.98
Cysteine	Cys C	121.16	240	v. sol.	1.92	10.70	8.37	5.15
Glutamic acid	Glu E	147.13	248	0.86	2.10	9.47	4.07	3.08
Glutamine	Gln Q	146.15	185	3.6[b]	2.17	9.13	—	5.65
Glycine	Gly G	75.07	292	25.0	2.35	9.78	—	6.06
Histidine	His H	155.16	287	4.16	1.80	9.33	6.04	7.64
Isoleucine	Ile I	131.18	285	4.12	2.32	9.76	—	6.04
Leucine	Leu L	131.18	294	2.43	2.33	9.74	—	6.04
Lysine	Lys K	146.19	224	66.6[b]	2.16	9.06	10.54	9.47
Methionine	Met M	149.21	281	5.14[b]	2.13	9.28	—	5.71
Phenylalanine	Phe F	165.19	283	2.96	2.20	9.31	—	5.76
Proline	Pro P	115.13	221	162.3	1.95	10.64	—	6.30
Serine	Ser S	105.09	228	25.0[b]	2.19	9.21	—	5.70
Threonine	Thr T	119.12	226	v. sol.	2.09	9.10	—	5.60
Tryptophan	Trp W	204.23	289	1.14	2.46	9.41	—	5.88
Tyrosine	Tyr Y	181.19	343	0.045	2.20	9.21	10.46	5.63
Valine	Val V	117.15	315	8.85	2.29	9.74	-	6.02

a: At these temperatures virtually all amino acids undergo some decomposition.

b: Temperatures: Arg 21°C, Gln, 18°C, Lys 20°C, Met 20°C, Ser 20°C.

Table 6: Salts: physical and chemical data

Compound	**Formula**	**Mol. wt.**	**Solubility (g/100 ml H_2O)** Cold(°C)	Hot (°C)
Ammonium acetate	CH_3COONH_4	77.08	148(4)	decomposes
Ammonium chloride	NH_4Cl	53.49	29.7 (0)	75.8(100)
Ammonium nitrate	NH_4NO_3	80.04	118.3(0)	871(100)
Ammonium persulfate (AMPS, Ammonium peroxydisulfate)	$(NH_4)_2S_2O_8$	228.20	58.2 (0)	v. sol.
Ammonium sulfate	$(NH_4)_2SO_4$	132.13	70.6 (0)	103.8(100)
Calcium chloride, dihydrate	$CaCl_2.2H_2O$	147.02	97.7 (0)	326(60)
Calcium chloride, hexahydrate	$CaCl_2.6H_2O$	219.08	279 (0)	536 (20)
Calcium hypochlorite	$Ca(OCl)_2$	142.99	sol.	sol.
Lithium chloride	LiCl	42.39	63.7 (0)	130 (95)
Magnesium acetate, tetrahydrate	$(CH_3COO)_2Mg.4H_2O$	214.40	120 (15)	v. sol.
Magnesium chloride, hexahydrate	$MgCl_2.6H_2O$	203.30	167 (0)	367 (100)
Magnesium nitrate, hexahydrate	$MgN_2O_6.6H_2O$	256.41	125 v.	sol.
Magnesium sulfate, heptahydrate	$MgSO_4.7H_2O$	246.47	71 (20)	91 (40)
Manganous chloride, anhydrous	$MnCl_2$	125.80	72.3(25)	123.8 (100)
Manganous chloride, tetrahydrate	$MnCl_2.4H_2O$	197.90	151(8)	656 (100)
Manganous sulfate, monohydrate	$MnSO_4.H_2O$	169.01	98.5 (48)	79.8 (100)
Manganous sulfate, heptahydrate	$MnSO_4.7H_2O$	223.06	172	—
Potassium acetate	CH_3COOK	98.14	253 (20)	492 (62)
Potassium chloride	KCl	74.55	34.7(20)	56.7 (100)
Potassium iodide	KI	166.00	127.5 (0)	208.0 (100)

Table 6 (Cont.)

Compound	Formula	Mol. wt.	Solubility (g/100 ml H_2O)	
			Cold (°C)	Hot (°C)
Potassium nitrate	KNO_3	101.10	13.3(0)	247(100)
Potassium permanganate	$KMnO_4$	158.03	6.4 (20)	25 (65)
Potassium sodium tartrate,	KOCOCH(OH)	282.22	26 (0)	66 (25)
tetrahydrate	$CH(OH)COONa.4H_2O$			
Potassium sulfate	K_2SO_4	174.25	12 (25)	24.1 (100)
Sodium chloride	NaCl	58.44	35.7 (0)	39.1 (100)
Sodium metabisulfite	$Na_2S_2O_5$	190.10	54 (20)	81.7 (100)
Sodium nitrate	$NaNO_3$	84.99	92.1 (25)	180.0 (100)
Sodium nitrite	$NaNO_2$	69.00	81.5 (15)	163 (100)
Sodium salicylate	$C_6H_4(OH).COONa$	160.10	111 (15)	125 (25)
Sodium succinate, hexahydrate	$(CH_2COONa)_2.6H_2O$	270.14	21.5 (0)	86.6(75)
Sodium sulfate, anhydrous	Na_2SO_4	142.04	4.7(0)	42.7(100)
Sodium sulfate, decahydrate	$Na_2SO_4.10H_2O$	322.19	11(0)	92.7(30)
Tetramethyl ammonium chloride	$(CH_3)_4NH_4Cl$	109.60	—	
Zinc chloride	$ZnCl_2$	136.29	432 (25)	615 (100)
Zinc sulfate, heptahydrate	$ZnSO_4.7H_2O$	287.54	96.5 (20)	663.6 (100)

Biochemical Calculations

1. Find the amount of solid NaOH required to prepare 300 ml of a 0.05 M solution. Express the concentration of this solution in terms of (a) g/litre (b)% (w/v) (c) mg % (d) N and (e) Osmolarity

Number of moles NaOH required = litres × M

i.e. 0.3 × 0.05 = 0.015 mole NaOH required

$$\text{Number of moles} = \frac{\text{wt. in g}}{\text{mol. wt.}}$$

$\therefore$ wt. in g = 0.015 × 40

wt. = 0.6 g

$\therefore$ 0.6 g must be dissolved in water and diluted to 300 ml.

(a) NaOH contains one OH/molecule

$\therefore$ $M = N$ and the solution is 0.05 N where N = normality = the no. of equivalents of solute per litre of solution.

The solution contains 0.6 g/300 ml i.e. 2 g/litre

(b) %(w/v) = g per 100 ml,

$\therefore$ 2 g/litre = 0.2 g/100 ml = 0.2%

(c) mg% = mg per 100 ml

$\therefore$ 0.2 g/100 ml = 200 mg/100ml = 200 mg%

(d) NaOH contains one OH per molecule.

$\therefore$ M = N and the solution is = 0.05 N

(e) NaOH yields 2 particles (Na^+ and OH^-)

$\therefore$ Osmolarity = 2 × M = 0.1 osmolar

Note:

i. % (w/v) = The weight in g of a solute per 100 ml of solution.

ii. mg% = The weight in milligram of a solute per 100 ml of solution.

iii. Generally $N = nM$ where N is the normality, M the molarity and n equals the number of electrons lost or gained per molecule (for oxidizing and reducing agents).

iv. Osmolarity = The molarity of a particle in a solution concentration based on degree of saturation.

2. Beginning with 350 ml of a 40% saturated ammonium sulfate solution at 0°C calculate (a) the weight of solid ammonium sulfate that must be added to bring the solution to 60% saturation and (b) the volume of saturated solution that must be added to attain 60% saturation. The specific volume of ammonium sulphate is 0.565 ml/g. The solubility of ammonium sulfate at 0°C is 706 g/1000 g water.

The volume of saturated solution at 0°C prepared by adding 706 g of ammonium sulfate (AS) to 1000 gm of H_2O is

$$1000 \text{ ml} + 706 \times (0.565) = 1399 \text{ ml}$$

The concentration of AS in the solution is $\frac{706}{1399} = 0.505$ g/ml = 505 g/litre

The molarity of the solution is $\frac{505}{132.14} = 3.82M$ as the mol. wt. of AS is 132.14.

(a) wt. in g = wt. of required solid ammonium sulfate

$$= \frac{505(S_2 - S_1)}{1 - 0.285S_2}$$ (where S_1 and S_2 are initial and final degrees of saturation)

$$= \frac{505 \times (0.6 - 0.4)}{1 - 0.285 \times 0.6}$$ (expressed as decimals)

$$= \frac{101.0}{1 - 0.1710} = \frac{101.0}{0.829}$$

$$= 121.8335$$

if the initial solution is 1000 ml.

$\therefore$ wt. in g for a saturated solution of 350 ml. $= \frac{121.8335}{1000} \times 350$

$$= 42.642 \text{ g}$$

(b) Vol. ml $= 350 \times \dfrac{(0.6 - 0.4)}{1 - 0.6}$

$= 350 \times \dfrac{0.2}{0.4} = 175$ ml

3. Calculate the molarity of pure ethanol. The density of ethanol is 0.789 g/ml. The mol. wt. of ethanol is 46.07.

Molarity of ethanol = no. of moles present in 1 litre of pure ethanol.

$$= \frac{\text{wt. in gm}}{\text{mol. wt.}} = \frac{\text{vol}_{\text{ml}} \times \rho_{\text{mg/ml}}\%\text{ as decimal}}{\text{mol. wt.}}$$

Where wt. in g = weight of pure substance in gm.

Vol_{ml} = volume in ml

% = fraction of total weight that is pure substance.

Here vol_{ml} = 1000

$\rho_{\text{mg/ml}}$ = 0.789 g/ml

and % = 100% = 1

$$\therefore \quad \frac{\text{wt. in g}}{\text{mol. wt.}} = \frac{1000 \times 0.789 \times 1}{46.07} = \frac{789}{46.07} = 17.126.$$

Molarity of ethanol = 17.126

Ionisation of strong acids and bases

4. Calculate the ionic strength of a 0.05 M solution $Fe_2(SO_4)_3$.

$$\text{Ionic strength} = \tfrac{\Gamma}{2} = \tfrac{1}{2}\sum M_i Z_i^2$$

where,

M_i = the molarity of the ion

Z_i = the net charge of the ion (irrespective of sign)

$\sum$ = a symbol meaning 'the sum of'

Ionic strength measures the concentration of charges in solution.

$$\frac{\Gamma}{2} = \frac{1}{2}\sum M_i Z_i^2 = \frac{1}{2}\left[M_{Fe^{3+}} Z^2_{Fe^{3+}} + \left(M_{SO_4^{2-}}\right) Z^2_{SO_4}\right]$$

The 0.05 M $Fe_2(SO_4)_3$ gives 0.1 M Fe^{3+} and 0.15 M SO_4^{2-}

$$\therefore \quad \frac{\Gamma}{2} = \frac{1}{2}\left[(0.1)(3)^2 + (0.15)(-2)^2\right]$$

$$= \frac{1}{2}[0.9 + 0.6] = \frac{1.5}{2} = 0.75$$

$$\therefore \quad \frac{\Gamma}{2} = 0.75$$

5. Explain how you will prepare 3 litres of 0.5 M HCl starting with a concentrated HCl solution (35% w/w HCl, Specific gravity = 1.15).

Number of moles = litres $\times M = 3 \times 0.5 = 1.5$

HCl needed 1.5 moles

wt. in. g = number of moles$\times$ mol. wt.= $1.5 \times 36.5 = 54.75$ g

Pure HCl needed $\longrightarrow$ 54.75 g.

The stock solution is only 35% HCl by weight

$$\therefore \text{ requirement of stock solution} = \frac{54.75}{0.35} = 156.43 \text{ g}$$

156.43 g stock solution is required.

If volume of stock solution is to be calculated then

$$\text{Vol}_{\text{ml}} = \frac{\text{wt. in g}}{\rho_{\text{g/ml}}} = \frac{156.43}{1.15} = 136.03$$

136.03 ml of stock solution is required

Hence we have to measure 136.03 ml of stock solution and dilute to 3 litres with water.

Strong Acids and Bases

6. What are the (a) H^+ ion concentration (b) pH and (c) OH^- ion concentration and (d) pOH of a 0.002 M solution of HCl?

(a) HCl being a strong inorganic acid it is 100% ionised in dilute solution. Hence when 0.002 M of HCl is introduced with one litre of H_2O it immediately dissociates with 0.002 M H^+ and 0.002 M Cl^-.

(b) $pH = -\log[H^+]$ [where $[H^+] = 0.002\ M = 2 \times 10^{-3}\ M$]

$$= -\log 2 \times 10^{-3} = \log \frac{1}{2 \times 10^{-3}}$$

$$= \log 0.5 \times 10^3 = \log 5 \times 10^2 = \log 5 + 2 \log 10$$

$$= 0.699 + 2 = 2.699$$

$pH = 2.699$

(c) $[H^+][OH^-] = K_w = 1 \times 10^{-14}$

$$[OH^-] = \frac{1 \times 10^{-14}}{[H^+]} = \frac{1 \times 10^{-14}}{2 \times 10^{-3}} = 0.5 \times 10^{-11}$$

$$[OH^-] = 0.5 \times 10^{-11} = 5 \times 10^{-12}$$

(d) $pOH = -\log[OH^-]$

$$= -\log(0.5 \times 10^{-11}) = -\log\left(\frac{1}{2} \times 10^{-11}\right) = \log \frac{2}{10^{-11}}$$

$$= \log 2 - \log 10^{-11} = \log 2 + 11 \log 10$$

$$= 0.3010 + 11$$

$pOH = 11.3010$

7. Calculate the pH, pOH and number of H^+ and OH^- ions per litre in the following solutions. (a) 10^{-4} M HNO_3 (b) 0.0025 M H_2SO_4 (c) 2.9×10^{-5} M NaOH

(a) $pH = -\log[H^+]$ where $[H^+] = 10^{-4}M$

$$= -\log 10^{-4} = \log 10^4 = 4$$

$pH = 4$

$pH + pOH = 14$

$pOH = 14 - 4 = 10$, $pOH =$ **10**

$H^+ = 10^{-4}M(10^{-4}\ g - ions/litre)$

$1g - ion/litre = 6.023 \times 10^{23}$ ions/litre

$\therefore\ 10^{-4}$ g ions/litre $= 6.023 \times 10^{23} 10^{-4}$ gm ions/litre

$= 6.023 \times 10^{19}$ ions/litre

$pOH = 10$

$$\therefore \quad [OH^-] = 10^{-10} \text{ M or } 10^{-10} \text{ g ions/litre}$$
$$= 6.023 \times 10^{23} \times 10^{-10} \text{ ions/litre}$$
$$= 6.023 \times 10^{13} \text{ ions/litre}$$

(b) $[H^+] = 0.0025 \text{ M} = 0.25 \times 10^{-2} \text{ M} = \frac{1}{4} \times 10^{-2} \text{ M}$

$$pH = -\log H^+$$
$$= -\log \frac{10^{-2}}{4}$$
$$= 2 + \log 4 = 2.6021$$

$$pH = 2.602$$

$$pOH = 14 - pH$$
$$= 11.398$$

$$H^+ = 2.5 \times 10^{-3} M$$
$$= 2.5 \times 10^{-3} \times 6.023 \times 10^{23} \text{ ions/litre}$$
$$= 1.51 \times 10^{21} \text{ ions/litre}$$

$$OH^- \text{ ions/l} = 6.023 \times 10^{23} \times \frac{1}{2.5} \times 10^{-11}$$
$$= 2.41 \times 10^{12}$$

H^+ ions/litre $= 1.51 \times 10^{21}$
OH^- ions/litre $= 2.41 \times 10^{12}$

(c) $pH = -\log[H^+] \qquad H^+ = 2.9 \times 10^{-5} M$

$$= -\log(2.9 \times 10^{-5})$$
$$= 5 - \log 2.9 = 5 - 0.4624 = 4.5376$$
$$= 4.54$$

$$pOH = 14 - 4.54 = 9.46$$

$$[H^+] \text{ ions/l} = 2.9 \times 10^{-5} \times 6.023 \times 10^{23}$$
$$= 17.47 \times 10^{18} = 1.75 \times 10^{19}$$

$$[H^+][OH^-] = 1 \times 10^{-14}$$

$$\text{OH}^- \text{ ions/l} = 6.023 \times 10^{23} \times \frac{1 \times 10^{-14}}{2.9 \times 10^{-5}}$$

$$= \frac{6.023}{2.9} 10^{14} \qquad [\text{H}^+] = 1.75 \times 10^{19} \text{ ions/litre}$$

$$= 2.1 \times 10^{14} \qquad [\text{OH}^-] = 2.1 \times 10^{14} \text{ ions/litre}$$

8. Obtain the α_{H^+} and γ_{H^+} in a 0.02 M HCl solution if the pH is 2.54

$$\text{pH} = 2.54$$

$$2.54 = \log \frac{1}{\alpha_{H^+}}$$

$$\therefore \quad \frac{1}{\alpha_{H^+}} = \text{anti log } 2.54 = 347$$

$$\therefore \quad \alpha_{H^+} = \frac{1}{347} = 2.9 \times 10^{-3}$$

$$\alpha_{H^+} = \gamma_{H^+}[\text{H}^+]$$

$$2.9 \times 10^{-3} = \gamma_{H^+}(0.02)$$

$$\gamma_{H^+} = \frac{0.0029}{0.02} = 0.145 \qquad \gamma_{H^+} = 0.145$$

9. In a 0.3 M solution KOH, γ_{OH^-} is 0.85. Calculate the pH and pOH of this solution.

$$\text{pOH} = -\log \gamma_{OH^-}[\text{OH}^-] \qquad [\text{H}^+] = 0.3\text{M}$$

$$= \log \frac{1}{\gamma_{OH^-}[\text{OH}^-]} \qquad = 3 \times 10^{-1} \text{ M}$$

$$= \log \frac{1}{0.85 \times 10^{-13}/3} \qquad [\text{H}^+][\text{OH}^-] = 1 \times 10^{-14}$$

$$= \log \frac{10^{13}}{0.85} \times 3 \qquad [\text{OH}^-] = \frac{10^{-14}}{3 \times 10^{-1}}$$

$$= \log (10^{13} \times 3.53) \qquad = \frac{10^{-13}}{3}$$

$$= 13.5478 = 13.55$$

$$\therefore \quad \text{pH} = 0.45$$

$$\boxed{\text{pOH} = 13.55; \quad \text{pH} = 0.45}$$

10. Find out how many millilitres of 0.24 M H_2SO_4 are required to neutralize $\frac{1}{3}$ of OH^- ions present in 750 ml of 0.36 N NaOH.

No.of moles of H^+ required = no.of moles (equivalents) of OH^- present

$\text{Litres} \times N_{acid} = \text{litres}_{base} \times N_{base}$

$H_2SO_4 = 0.24 \text{ M} = 0.48 \text{ N}$

$\text{Litres} \times N = \text{litres} \times 0.48 = \frac{1}{3} \times 0.750 \times 0.36$

$$\therefore \text{litres}_{acid} = \frac{1}{3} \times \frac{0.750 \times 0.36}{0.48} = 0.1875 \text{ litre}$$

Requirement of acid = 187.5 ml.

Weak Acids and Bases

11. The weak acid HA is 1.5% dissociated in a 0.5 M solution. Calculate (a) the Ka (b) the pH of the solution (c) the amount of 0.3 N KOH required to neutralize 500 ml of the weak acid solution.

	HA	$\rightleftharpoons$	H^+	$+ A^-$
Starting	0.5		0	0

Change – (1.5% of 0.5 M) $= \frac{1.5}{100} \times 0.5$

$= -0.75 \times 10^{-2}$ M

$= -7.5 \times 10^{-3}$ M $\quad + \quad 7.5 \times 10^{-3}$M $\quad + \quad 7.5 \times 10^{-3}$M

HA $\quad\quad H^+ \quad\quad A^-$

Equilibrium

$0.5 - 7.5 \times 10^{-3}$ M; $\quad 7.5 \times 10^{-3}$ M; $\quad 7.5 \times 10^{-3}$M

$$\text{(a)}\ K_a = \frac{[H^+][A^-]}{[HA]} = \frac{(7.5 \times 10^{-3})(7.5 \times 10^{-3})}{0.5 - 7.5 \times 10^{-3}}$$

$$= \frac{(7.5 \times 10^{-3})(7.5 \times 10^{-3})}{0.5} = 112.5 \times 10^{-6}$$

$$= 1.125 \times 10^{-8} = 1.13 \times 10^{-8}$$

$$K_a = 1.13 \times 10^{-8}$$

(The term 7.5×10^{-3} in the denominator being very small, is omitted.)

(b) $pH = \log \frac{1}{[H^+]} = \log \frac{1}{7.5 \times 10^{-3}} = 3 - \log 7.5$

$= 3 - 0.8751 = 2.1249$

$= 2.12 \qquad pH = 2.12$

(c) No. of moles of OH^- required = Total no. of moles H^+ available

$\text{litre}_{\text{base}} \times N_{\text{base}} = \text{litre}_{\text{acid}} \times N_{\text{acid}}$

HA is monoprotic $\quad \therefore \quad N = M$ (in general N = nM - here $n = 1$)

$\text{litres}_{\text{base}} \times 1.5 = 0.5 \times 0.3$

$$\text{litre}_{\text{base}} = \frac{0.5 \times 0.3}{1.5} = \frac{0.15}{1.5} = 100 \text{ ml}$$

$\therefore$ 100 ml of 0.3 N KOH is required for neutralization.

12. The K_a for a weak acid HA is 4×10^{-4}. Find out the (a) pH (b) degree of dissociation of the acid in a 0.25 M solution (c) also calculate $_pK_a$ and $_pK_b$.

Let x = M that dissociates so that x = M of H^+ and M of A^- produced.

HA	$\rightleftharpoons$	H^+	+	A^-
0.25 M		0		0
$- x$ M	+	x M	+	x M
Equilibrium				
$2.5 \times 10^{-1} - x$ M		x M		x M

$$K_a = \frac{[H^+][A^-]}{[HA]} = \frac{x.x}{2.5 \times 10^{-1} - x} = 4 \times 10^{-4}$$

Assuming x to be very small this reduces to

$$\frac{x^2}{2.5 \times 10^{-1}} = 4 \times 10^{-4}$$

$$x^2 = 10 \times 10^{-1} \times 10^{-4} = 10^{-4}$$

$$x = 10^{-2}$$

$$H^+ = 10^{-2}M$$

Also $[H^+] = \sqrt{K_a[HA]}$

$$= \sqrt{(4 \times 10^{-4}(0.25)} = 2 \times 10^{-2} \times 0.5 = 10^{-2}\text{M}$$

(a) $pH = \log \frac{1}{[H^+]}$

$= \log \frac{1}{10^{-2}} = 2$

$pH = 2$

(b) Degree of dissociation $= \dfrac{[H^+]}{[HA]_{orig}} \times 100$

$= \dfrac{10^{-2}}{0.25} \times 100 = \dfrac{1}{0.25} = 4\%$

(c) $pK_a = -\log\ K_a = -\log 4 \times 10^{-4}$

$= 4 - \log 4 = 4 - 0.6021$

$= 3.3979$

$pK_a + pK_b = 14$

$pK_b = 14 - 3.3979$

$= 10.6021$

13. What is the H^+ ion concentration in a 0.04 M solution of a fairly strong acid HA where $K_a = 5 \times 10^{-2}$? Find also the dissociation of the acid.

Let x = M of HA dissociated

= M of H^+ produced

	HA	$\rightleftharpoons H^+$	$+ A^-$
	0.04 M	0	0
	$-x$ M	$+ x$ M	$+ x$ M
Equilibrium	$0.04 - x$ M	x M	x M

$$Ka = \frac{[H^+][A^-]}{[HA]}$$

i.e. $5 \times 10^{-2} = \dfrac{(x)(x)}{0.04 - x}$

$x^2 = 5 \times 10^{-2}(0.04 - x)$

$x^2 + 5 \times 10^{-2}x - 20 \times 10^{-4} = 0$

$$x = \frac{-5 \times 10^{-2} \pm \sqrt{25 \times 10^{-4} + 80 \times 10^{-4}}}{2}$$
$$= \frac{(-5 \pm \sqrt{105})10^{-2}}{2} = \frac{-5 \pm 10.25}{2} \times 10^{-2}$$
$$= \frac{5.25}{2} \times 10^{-2} \quad \text{(considering only the +ve value)}$$
$$= 2.625 \times 10^{-2} = 2.63 \times 10^{-2} \text{ M}$$

$[H^+] = 2.63 \times 10^{-2}$ M

When,

$x^2 + bx + c = 0$

$x = \frac{-b \pm \sqrt{b^2 - 4c}}{2}$

Here,

$b = 5 \times 10^{-2}$;

$c = -20 \times 10^{-4}$

$$\text{Degree of dissociation} = \frac{[H^+]}{[HA]_{orig}} \times 100 = \frac{2.63 \times 10^{-2}}{0.04} \times 100 = 65.75\%$$

Buffer

14. Describe the preparation of 4 litres of a 0.3 M acetate buffer, pH 4.58 starting from solid sodium acetate trihydrate (mol. wt. 136) and a 2.5 M solution of acetic acid.

We will first calculate the molarities of OAc^- and HOAc present.

pH = 4.58

$[H^+] = 10^{-4.58}$

(Ka for acetic acid = 1.70×10^{-5})

$$Ka = \frac{[H^+][OAc^-]}{[HOAc]}$$

$$1.7 \times 10^{-5} = \frac{10^{-4.58} \times y}{0.3 - y} \qquad \text{where} \quad y = [OAc^-]$$

$$y(10^{-4.58} + 1.7 \times 10^{-5}) = 0.3 \times 1.7 \times 10^{-5}$$

$$y = \frac{0.51 \times 10^{-5}}{10^{-5}(10^{0.42} + 1.7)}$$
$$= \frac{0.51}{4.33} = 0.118 \text{ M}$$

$[OAc^-] = 0.118$ M

$[HOAc] = 0.3 - 0.118 = 0.182$ M

$$4 \text{ litres} \times 0.3 \text{ M} = 1.2 \text{ (mole) (total of HOAc}^- \text{ and OAc)}$$

$$\therefore \quad 4 \text{ litres} \times 0.118 = 0.472 \text{ m OAc}^-$$

$$4 \text{ litres} \times 0.182 = 0.728 \text{ m HOAc}^-$$

$$\text{no. of moles} = \text{wt. in g/m wt.} = 0.472$$

$$\therefore \quad \text{wt. in g} = (0.472)(\text{mole. wt.}) = (0.472)(136)$$

$$= 64.192 \text{ g}$$

$$\text{wt. in g} = 64.2 \text{ g}$$

The 0.728 mole of HOAC comes from a 2.5 M stock solution

$$\text{no. of moles} = \text{litres} \times \text{M}$$

$$0.728 = \text{litres} \times 2.5$$

$$\text{litres} = \frac{0.728}{2.5} = 0.2912 = 291.2 \text{ ml}$$

Therefore 64.2 g of sodium acetate and 291.2 ml of acetate is made up to 4 litres.

15. Find the concentrations of HOAc and OAc$^-$ in a 0.4 M acetate buffer, pH 5.48. The Ka for acetic acid is 1.70×10^{-5}(pKa = 4.77).

The 0.4 M acetate contains a total of 0.4 mole of acetate per litre. Part of this is in the conjugate acid form HOAc and some in the conjugate base form OAc$^-$. The concentration of each, we find by using the Ka expression.

$$\text{Ka} = \frac{[\text{H}^+][\text{OAc}^-]}{[\text{HOAc}]}$$

Let $t = M$ of OAc$^-$

$$\therefore \quad 0.4 - t = \text{M of HOAc}$$

$$\text{pH} = 5.48$$

$$\therefore \quad [\text{H}^+] = 10^{-5.48}$$

$$1.70 \times 10^{-5} = \frac{10^{-5.48} t}{0.4 - t}$$

$$= \frac{10^{-6} \times 10^{0.52} t}{0.4 - t}$$

$$6.8 \times 10^{-6} - 1.7 \times 10^{-5}t = 10^{-6} \times 10^{0.52}t$$

$$6.8 \times 10^{-6} = 10^{-6}(17 + 10^{0.52})\ t$$

$$t = \frac{6.8}{17 + 3.312}$$

$$= \frac{6.8}{20.312} = 0.335$$

$$[\text{OAc}^-] = 0.335;\ [\text{HOAc}] = 0.4 - 0.335;\ [\text{HOAc}] = 0.065\ \text{M}$$

Alternately we have

$$\text{pH} = \text{pKa} + \log \frac{[\text{OAc}^-]}{[\text{HOAc}]}$$

Let, $t = M$ of OAc^-

$\therefore\ 0.4 - t = M$ of HOAc

$$5.48 = 4.77 + \log \frac{t}{0.4 - t}$$

$$0.71 = \log \frac{t}{0.4 - t}$$

$$\frac{t}{0.4 - t} = 5.129 \quad \text{(antilog of 0.71)}$$

$$6.129t = 0.4 \times 5.129$$

$$t = \frac{2.052}{6.129} = 0.335$$

i. e. $[\text{OAc}] = 0.335$

$[\text{HOAc}] = 0.4 - 0.335$

$\therefore\ [\text{HOAc}] = 0.065$

16. Calculate the pH of a 0.30 M solution of an amine with a pKa of 8.7.

$$\text{pK}_a + \text{pK}_b = 14$$

$$\text{pK}_b = 14 - \text{pKa}$$

$$= 14 - 8.7$$

$$= 5.3$$

$$\text{p}[\text{RNH}_2] = -\log[\text{RNH}_2]$$

$$= -\log 30 \times 10^{-2} = 2 - \log 30$$

$$= 2 - 1.4771 = 0.5229 = 0.52$$

$$\text{pOH} = \frac{\text{pK}_b + \text{p}[\text{RNH}_2]}{2} = \frac{5.3 + 0.52}{2} = \frac{5.82}{2} = 2.91$$

$$\text{pH} = 14 - \text{pOH} = 14 - 2.91 = 11.09$$

$$\text{pH} = 11.09$$

17. What is the concentration of dissolved CO_2 in equilibrium with an atmosphere containing a partial pressure of CO_2 of 42 mm Hg?

Now $[CO_2]$ dissolved $= k \times pCO_2$ [where $k = 3.01 \times 10^{-5}$]

$= (3.01 \times 10^{-5}) \times 42 = 126.42 \times 10^{-5}$

$= 1.26 \times 10^{-3}$ M $= 1.3 \times 10^{-3}$ M

18. How many different linear α-linked pentapeptides can be made from 15 different L-α-amino acids (a) using any amino acid for any of the five positions (repetitions allowed) (b) using any amino acid only once in the chain? (c) Find the no.of qualitatively different amino acid analysis patterns possible.

(a) Since repetitions are used for making linear α-linked pentapeptides using the 15 different L-amino acids, each of the five positions can be occupied by any one of the 15 amino acids and there are 15 ways for each position.

Hence the required no. of pentapeptides $= 15 \times 15 \times 15 \times 15 \times 15$

$= 15^5 = 759375$

(b) The first link can be occupied by any one of the 15 amino acids in 15 ways. Since repetitions are not allowed the second place is filled by any one of the remaining 14 amino acids, the third by 13 and so on.

Hence the required number $= 15 \times 14 \times 13 \times 12 \times 11$

$= 360360$

(c) No. of qualitatively different amino acid patterns = sum of all combinations of 15 objects taken 1,2,3,4 and 5 at a time

$$= {}^{15}C_1 + {}^{15}C_2 + {}^{15}C_3 + {}^{15}C_4 + {}^{15}C_5$$

$$= 15 + \frac{15.14}{1.2} + \frac{15.14.13}{1.2.3} + \frac{15.14.13.12}{1.2.3.4} + \frac{15.14.13.12.11}{1.2.3.4.5}$$

$$= 15 + 105 + 455 + 1365 + 3003 = 4943$$

Note: $^{15}C_2 = \dfrac{15!}{2!(15-2)!}$ where 15! stands for the product of numbers from 15 and numbers less than 15. i. e. 15! = 15 . 14 . 13 ... 3.2.1.

since $^{n}C_r = \dfrac{n!}{r!(n-r)!}$

Also $^{15}C_2 = \dfrac{15!}{2!.13!} = \dfrac{15.14.(13.12\ldots1)}{1.2(13!)} = \dfrac{(15.14)13!}{1.2.13!} = \dfrac{15.14}{1.2}$

Similarly $^{15}C_3 = \dfrac{15.14.13}{1.2.3}$ and so on.

19. (a) Calculate the axial length of an α-helix containing 145 amino acid residue (b) How long would the chain be if it were fully extended? (c) Approximately what is the mol. wt. of the protein?

(a) The helix rises 5.4 Å for every 3.6 residues. $\therefore$ $\dfrac{5.4}{3.6} = \dfrac{\text{length } \ell}{145}$

$$\ell = \frac{5.4 \times 145}{3.6} = 217.5\text{Å}$$

The axial length of the α–helix is 217.5 Å.

(b) In a fully extended chain the distance between residues is 3.6 Å.

$\therefore$ Total length of the polypeptide chain $= 145 \times 3.6 = 522$Å

(c) The average mol. wt. of an amino acid residue is 120. $\therefore$ The mol. wt. of the protein containing 145 amino acid residues $= 120 \times 145 = 17400$

20. Calculate the width of the peptide plane i. e (a) The distance 'a' between the α-carbon and the oxygen and (b) The distance 'b' between the H and α-carbon atom

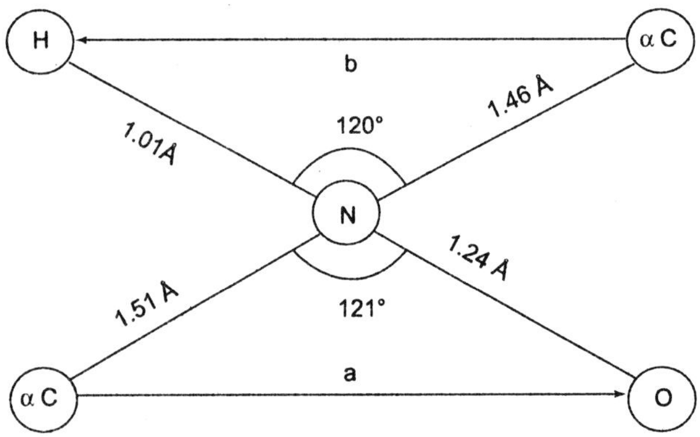

Fig. 2-7

From Δ H αCN

$$b^2 = 1.01^2 + 1.46^2 - 2 \times 1.01 \times 1.46 \times \cos 120°$$
$$= 1.02 + 2.132 + 1.4746$$
$$= 4.6266$$
$$b = 2.151 \text{Å}$$

From Δ NαCO,

$$a^2 = 1.51^2 + 1.24^2 - 2 \times 1.51 \times 1.24 \cos 121°$$
$$= 2.2801 + 1.5376 - 3.7448(-.515)$$
$$= 2.2801 + 1.5376 + 1.929$$
$$= 5.7467$$
$$a = 2.397 \text{Å}$$

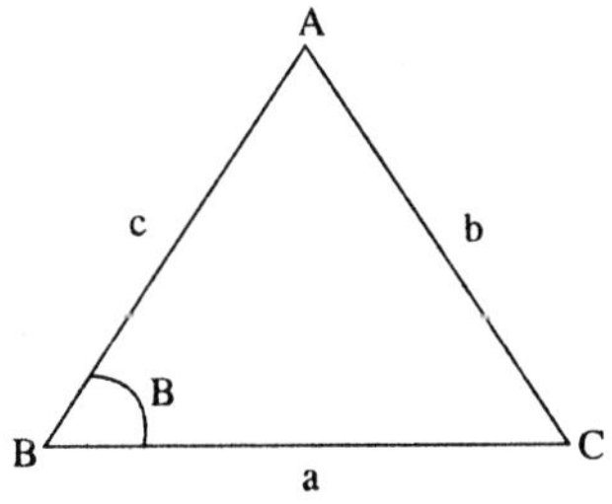

In △ ABC;

$$b^2 = a^2 + c^2 - 2ac \cos B$$

21. (a) At 10°C the osmotic pressure of a protein solution (6 mg/ml) was 5.628×10^{-3} atm. Calculate mol. wt. (b) An osmotic pressure of 2.766×10^{-3} was found at 3mg/ml atm. Calculate accurate value for mol. wt. R = 0.0821 litre × atm × mole^{-1} degree $^{-1}$

C	π	π/C
6 mg/ml	5.628×10^{-3}	9.38×10^{-4}
3 mg/ml	2.766×10^{-3}	9.22×10^{-4}
∴ at 0 mg/ml	—	9.06×10^{-4} (assuming there is a uniform decrease of $\frac{\pi}{c}$)

$$\text{mol. wt.} = \frac{RT}{\pi/C}$$

Here R = 0.0821 and T = absolute temperature = 283

$$\text{(a) mol wt.} = \frac{0.0821 \times 283}{9.38 \times 10^{-4}}$$
$$= \frac{23.2343}{9.38} \times 10^4$$
$$= 2.4770042 \times 10^4$$
$$= 24770$$

(b) From the table the value of π/C decreases by 0.16×10^{-4} $(9.38 - 9.22 = 0.16)$ for a decrease of 3 mg/ml of C. Hence at infinite dilution we can take π/C as 9.06×10^{-4}.

$$\text{Hence mol wt.} = \frac{0.0821 \times 283}{9.06 \times 10^{-4}}$$

$$= \frac{23.2343}{9.06} \times 10^4$$

$$= 2.56449 \times 10^4$$

$$= 25645.$$

22. Calculate the average mol. wt. of the triglyceride present in the given fat sample of KOH. The saponification number of the fat sample is 230. The molecular weight of KOH is 56.

For each mole of a triglyceride it takes 3 moles of KOH.

$$\text{Required mol. wt.} = \frac{3 \times 56 \times 1000}{\text{saponification number}}$$

$$= \frac{3 \times 56 \times 1000}{230}$$

Mol. wt. of triglyceride $= 7.304 \times 100 = 730.4$

23. A lipoprotein contains 70% protein and 30% lipid by weight. Calculate its specific volume and density given that the average densities of protein and lipid are 1.30 g/cm^3 and 0.92 g/cm^3 respectively.

$$\bar{\nu}_{\text{lipoprotein}} = \sum(\text{wt. fraction of components}) \times (\bar{\nu} \text{ of components})$$

$$= 0.7 \times \frac{1}{1.30} + 0.3 \times \frac{1}{0.92}$$

$$= 0.5385 + 0.3261$$

$$= 0.8646 = 0.865 \text{ cm}^3/\text{g}$$

$$\rho = \frac{1}{\bar{v}} = \frac{1}{0.865} = 1.156 \text{ g/cm}^3$$

24. In E. coli the ribosome is found to be 8% of cell volume. Each ribosome is a sphere of 200 Å in diameter, Calculate the number of ribosomes in an E. coli cell. Assume that E. coli is a cylinder 1μ in diameter and 2μ long.

The volume of an E. coli cell (assuming it to be a cylinder) is given by $V = \pi r^2 \ell$

where, $r = \frac{1}{2}\mu = 0.5 \times 10^{-4}$ cm

$$= 5 \times 10^{-5} \text{cm}$$

$$\ell = 2\mu = 2 \times 10^{-4} \text{ cm}$$

$$V = (3.14)(0.5 \times 10^{-4})^2(2 \times 10^{-4}) \text{ cm}^3$$

$$= 1.57 \times 10^{-12} \text{ cm}^3$$

$$8\% \text{ of this volume} = 1.57 \times 10^{-12} \times \frac{8}{100}$$

$$= 1.256 \times 10^{-13} \text{cm}^3$$

Each ribosome is a sphere of radius $\left(\frac{200 \text{ Å}}{2}\right) = 100 \times 10^{-8}$

$$= 10^{-6} \text{cm}$$

$$\therefore \quad \text{Volume of each ribosome} = \frac{4}{3}\pi r^3 = \frac{4}{3} \times 3.14 \times (10^{-6})^3$$

$$= \frac{4}{3} \times 3.14(10^{-6})^3$$

$$= \frac{4}{3} \times 3.14 \times 10^{-18}$$

$$\therefore \quad \text{No. of ribosome} = \frac{1.256 \times 10^{-13}}{(\frac{4}{3}) \times 3.14 \times 10^{-18}}$$

$$= \frac{1.256 \times 3 \times 10^5}{4 \times 3.14}$$

$$= 0.3 \times 10^5 = 30,000$$

So, the no. of Ribosome in E. coli is 30,000.

25. How many different L. triglycerides can be made from glycerol and 5 different fatty acids? How many triglycerides of quantitatively different compositions can be made from glycerol and five different fatty acids?

(a) The no. of different L. triglycerides that can be made from glycerol and 5 different fatty acids $= n^r = 5^3 = 125$

(b) No. of triglycerides of quantitatively different compositions from glycerol and 5 different fatty acids = ${}_nN_r$ where $n = 5$, $r = 3$

$$= \frac{(n+r-1)!}{r!(n-1)!} = \frac{(8-1)!}{3!4!} = \frac{7!}{3!4!}$$

$$= \frac{1 \times 2 \times 3 \times 4 \times 5 \times 6 \times 7}{(1 \times 2 \times 3)(1 \times 2 \times 3 \times 4)} = 35$$

26. Find the mol. wt. of a protein coded for by an mRNA containing 1200 nucleotides.

1200 nucleotides have $\frac{1200}{3}$ amino acids.

The average mol. wt. of one amino acid residue is 120.

$$\therefore \quad \text{The mol. wt.} = \frac{1200}{3} \times 120 = 48,000$$

27. Amino acid analysis of 15 mg of the purified protein yielded 83 μg of tryptophan. What is the minimum molecular weight of the protein? (The mol. wt. of tryptophan is 204.1)

The minimum molecular weight of the protein is calculated as follows.

$$\frac{15 \times 10^{-3} \text{ protein}}{83 \times 10^{-6} \text{ g tryptophan}} = \frac{\text{mol. wt.}}{204.1}$$

$$\therefore \quad MW_{min} = \frac{204.1 \times 15}{83} \times 10^3$$

$$= 36.8855 \times 10^3$$

$$= 36886$$

3

Biochemical Energetics

Living organisms utilize the chemical energy in their environment to carry out their biochemical activities. These processes are subject to the basic laws of thermodynamics. So it is necessary to define and understand a few fundamental terms of thermodynamics, a science that relates the energy changes that occur in chemical and physical processes. The First Law of Thermodynamics states that the energy can neither be created nor destroyed. But in a given process, one form of energy may be converted into another form while the total energy of the system and its surroundings remain constant. This law only states about conservation of energy. The second law of Thermodynamics states that all natural processes proceed in a direction that leads to a minimum potential energy level i.e. towards equilibrium. The naturally occurring processes are the spontaneous reactions which release energy as they progress towards equilibrium. Example for a spontaneous reaction is the gas diffusion from a region of high pressure and concentration to a region of low pressure and concentration. It is important to understand that in all the spontaneous reactions energy is conserved but the energy is distributed in a qualitatively different ways. The second law highlights the unidirectionality of spontaneous processes through the term *Entropy*. Entropy (S) is defined as the measure of the randomness or orderliness of the energy and matter in a system. The more random, disordered a system, the higher its entropy. In a more random, disordered and disorganized system the entropy increases while in organized, orderly and constrained system, the entropy decreases. So increase in entropy shows a loss of organization and a decrease in the potential to do further work.

According to the third law, the entropy of the perfect crystals of a compound rises with the increase in temperature from a minimum or zero value at the zero of the Kelvin scale (0 K or −273°C). In other words, at absolute zero, where all random motion ceases, the entropy of a perfect crystal of every substance is zero, where all the atoms are maximally ordered.

The living body consists of a mass of chemical substances organized in an orderly manner in cells, tissues and organs. As per the thermodynamic laws, the living organisms constantly tend to move spontaneously towards increasing entropy. But the total energy of the living system receiving the energy plus that of the system providing the energy remains constant. However, the entropy of just the system receiving the energy may increase, decrease or remain constant. In general, a living cell consumes raw materials at high entropy state from its environment and orders these materials in to a lower entropy state at the expense of the environment, which also supplies chemical energy at a low entropy state. In biological systems, it is impractical to determine the entropy of a system by counting the number of ways it can assume its most probable state. Therefore a state function that predicts whether or not a given process is spontaneous was formulated by J. Williard Gibbs in 1978. This thermodynamic functions of state that are more easily measured or calculated is called the change in free energy. The free energy change is the measure of the maximum useful work that a reaction could perform at constant temperature and pressure and depends on the displacement of the system from equilibrium.

At constant temperature and pressure, the energy difference is called the free energy difference. This is also known as Gibbs free energy change (ΔG) which is the maximum potential of a reaction for performing useful work. ΔG is defined as the free energy content of the product minus the free energy content of the reactants. The exergonic reactions have negative ΔG values while endergonic reactions have positive ΔG values.

Example

Endergonic or Synthetic Reaction: $A + B \longrightarrow A \sim B$; $\Delta G = +7$ kcal/mole

Exergonic or Degradation Reaction: $A \sim B \longrightarrow A + B$; $\Delta G = -7$ kcal/mole

The relationship between the amount of energy released and the concentrations of the substrate and products in a biochemical reaction is given in the following equation.

$$\Delta G = -2.3RT \log K_{\text{eq}} + 2.3RT \log \frac{[P_1]^a[P_2]^b}{[S_1]^c[S_2]^d}$$

where,

R	$=$ Gas constant (1.987 cal, $\times$ mole^{-1} °K^{-1})
T	$=$ Absolute temperature °K
$[S][P]$	$=$ Actual concentration of the substrate and the products
a, b, c, d	$=$ Coefficients of P, S etc. in balanced chemical reaction.

At equilibrium $\frac{[P]}{[S]}$ ratio $= K_{eq}$

Hence

$$\Delta G = 0$$
$$\text{At } 25\ ^\circ\text{C},\ 2.303\ RT = 1364$$
$$37\ ^\circ\text{C},\ 2.303\ RT = 1419$$

The ΔG under standard state condition is written as ΔG°

$$\begin{aligned}\Delta G &= -1364 \log K_{eq} + 1363 \log 1 \\ &= -1364 \log K_{eq} + 0 \\ \Delta G^\circ &= -1364 \log K_{eq} \text{ at } 25^\circ \text{ C}\end{aligned}$$

So the actual ΔG under any particular set of concentration condition can be calculated from

$$\Delta G = \Delta G^\circ + 1364 \log \left\{ \frac{[P_1]^a [P_2]^b \ldots}{[S_1]^c [S_2]^d \ \ldots} \right\} \text{ at } 25^\circ\text{C}$$

The ΔG of a reaction indicates only the difference between the free energy contents of the products and the original substrates. This difference must be the same regardless of the path taken or the number of steps involved. A negative ΔG means that the reaction is spontaneous and will proceed from left to right toward a state of minimum energy. The overall ΔG and the equilibrium constant K_{eq} will remain the same for a substrate and product irrespective of its reaction sequence.

Biochemical Calculations

1. Calculate the equilibrium concentration of all components of the isocitritate reaction (isocitritate $\rightleftharpoons$ glyoxylate + succinate) when the concentration of isocitritate is (a) 3 M (b) 0.3 M, (c) 0.03 M (d) 3×10^{-3} M. The $\Delta G'$ for the isocitritate reaction is +2110 cal/mole.

(a) Let

$$\begin{aligned} y &= \text{M glyoxylate concentration} \\ y &= \text{M succinate conc.} \\ 1-y &= \text{M isocitritate that remains} \end{aligned}$$

isocitritate $\rightleftharpoons$ glyoxylate + succinate

Start	1	0	0
Change	$-y$	$+y$	$+y$
Equilibrium	$1-y$	y	y

$$\Delta G' = -1364 \log K'_{\text{eq}}$$

$$2110 = -1364 \log K'_{\text{eq}}$$

$$\log K'_{\text{eq}} = -\frac{2110}{1364} = -1.547$$

$$K'_{\text{eq}} = 0.0284 \text{ (antilog)}$$

$$K'_{\text{eq}} = \frac{y \times y}{3-y} = 0.0284$$

Assuming that y is small compared to 3 M we ignore it in the denominator

$$\begin{aligned} 0.0284 &= \frac{y^2}{3} \\ y^2 &= 0.0852 \\ y &= 0.292 \\ 1-y &= 0.708 \end{aligned}$$

isocitritate = 0.708 M
glyoxylate = 0.292 M
succinate = 0.292 M

(b) Here the initial concentration of isocitritate is 0.3 M. Proceeding as above

$$0.0284 = \frac{y^2}{0.3-y}$$

$$y^2 + 0.0284y - 0.00852 = 0$$

$$y = \frac{-0.0284 \pm \sqrt{0.0284^2 + 4 \times 0.00852}}{2}$$

$$= 0.0793$$

$$0.3 - y = 0.2207$$

isocitritate = 0.2207 M
glyoxylate = 0.0793 M
succinate = 0.0793 M

(c) $$0.0284 - \frac{y^2}{0.03 - y}$$

$$y^2 + 0.0284y - 0.03 \times 0.0284 = 0$$

$$y = \frac{-0.0284 \pm \sqrt{0.0284^2 + 4 \times 0.000852}}{2}$$

$$= 0.0183$$

$$0.03 - y = 0.0117$$

isocitritate = 0.0117 M
glyoxylate = 0.0183 M
succinate = 0.0183 M

(d) $$0.0284 = \frac{y^2}{0.003 - y}$$

$$y^2 + 0.0284y - 0.003 \times 0.0284 = 0$$

$$y = \frac{-0.0284 \pm \sqrt{0.0284^2 + 4 \times 0.0000852}}{2}$$

$$= 0.0027$$

$$0.003 - y = 0.003 - 0.0027 = 0.0003$$

isocitritate = 0.0003 M
glyoxylate = 0.0027 M
succinate = 0.0027 M

2. Arrive at the reduction potential of the flavoprotein/flavoprotein - H_2 half reaction when the oxidized/reduced concentration (activity) ratios are (a) 10^{-3} (b) 0.3 (c) 2 (d) 4 (e) 30 (f) 500, $E_0' = -0.06v$

$$E = E_0' + \frac{0.059}{n} \log \frac{\text{Oxidized form}}{\text{Reduced form}}$$

(a) $$E = -0.06 + \frac{0.059}{2} \log 10^{-3} = -0.06 + 0.0295 \times (-3)$$

$$= -0.1485, \quad E = -0.149v$$

(b) $$E = -0.06 + \frac{0.059}{2} \log 0.3 = -0.06 + 0.0295 \times \bar{1}.4771$$

$$= -0.06 + (0.0295) \times (-0.5229) = -0.06 - 0.01543$$

$$= -0.07543, \quad \therefore \quad E = -0.075v$$

$$\bar{1}.4771 = -1 + .4771$$

$$= -0.5229$$

(c) $E = -0.06 + \frac{0.059}{2} \log 2$

$= -0.06 + 0.0295 \times 0.3010 = 0.00888 - 0.06$

$= -0.05112 = -0.051v$

(d) $E = -0.06 + 0.0295 \times \log 4$

$= -0.06 + 0.0295 \times 0.6021 = -0.06 + 0.01776$

$= -0.04224 = -0.0422v$

(e) $E = -0.06 + 0.0295 \times \log 30$

$= -0.06 + 0.0295 \times 1.4771 = -0.06 + 0.04357$

$= -0.01643 = -0.0164v$

(f) $E = -0.06 + 0.0295 \times \log 500$

$= -0.06 + 0.0295 \times 2.6990 = -0.06 + 0.07962$

$= +0.01962 = +0.0196v$

3. The K'_{eq} for the hydrolysis reaction glucose-6-phosphate + $H_2O \rightleftharpoons$ glucose + P is 199.8. Calculate the $\Delta G'$ for the hydrolysis reaction.

$K'_{eq} = 199.8$

$\Delta G' = -1364 \log 199.8 = -1364 \times 2.3006$

$= -3138.02 = -3138$ cal/mole

4. Calculate the energy in one einstein of photons for light of wavelengths (a) 360 nm and (b) 650 nm.

1 nm $= 10^{-9}$ m $= 10^{-7}$ cm

360 nm $= 360 \times 10^{-7}$ cm $= \lambda$

650 nm $= 650 \times 10^{-7}$ cm $= 6.5 \times 10^{-5}$ cm

Each photon or quantum of light has an energy of $h\nu$

h = Plancks constant $= 6.627 \times 10^{-27}$ erg × sec

$= 1.58 \times 10^{-34}$ cal × sec

and ν = the frequency of light in sec^{-1}

The energy in one einstein (one mole) of photon is given by

$$\xi = Nh\nu = Nh\frac{c}{\lambda} \text{ cal/einstein}$$

where N = Avogadro's number = 6.023×10^{23} photons/einstein

λ = the wave length of light in cm/sec

c = the speed of light in cm/sec

$= 3 \times 10^{10}$ cm/sec if λ is given in cms

$$\xi = \frac{6.023 \times 10^{23} \times 1.58 \times 10^{-34} \times 3 \times 10^{10}}{\lambda}$$

$$= \frac{28.549 \times 10^{-1}}{\lambda} = \frac{2.855}{\lambda} \text{ cal/einstein}$$

(a) Let ξ_1 be the energy corresponding to a wavelength of 360 nm, i.e. 360×10^{-7} cm

$$\xi_1 = \frac{2.855}{\lambda} \text{ cal/einstein} = \frac{2.855}{3.6 \times 10^{-5}}$$

$$= 0.7930555 \times 10^5 = 79306 \text{ cal/einstein}$$

(b) Let ξ_2 be the energy corresponding to a wavelength of 650 nm, i.e. 650×10^{-7} cm

$$\xi_2 = \frac{2.855}{6.5 \times 10^{-5}} = 0.4392307 \times 10^5$$

$$= 43923 \text{ cal/einstein}$$

5. Plasma contains about 0.1 M Cl^-. Kidney tissue contains about 0.04 M Cl^-. Calculate (a) the ΔG for transport of Cl^- from plasma into kidney cells and (b) the energy expended by kidney cells in transporting Cl^- outward against the concentration gradient.

(a) $$\Delta G = 1362 \log \frac{C_2}{C_1} = 1362 \log \frac{0.04}{0.1} = 1362 \log 0.4$$

$$= 1362 \times (\bar{1}.6021) = 1362 \times (-0.3979)$$

$$= -542$$

(b) $$\Delta G = 1362 \log \frac{0.1}{0.04} = 1362(\log 1 - \log 0.4)$$

$$= 1362 \times (0 - \bar{1}.6021) = 1362 \times 0.3979$$

$$= +542$$

6. The K'_{eq} of a deamination reaction at 25°C is 175. The K'_{eq} at 37°C is 85 (a) calculate ΔH and ΔG; $T\Delta S$ and ΔS at 37°C (b) what is K'_{eq} at 32°C?

$$\text{(a) } \log\left(\frac{K'_{eq.37°C}}{K'_{eq.25°C}}\right) = \frac{\Delta H}{2.3R}\frac{(T_2 - T_1)}{T_2T_1}$$

where T_1, T_2 represent absolute temperatures corresponding to 25°C, 37°C i.e. absolute temp 298 and 310 respectively.

$$\Delta H = (2.3R)\left(\log\frac{K'_{eq.37°C}}{K'_{eq.25°C}}\right)\frac{T_2T_1}{T_2 - T_1}$$

$$= 2.3 \times 1.987\left(\log\frac{85}{175}\right)\frac{298 \times 310}{12}$$

$$= 2.3 \times 1.987(-0.3136)\frac{298 \times 310}{12}$$

$$= -11033.12 \text{ cal/mole}$$

$$\Delta G'_{37°C} = -2.303 \text{ RT} \log K'_{eq37°C} = -2.303 \times 1.987 \times 310 \times \log 85$$

$$= -2.303 \times 1.987 \times 310 \times 1.9294 = -2737 \text{ cal/mole}$$

$$T\Delta S = \Delta H - \Delta G'$$

$$= -11033.12 + 2737 = -8296 \text{ cal/mole}$$

$$\Delta S = \frac{-8296}{T} = \frac{-8296}{310} = -26.76$$

$$= -26.8 \text{ e.u. (entropy units)} \quad \{1 \text{ e.u.} = 1 \text{ cal} \times \text{ mole}^{-1} \times \text{degree}^{-1}\}$$

$$\log\frac{K'_{eq.37°C}}{K'_{eq.32°C}} = \frac{-11033}{2.303 \times 1.987} \times \frac{5}{310 \times 305} = -0.1275$$

$$\log K'_{eq.37°C} - \log K'_{eq.32°C} = -0.1275$$

$$\log K'_{eq.32°C} = 0.1275 + \log K'_{eq.37°C}$$

$$= 0.1275 + 1.9294 = 2.0569$$

$$\text{or,} \quad K'_{eq.32°C} = 114$$

$$\therefore \quad K'_{eq.\text{ at }32°C} = 114$$

7. Calculate the number of moles of ATP that could be synthesized by a photosynthetic organism upon absorption of 2 einstein of orange light of 600 nm? (b) How many molecules of ATP could be produced from one photon? (c) What is the overall efficiency of energy conversion if one mole of ATP is formed per three equivalents of electron excited by orange light (i.e. per 3 einsteins of photons)?

(a) $$\xi = \frac{2.855}{\lambda} = \frac{2.855}{6 \times 10^{-5}}$$

$$= 0.47583 \times 10^5$$

$$= 47583$$

$$\text{No. of moles of ATP} = \frac{47583}{7700}$$

$$= 6.17 \approx 6 \text{ moles of ATP/einstein}$$

(The synthesis of ATP under standard conditions require 7700 cal/mole.)

(b) The ratio of photons/einstein is the same as the ratio of molecules/mole. Hence 6 molecules of ATP/photon could be produced.

(c) Assuming that 3 photons must be absorbed in order to excite three electron to a sufficient energy level.

$$\xi \text{ input} = 3 \times 47583$$

$$= 142749 \text{ cal}$$

$$\xi \text{ conserved} = 1 \times 7700 \text{ cal}$$

$$\text{efficiency} = \frac{7700}{3 \times 47583} \times 100$$

$$= 5.4\%$$

8. The velocity of reaction A + B $\longrightarrow$ C + D at 40°C is three times as great as the velocity at 25°C. Calculate the activation energy.

$$\log \frac{K_2}{K_1} = \frac{E_a}{2.3R} \frac{T_2 - T_1}{T_1 T_2}$$

$$E_a = \left(\log \frac{K_2}{K_1}\right) \frac{2.3R \times T_2 T_1}{T_2 - T_1}$$

$$= (\log 3) 2.3 \times 1.987 \times \frac{298 \times 313}{15}$$

$$= 0.4771 \times \frac{2.3 \times 1.987 \times 298 \times 313}{15}$$

$$= 13558.275$$

$$E_a = 13558 \text{ cal/mole}$$

9. Calculate the E_0' values at pH 10.0 for (a) the acetaldehyde/ethanol half reaction and (b) the Metmyoglobin Fe^{+3}/myoglobin Fe^{+2} half reaction. (The E_0' values for pH 7 are given as $-0.163v$ and $+0.046v$ respectively).

(a) $E'_{0\,\text{pH}_{10}} = E'_{0\,\text{pH}_7} - 0.059 \times (10 - 7) = E'_{0\,\text{pH}_7} - 0.059 \times 3$

$$= -0.163 - 0.177 = -0.34 = -0.34v$$

(b) $E'_{0\,\text{pH}_{10}} = E'_{0\,\text{pH}_7}$ (since H^+ is not involved in the reaction)

$$= 0.046v$$

$$E'_{0\,\text{pH}_{10}} = 0.046v$$

4

Enzyme

The term enzyme kinetics implies a study of the speed, rate or velocity of an enzyme catalysed reaction, and of the various factors which may affect this. The most important factors are enzyme concentration, ligand concentration like substrates, products, inhibitors, activators etc., pH, ionic strength, and temperature. The study of enzyme kinetics is a knowledge of the way in which reaction velocity is altered by changes in the concentration of the enzyme's substrate and of the simple mathematics underlying this. The reaction carried out by an enzyme can be represented by the following equation:

$$S \xrightarrow{E} P$$

in which: S is the substrate product and E is the enzyme. In a normal, non-catalysed chemical reaction the velocity of the reaction would be directly proportional to the concentration of substrate. For a non-catalysed reaction, then a graph of velocity against substrate concentration would be a straight line. In an enzyme catalysed reaction the same type of experiment, measuring reaction velocity at various different concentrations, does not give a straight line but a curve.

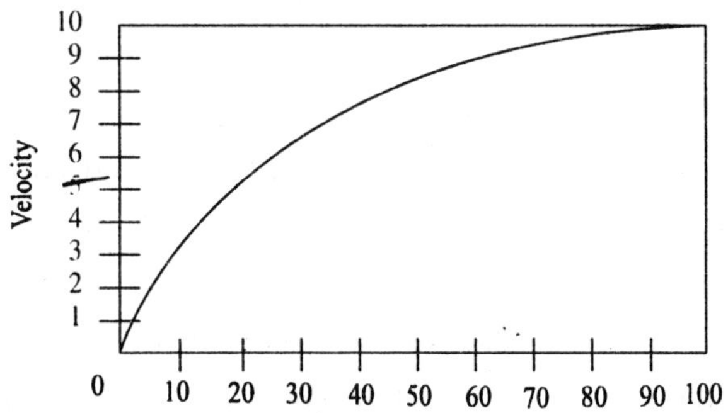

Fig. 4-1

It is described mathematically as a rectangular hyperbola and its equation is slightly more complex. Our knowledge of the equation of the $\frac{V}{[A]}$ curve is based on pioneering work carried out by Michaelis and Menton and by Briggs and Haldane, and is usually called the Michaelis equation after the first of these workers. It can be written in a number of different ways, but probably the commonest form is:

$$\nu = \frac{V_{\max} a}{K_m + a} \qquad \text{or} \qquad \nu = \frac{V_{\max}[s]}{K_m + [s]}$$

The two variables in this equation are the reaction velocity v, and the concentration of substrate, 'a' $V_{\max}$ and K_m are two constants. $V_{\max}$ is the reaction velocity at very high, saturating, concentrations of substrate. $V_{\max}$ is directly proportional to the concentration of enzyme but K_m is independent of enzyme concentration.

Enzymes as biological catalysts

Activation energy: The fact that a reaction has a very large negative free energy change does not mean that it will proceed at a rapid rate. A negative ΔG simply means that the existing $\frac{[P]}{[S]}$ ratio is smaller than the equilibrium ratio. Before substrate S can become a molecule of product P it must possess a certain minimum energy in order to pass into a transition state $S \ldots P$. The amount of energy required is called the activation energy. The transition state represents a halfway point where the bonds of S are distorted sufficiently so that conversion to P becomes possible. The rate of the reaction $S \rightarrow P$ depends on the number of molecules of S that enter the transition state per unit time.

There are two ways of increasing the reaction rate. One is the raise the temperature. The other is to lower the activation energy. At life temperatures, a few or no reactions of intermediary metabolism would occur at a rate sufficient to permit cell growth and maintenance. But living cells can exist under relatively mild environmental conditions. They possess enzymes, the biological catalyst which selectively lower the energies of activation of vital chemical reactions.

A number of factors are responsible to lower the activation energy of an enzyme. Most of the enzyme-catalyzed reactions proceed via recognized organic reaction mechanisms in which the enzyme provides the catalytic groups. In such reactions the rate enhancement by an enzyme stems from proximity and orientation factors. For a substrate molecule and a catalytic group to react they must get close enough to each other and the approach must occur at the proper angle. In solution, the random motion of the two molecules would yield a low probability of an effective collision. When the two molecules are absorbed onto the active site of the enzyme where one of the reactants is the substrate and the other reactant is a R-group of the active site, then both the intermolecular distance and orientation may be optimized. The effective concentration of a substrate in the volume of the active site is much greater than the concentration in the solution from which the substrate was absorbed. Koshland proposed that the active sites of enzymes are so constructed that they align the oribitals of the substrate and catalytic groups optimally to enter the transition state. This concept is called orbitial steering. Another qualitatively similar concept of steropopulation control discussed by Wilstien and Cohen point out to the combined effect of multipoint attachment and the precise fit of the substrate into the active site with unique conformation. The substrates confined to the active site of an enzyme have a relatively long residence time. As a consequence of this substrate anchoring the number of substrate molecules attaining the transition state per unit time may be increased tremendously. Several other workers have suggested that the substrates are distorted upon binding to the enzyme. This is called as rack mechanism which assumes that the substrate fits loosely into the activesite, but the bonds that are formed between the enzyme and the substrate are so strong that a susceptible bond within the substrate is distorted producing the activated transition state. In this mechanism a portion of the activation energy is provided directly by the binding forces between enzyme and substrate.

Table 1: Enzymes active on carbohydrates and sugars

Enzyme	Source	EC number	Mol. wt	Optimum pH	Specificity
β-N-acetyl-D-glucosaminidase	Beef kidney	3.2.1.30	155000	4.5	Broad specificity
N-acetylneuraminic acid aldolase	E. coli	4.1.3.3	–	–	Cleaves N-acetylneuraminate
α-amylase	Bacillus subtilis	3.2.1.1	97000	5.7	α-1, 4-glucosidic links
	Pig pancreas	–	50000	7.0	
β-amylase	Various plants	3.2.1.2	206 000	4.8	α-1, 4-glucosidic links
Amyloglucosidase	Aspergillus niger	3.2.1.3	97 000	4.7	α-1, 4 and α-1,6-glucan links at non-reducing terminus
Cellulase	Various fungi	3.2.1.4	various	4.8	β-1, 4-glucan links in cellulose
Chitinase	Streptomyces antibioticus	3.2.1.14	–	–	α-1, 4-acetamido-2-deoxy-D-glucoside links in chitin and chitosan
Endo-α-N-acetyl Galactosaminidase	Diplococcus pneumoniae	3.2.1.97	–	–	Gal-β1, 3-GalNAc links in glycoproteins
Endo-β-galactosidase	Bacteroides fragilis	3.2.1.103	32 000	5.8	β-1, 4-galactosidic links in GlcNAc-Gal-GlcNAc
Endoglycosidase D	Diplococcus pneumoniae	3.2.1.96	280 000	6.5	Cleaves various high mannose glycans
Endoglycosidase F	Flavobacterium meningosepticum	3.2.1.96	32 200	5.7	Cleaves various high mannose glycans

Table 1: Contd.

Enzyme	Source	EC number	Mol. wt	Optimum pH	Specificity
Endoglycosidase H	Streptomyces sp.	3.2.1.96	29 000	5.5	Cleaves various high mannose glycans
β-fructosidase	S. cerevisiae	3.2.1.26	270 000	4.6	β-D-fructofuranosides
α-fucosidase	Beef kidney	3.2.1.51	217 000	5.0	α-L-fucopyranosides
α-galactosidase	Various	3.2.1.22	329 000	7.2	Terminal α-galactosyl residues from α-D-galactopyranosides
β-galactosidase	Various	3.2.1.23	540 000	7.0	E.coli enzyme hydrolyses β-D-galactopyranosides, beef enzyme removes terminal Gal from Gal-β 1,3-GlcNAc, Gal-β 1, 4-GlcNAc or Gal-β, 3-GalNAc
α-glucosidase	S. cerevisiae	3.2.1.20	68 500	7.2	Broad specificity for α-glucosides
β-glucosidase	Almonds	3.2.1.21	67 000	5.0	Broad specificity for β-D-glucosides
β-glucuronidase	E. coli	3.2.1.31	–	6.2	Broad specificity for β-D-glucosides
Glycopeptidase F	Flavobacterium meningosepticum	3.2.2.18	35 500	7.5	Glycoproteins between Asn and GlcNAc meningosepticum
Hyaluronidase	Mammalian testes	3.2.1.35	55 000	5.2	1,4 links between N-acetylhexosamine and D-glucuronatein hyaluronic acid, chondroitin, etc.
Hyaluronidase	Leech	3.2.1.35	–	–	α-glucuronidase specific for hyaluronic acid

Table 1: Contd.

Enzyme	Source	EC number	Mol. wt	Optimum pH	Specificity
Laminarinase	Spisula solidissima	3.2.1.6	–	–	β-1,3 and β-1,4 links adjacent to α-1,3-link
Lysozyme (muramidase)	Chicken egg white	3.2.1.7	14 400	6.5	β-1,4 links between GlcNAc and N-acetyl-muramic acid in peptidoglycans
α-mannosidase	Jack bean	3.2.1.24	–	4.0	Broad for α-mannosides
Neuraminidase (sialidase)	Arthrobacter ureafaciens	3.2.1.18	90 000	5.2	NeuAc-α2,6-Gal,NeuAc-α2,
	Clostridium perfringenes	3.2.1.18	60 000	5.0	6-GlcNAc or NeuAc-α2,3-Gal
Novozym	Trichoderma harzianum	–	–	–	Multienzyme (cellulase, chitinase, laminarinase, proteinase, xylanase) active against yeast cell walls
Novozym	Aspergillus aculeatus	–	–	–	Multienzyme (pectolytic, cellulolytic, hemicellulolytic, proteolytic, saccharifying) active against plant cell walls
Pullulanase	Aerobacter aerogenes	3.2.1.41	145 000	5.0	α-1,6-glucan links

a = Abbreviations: Endo, endoglycosidase; Exo, exoglycosidase.

b = maltase.

Table 2: Proteases

Enzyme	Source	EC number	Mol. wt	Optimum pH	Specificity
Alkaline protease	Streptomyces griseus	-	-	-	Broad specificity
Aminopeptidase M	Pig kidney	3.4.11.2	280 000	7.2	Cleaves N-terminal residue from di-and tripeptides
Ancrod	Agkistrodon	3.4.21.28	38 000	7.5	Arg-X, especially arg-gly
Bromelain	Pineapple	3.4.22.3	28 000	8.2	Lys-X, ala-X, tyr-X
Carboxypeptidase A	Bovine pancreas	3.4.17.1	34 500	8.0	Cleaves C-terminal residue, preference for aromatic and aliphatic R-groups
Carboxypeptidase B	Pig pancreas	3.4.17.2	34 300	8.0	Cleaves C-terminal arg, lys or ornithine
Carboxypeptidase P	Penicillium janthinellum	3.4.16.1	51 000	3.7	Cleaves C-terminal residue
Carboxypeptidase Y	S. cerevisiae	3.4.16.1	61 000	6.0	Cleaves C-terminal residue, preference for aromatic and aliphatic R-groups
Cathepsin C	Bovine spleen	3.4.14.1	210 000	5.0	Cleaves N-terminal dipeptide, blocked by lys, arg and pro
Chymotrypsin A4 (α Chymotrypsin)	Bovine pancreas	3.4.21.1	25 000	8.0	Tyr-X, phe-X, trp-X; lower specificity for leu-X, met-X, ala-X, asp-X and glu-X
Clostripain (clostridiopeptidase B)	Clostridium histolyticum	2.3.22.8	50 000	7.6	Arg-X

Table 2: Contd.

Enzyme	Source	EC number	Mol. wt	Optimum ph	Specificity
Collagenase	Achromobacter isophagus	3.4.24.8	70 000	7.2	Predominantly X-gly of pro-X-gly-pro
Collagenase	Clostridium histolyticum	3.4.24.3	105 000	7.0	Predominantly X-gly of pro-X-gly-pro
Collagenase Leech	–	–	–	–	Degrades collagen to peptides
Dispase	Bacillus polymyxa	3.4.24.4	35 900	8.5	Non-specific
Elastase	Pig pancreas	3.4.21.36	25 900	8.5	Ala-X, gly-X; lower specificity for val-X, leu-X,ile-X and ser-X
Endoproteinase Arg-C submaxillary protease	Mouse submaxillary gland	3.4.21.40	25 000	8.2	Arg-X (submaxillary protease)
Endoproteinase Glu-C (V8 protease)	Staphylococcus aureus	3.4.21.19	12 000	4.0 & 7.8	Glu-X, lower specificity for asp-X
Endoproteinase Lys-C	Lysobacter enzymogenes	3.4.99.30	33 000	8.6	Lys-X
Factor Xa	Bovine plasma	3.4.21.6	55 000	8.3	Arg-X of gly-arg-X
Ficin	Fig tree	3.4.22.3	23 800	6.4	Bonds involving uncharged and/or aromatic amino acids
Kallikrein	Pig pancreas	3.4.21.8	28 000	7.5	Arg-X, preferentially of phe-arg-X or leu-arg-X
Leucine amino-peptidase	Pig kidney	3.4.11.1	300 000	9.2	Cleaves N-terminal residue

Table 2: Contd.

Enzyme	Source	EC number	Mol. wt	Optimum pH	Specificity
Papain	Carica papaya	3.4.22.2	23 000	6.5	Bonds involving arg, lys, glu, his, gly and/or tyr
Pepsin	Pig stomach	3.4.23.1	34 500	1.0	Broad specificity but preference for phe-X, met-X, leu-X and trp-X where **X** is another hydrophobic residue
Plasmin	Bovine plasma	3.4.21.7	85 000	8.9	Arg-X, lys-X
Pronase	Streptomyces griseus	mixture	–	~ 7.0	Broad spectrum
Proteinase K	Tritirachium album	3.4.21.14	27 000	~ 8.0	Broad spectrum
Pyroglutamate aminopeptidase	Calf liver	3.4.11.8	75 000	8.0	Cleavage of N-terminal pyroglutamate residue
Subtilisin	Bacillus subtilis	3.4.21.14	27 600	~ 8.0	Broad spectrum
Thermolysin	Bacillus thermoproteolyticus	3.4.24.4	35 000	8.0	X-leu, X-phe and other non-polar residues
Thrombin	Bovine plasma	3.4.21.5	34 000	8.6	Arg-X
Trypsin	Bovine pancreas	3.4.21.4	23 500	8.0	Arg-X, lys-X

Table 3: Proteins: physical, chemical and biological data

Protein	Source	Approx. mol. wt	Form	Use
Albumin (BSA)	Bovine serum	67 000	mixed	Used to stabilize enzymes during purification and dilution.
Avidin	Egg white	66 000	tetramer	Binds biotin and biotinylated derivatives, typically 10-15 μg d-biotin per mg protein.
Protein A	Staphylococcus aureus	42 000	single polypeptide	Binds the Fc fragment of IgG, typically 12-14 mg of human IgG per mg protein.
Streptavidin	Streptomyces avidini	58 000	tetramer	Binds biotin and biotinylated derivatives, typically 10-15 μg d-biotin per mg protein; less non-specific binding than egg white avidin.

Table 4: Enzyme substrates: physical, chemical and biological data

Substrate	Mol. wt	Assay[a]
Alkaline phosphatase		
5-bromo-4-chloro-3-indolyl phosphate (X-phos, BCIP)	433.6 (p-toluidine salt)	C
4-methylumbelliferyl phosphate (4-MUP)	300.1 (.2Na)	F
Nitroblue tetrazolium[b]	817.7	C
p-nitrophenylphosphate (PNPP)	371.1 (.2Na)	–
β-galactosidase		
5-bromo-4-chloro-3-indolyl- β-D-galactopyranoside (X-gal)	408.6	C
5-bromo-3-indolyl-β-D-galactopyranoside (Indigal)	374.0	C
Halogenated indolyl-β-D-galactopyranoside (Bluo-gal)[c]	353.3	C

Table 4: Contd.

β-galactosidase		
methylumbelliferyl-α-D-galactopyranoside	338.3	F
o-nitrophenyl-β-D-galactopyranoside (ONPG)	301.2	-
p-nitrophenyl-β-D-galactopyranoside (PNPG)	301.2	S
Isopropyl-β-D-thiogalactopyranoside (IPTG)	238.3	I
β-glucuronidase		
5-bromo-4-chloro-3-indolyl-β-D-glucuronic acid (X-glu)	422.6	H
4-methylumbelliferyl-β-D-glucuronide (MUG)	352.3	F
p-nitrophenyl-β-D-glucuronide (PNPGlu)	315.2	S
resorufin glucuronide	–	H
trifluoromethylumbelliferyl glucuronide	–	H
Peroxidase		
2, 2′-azino-di(3-ethylbenz-thiazoline sulfonic acid) (ABTS)	548.7	C
4-chloro-1-naphthol (4CN)	178.6	C
3′, 3-diaminobenzidine tetrahydrochloride (DAB)	360.1	C
o-phenyldiamine dihydrochloride (OPD)	181.1	C
3, 3′, 5, 5′-tetramethylbenzidine (TMB)	240.4	C

[a]C = colorimetric, F = fluorometric, S = spectrophotometric, I = inducer, H = histochemical.

[b]Nitroblue tetrazolin = 3, 3′-(3, 3′-dimethoxy-4, 4′-biphenylylene) -bis-2-(p-nitrophenyl)-5-phenyl-2H-tetrazolium chloride.

Biochemical Calculations

1. The following table gives the concentration velocity for an enzyme catalyzing a reaction $S \longrightarrow P$. (a) Calculate K_m and $V_{\max}$(b) show that the enzyme obeys hyperbolic saturation kinetics (c) also find the first order rate constant for the enzyme concentration.

[S] M	v n moles × litre^{-1} × min^{-1}
2.2×10^{-6}	25
3.5×10^{-6}	32
4.0×10^{-6}	40
1.0×10^{-5}	60
3.0×10^{-5}	80
4.0×10^{-5}	95
2.0×10^{-3}	108
3.0×10^{-3}	119
1.0×10^{-2}	120

As per the above values the change in v is minimal for changes in $[S]$ above 3×10^{-3} M. Hence in the region of 10^{-3} M to 10^{-2} M, v must be very close to $V_{\max}$.

$$V_{\max} = \text{n moles} \times \text{ litre}^{-1} \times \text{min}^{-1}$$

We calculate K_m by taking any v and the corresponding $[S]$.

$$\frac{v}{V_{\max}} = \frac{[S]}{K_m + [S]}$$

$$\frac{60}{120} = \frac{1 \times 10^{-5}}{K_m + 10^{-5}}$$

$$\frac{1}{2} = \frac{10^{-5}}{K_m + 10^{-5}}$$

$$K_m + 10^{-5} = 2 \times 10^{-5}$$

$$K_m = 10^{-5} \text{ M}$$

Since any set of data gives the same answer, the enzyme obeys the Henie-Michaelis Menton equation.

When $[S] = 1 \times 10^{-5} = K_m$

$v = 60 = 0.5 \times V_{\text{max}}$

When $[S] = 2 \times 10^{-5} = 2K_m$

$$\frac{V}{120} = \frac{2 \times 10^{-5}}{10^{-5} + 2 \times 10^{-5}} = \frac{2}{3}$$

$$V = \frac{2}{3}120 = 80 = \frac{2}{3}V_{\text{max}}$$

$$\therefore \quad V \neq V_{\text{max}}$$

Thus the relationship is hyperbolic and not linear.

We calculate the first order rate constant from the equations

$$v = \frac{V_{\text{max}}}{K_m} \times [S] = K[S]$$

where,

K = the first order rate constant expressed as units of min^{-1} while v is expressed in moles $\times$ $\text{litre}^{-1} \times \text{min}^{-1}$ and K_m is expressed in moles $\times$ litre.

V_{max} = 120 m moles $\times$ $\text{litre}^{-1} \times \text{min}^{-1}$

= 120×10^{-9} moles $\times$ $\text{litre}^{-1} \times \text{min}^{-1}$

K_m = 10^{-5} M

$$K = \frac{V_{\text{max}}}{K_m} \text{ gives}$$

$$K = \frac{120 \times 10^{-9}}{10^{-5}} = 120 \times 10^{-4} = 0.012 \text{ min}^{-1}$$

$$K = 0.012 \text{ min}^{-1}$$

2. An enzyme with a K_m of 2.5×10^{-4} M was assayed at the following substrate concentrations (a) 4×10^{-7} M (b) 5.4×10^{-5} M (c) 10^{-4} M (d) 3×10^{-3} M (e) 0.08 M. The velocity observed at 0.08 M was 132 n moles $\times$ min^{-1}. Calculate the initial velocities at the other substrate concentrations.

$$v_5 = 132 = \frac{(0.08) \times V_{\text{max}}}{2.5 \times 10^{-4} + 0.08}$$

$$= \frac{0.08 \times 10^4}{802.5} \times V_{\text{max}} \qquad \therefore \quad V_{\text{max}} = \frac{802.5 \times 132}{800} = 132.4$$

$$v_4 = \frac{3 \times 10^{-3} \times V_{\max}}{2.5 \times 10^{-4} + 3 \times 10^{-3}}$$

$$= \frac{3}{3.25} \times 132.4$$

$$= 121.8$$

$$v_3 = \frac{10^{-4}}{2.5 \times 10^{-4} + 10^{-4}} \times V_{\max}$$

$$= \frac{1}{2.5 + 1} \times 132.4$$

$$= \frac{1}{3.5} \times 132.4 = 37.83$$

$$v_2 = \frac{5.4 \times 10^{-5}}{2.5 \times 10^{-4} + 5.4 \times 10^{-5}} V_{\max} = \frac{5.4}{30.4} \times 132.4 = 23.52$$

$$v_1 = \frac{4 \times 10^{-7}}{2.5 \times 10^{-4} + 4 \times 10^{-7}} V_{\max} = \frac{4}{2504} \times 132.4 = 0.212$$

The velocity values are

$v_1 = 0.212$ n moles $\times$ litre^{-1} $\times$ min^{-1}

$v_2 = 23.52$ n moles $\times$ litre^{-1} $\times$ min^{-1}

$v_3 = 37.83$ n moles $\times$ litre^{-1} $\times$ min^{-1}

$v_4 = 121.8$ n moles $\times$ litre^{-1} $\times$ min^{-1}

$v_5 = 132.4$ n moles $\times$ litre^{-1} $\times$ min^{-1}

3. Find v_i and the degree of inhibition caused by a competitive inhibitor under the following conditions (a) [S] $= 3 \times 10^{-3}$ M and $I = 4 \times 10^{-3}$ M (b) $[S] = 2 \times 10^{-4}$ M and $I = 3 \times 10^{-3}$ M (c) $[S] = 5.5 \times 10^{-3}$ M and $I = 2 \times 10^{-5}$ M. Assume that K$_m = 3 \times 10^{-3}$ M, K$_i = 2 \times 10^{-4}$ M and $V_{\max} = 250$ nmoles$\times$ litre^{-1}$\times$ min^{-1}

$$v_i = \frac{[S]V_{\max}}{K_m\left(1 + \frac{[1]}{K_i}\right) + [S]}$$

$$v_0 = \frac{[S]V_{\max}}{K_m + [S]}.$$

$$\text{(a)}\quad v_i = \frac{3\times10^{-3}\times250}{3\times10^{-3}\left(1+\frac{4\times10^{-3}}{2\times10^{-4}}\right)+3\times10^{-3}}$$

$$= \frac{250}{(1+20)+1} = \frac{250}{22} = 11.4$$

$$v_0 = \frac{3\times10^{-3}\times250}{3\times10^{-3}+3\times10^{-3}} = \frac{250}{2} = 125 \text{ n moles.}$$

$$a = \frac{v_i}{v_0} = \frac{11.4}{125} = 0.0912 = 0.09 = 9\%$$

$$\therefore\quad i = 1-0.09 = 0.91 \longrightarrow 91\%$$

$v_i \doteqdot 11.4$ n moles $\times$ litre$^{-1}\times$ min^{-1}, Degree of inhibition $= 91\%$

$$\text{(b)}\quad v_i = \frac{2\times10^{-4}\times250}{3\times10^{-3}\left(1+\frac{3\times10^{-3}}{2\times10^{-4}}\right)+2\times10^{-4}}$$

$$= \frac{0.2\times250}{3(1+15)+0.2} = \frac{50}{48.2} = 1.04$$

$$v_0 = \frac{2\times10^{-4}\times250}{3\times10^{-3}+2\times10^{-4}} = \frac{500}{32} = 15.63$$

$$a = \frac{v_i}{v_0} = \frac{1.04}{15.6} = 0.066 = 0.07$$

$$i = 1-0.07 = 0.93 \longrightarrow 93\%$$

$v_i = 1.04$ n moles/lit^{-1}/min^{-1}, degree of inhibition $= 93\%$

$$\text{(c)}\quad v_i = \frac{5.5\times10^{-3}\times250}{3\times10^{-3}\left(1+\frac{2\times10^{-3}}{2\times10^{-4}}\right)+5.5\times10^{-3}}$$

$$= \frac{5.5\times250}{3\times11+5.5} = \frac{1375}{38.5} = 35.7$$

$$v_0 = \frac{5.5\times10^{-3}\times250}{3\times10^{-3}+5.5\times10^{-3}} = \frac{1375}{8.5} = 161.8$$

$$a = \frac{v_i}{v_0} = \frac{35.7}{161.8} = 0.22$$

$$i = 1-a = 1-0.22 = 0.78 \rightarrow 78\%$$

$v_i = 35.7$ n moles $\times$ litre$^{-1}\times$ min^{-1}

degree of inhibition $= 78\%$.

4. What is the K_i for a non-competitive inhibitor if 3×10^{-4} n[I] yields 70% inhibition of enzyme catalyzed reaction?

For a non-competitive inhibition: $i = \dfrac{[I]}{K_i + [I]}$

$$i = 70\% = 0.7$$

$$0.7 = \frac{3 \times 10^{-4}}{K_i + 3 \times 10^{-4}}$$

$$K_i = \frac{3 \times 10^{-4}(1 - 0.7)}{0.7} = \frac{3}{7} \times 3 \times 10^{-4}$$

$$= 1.286 \times 10^{-4} \qquad K_i = 1.29 \times 10^{-4} M$$

5. Find (a) the velocity and (b) the degree of inhibition of an enzyme-catalyzed reaction in the presence of 5×10^{-4} M substrate ($K_m = 2 \times 10^{-3}$M) and 3×10^{-4} M non competitive inhibitor ($K_i = 3 \times 10^{-5} M$). $V_{\max}$ 475 n moles × litre^{-1}× min^{-1}.

$$v_i = \frac{[S]V_{\max}}{([S] + K_m)\left(1 + \frac{[I]}{K_i}\right)}$$

$$= \frac{5 \times 10^{-4} \times 475}{(5 \times 10^{-4} + 2 \times 10^{-3})\left(1 + \frac{3\times10^{-4}}{3\times10^{-5}}\right)}$$

$$v_i = \frac{5 \times 475}{25 \times 11} = 8.64$$

$v_i = 8.64$ n moles × litre^{-1}× min^{-1}

degree of inhibition $i = 1 - a$, where $a = \dfrac{v_i}{v_0}$

$$v_0 = \frac{[S]V_{\max}}{K_m + [S]} = \frac{5 \times 10^{-4} \times 475}{2 \times 10^{-3} + 5 \times 10^{-4}}$$

$$= \frac{5 \times 475}{25} = 95$$

$$a = \frac{v_i}{v_0} = \frac{8.64}{95}$$

$$= 0.0909 = 0.091$$

$i = 1 - a = 0.909 = 0.91 \qquad i = 91\%$

6. How will you relate between the substrate concentration required in the presence of a competitive inhibitor $[S]_i$, to the substrate concentration required in the absence of inhibitor $[S]_0$ in order to observe a given velocity?

For a given velocity v_i

$$v_i = \frac{[S]_0 V_{\max}}{K_m + [S]_0} \quad \text{in the absence of inhibitor}$$

$$= \frac{[S]_i V_{\max}}{[S]_i + K_m\left(1 + \frac{[I]}{K_i}\right)} \quad \text{in the presence of a competitive inhibitor}$$

Hence,

$$[S]_0\left\{[S]_i + K_m\left(I + \frac{[I]}{Ki}\right)\right\} = [S]_i\{K_m + [S]_0\}$$

$$[S]_0[S]_i + [S]_0 K_m + [S]_0\frac{K_m[I]}{K_i} = [S]_i K_m + [S]_i[S]_0$$

$$[S]_0\left\{K_m + \frac{[I]K_m}{K_i}\right\} = [S]_i K_m$$

$$[S]_i = \left\{1 + \frac{[I]}{Ki}\right\}[S]_0$$

7. The effect of temperature on the rate of reaction is expressed as Q_{10} for a temperature difference of 10°C. $V_{\max}$ at 24°C and 37°C was 150 nmoles × litre × min^{-1} and 450 n moles × litre^{-1} × min^{-1} respectively. Calculate the activation energy and (b) Q_{10} value between 25°C and 35°C.

(a) T_1 = 24°C = 297 K, T_2 = 37°C = 310 K, R = 1.98

$$\log\frac{V_{\max 2}}{V_{\max 1}} = \frac{E_a}{2.3 \times R} \times \frac{T_2 - T_1}{T_2 T_1}$$

$$\log\frac{450}{150} = \frac{E_a}{2.3 \times 1.98} \times \frac{(310 - 297)}{310 \times 297}$$

$$E_a = \frac{(\log 3)(2.3)(1.98)(310)(297)}{13}$$

$$= \frac{0.4771 \times 2.3 \times 1.98 \times 310 \times 297}{13}$$

$$= 15387.8 \text{ cal. mole}$$

(b) For $T_1 = 25°C = 298$ K, $T_2 = 35°C = 308$ K

$$\log \frac{V_{\max_2}}{V_{\max_1}} = \frac{15382.8}{2.3 \times R} \frac{(308 - 298)}{308 \times 298} = 0.3681$$

$$\frac{V_{\max_2}}{V_{\max_1}} = \text{antilog } 0.3681 = 2.334$$

$$Q_{10} = 2.334$$

> **Note:** Q_{10} is the factor by which rate of the reaction increases for rise 10°C in temperature.

8. The specific activity of an enzyme is 125 units/mg protein (a) find the turn over number if mol. wt = 360000 (b) find the time required for one catalytic cycle.

$$\begin{aligned}
\text{Specific activity S.A.} &= 125 \text{ units/mg protein} \\
&= 125 \times 10^3 \text{ units/g protein} \\
&= 1.25 \times 10^5 \text{ units/g protein} \\
&= 1.25 \times 10^5 \text{ units/g protein} \times 3.6 \times 10^5 \text{ g/mole} \\
&= 4.5 \times 10^{10} \text{ units/mole enzyme}^{-1}
\end{aligned}$$

$$\begin{aligned}
\text{Turnover number} &= 4.5 \times 10^{10} \mu \text{ moles} \times \text{min}^{-1} \\
&\quad \times \text{mole enzyme}^{-1} \times 10^{-6} \text{ moles} \times \mu \text{ mole}^{-1}
\end{aligned}$$

(a) Turnover number $= 4.5 \times 10^4 \text{ min}^{-1}$

(b) The time required for catalytic cycle $= \dfrac{1}{4.5 \times 10^4}$

$$= 0.22 \times 10^{-4} = 2.2 \times 10^{-5}$$

The time required $= 2.2 \times 10^{-5}$ min

9. (a) What is the n_{app} value of an enzyme if $\dfrac{[S]_{0.9}}{[S]_{0.1}}$ is 7.5?

n is the number of substrate binding site per molecule of enzymes. Since the calculated value of n will be less the actual no. of sites, it is expressed as an approximate value, i.e. n_{app}.

(b) Calculate the $\frac{[S]_{0.9}}{[S]_{0.1}}$ ratio of an allosteric enzyme where $n_{\text{app}} = 3.5$

(a) $\frac{[S]_{0.9}}{[S]_{0.1}} = \sqrt[n]{81}$

$$n_{\text{app}} = \frac{\log 81}{\log \frac{[S]_{0.9}}{[S]_{0.1}}} = \frac{\log 3^4}{\log 7.5} = \frac{1.9085}{0.8751}$$

$$= 2.181 = 2.18$$

$$n_{\text{app}} = 2.18$$

(b) If $n_{\text{app}} = 3.5 = \frac{\log 81}{\log \frac{[S]_{0.9}}{[S]_{0.1}}}$

$$\log \frac{[S]_{0.9}}{[S]_{0.1}} = \frac{1}{3.5} \log 81 = \frac{1.9085}{3.5} = 0.55$$

$$\frac{[S]_{0.9}}{[S]_{0.1}} = 3.55 \text{ (antilog of 0.55)}$$

10. Calculate the value of L for an allosteric dimer (n = 2) if $\frac{V}{V_{max}} = 0.45$ when $\frac{[S]}{K_S} = 6$. Assume C = 0. The equilibrium constant for the transition R_0 to T_0 is represented as L.

$$\frac{v}{V_{\max}} = \frac{\frac{[S]}{K_S}\left(1 + \frac{[S]}{K_S}\right)}{L + \left(1 + \frac{[S]}{K_S}\right)^2}$$

where, $K_s = E + S \rightleftharpoons ES$

$$0.45 = \frac{6(1+6)}{L + (1+6)^2} = \frac{42}{L + 49}$$

$$L = \frac{42 - 0.45 \times 49}{0.45} = \frac{19.95}{0.45} = 44.3$$

5

Spectrophotometry

One of the most valuable analytical tools available to biochemists is spectrophotometry. Unknown compounds may be identified by their characteristic absorption spectra in the ultraviolet, visible, or infrared. Many molecules absorb ultraviolet or visible light. The absorbance of a solution increases as attenuation of the beam increases. Absorbance is directly proportional to the path length, b, and the concentration, c, of the absorbing species. Beer's Law states that

$$A = \varepsilon bc$$

where ε is a constant of proportionality, called the absorbtivity.

Different molecules absorb radiation of different wavelengths. An absorption spectrum will show a number of absorption bands corresponding to structural groups within the molecule. For example, the absorption that is observed in the UV region for the carbonyl group in acetone is of the same wavelength as the absorption from the carbonyl group in diethyl ketone.

5.1 Electronic Transitions

The absorption of UV or visible radiation corresponds to the excitation of outer electrons. There are three types of electronic transition which can be considered:

1. Transitions involving π, σ, and n electrons.
2. Transitions involving charge-transfer electrons.
3. Transitions involving d and f electrons.

When an atom or molecule absorbs energy, electrons are promoted from their ground state to an excited state. In a molecule, the atoms can rotate and vibrate with respect to each other. These vibrations and rotations also have discrete energy levels, which can be considered as being packed on top of each electronic level.

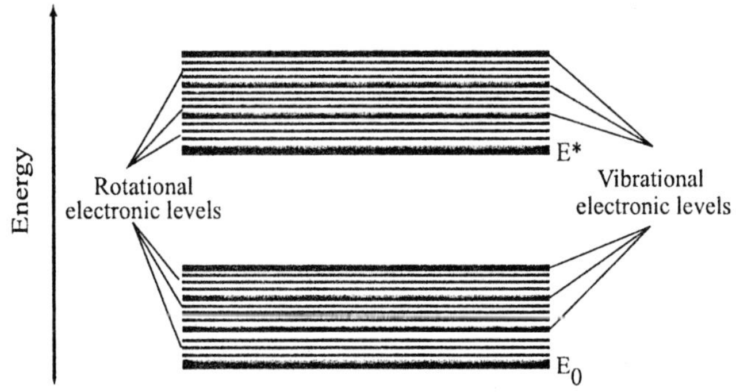

Fig. 5-1

5.2 Absorbing Species Containing π, σ, and n Electrons

Absorption of ultraviolet and visible radiation in organic molecules is restricted to certain functional groups (chromophores) that contain valence electrons of low excitation energy. The spectrum of a molecule containing these chromophores is complex. This is because the superposition of rotational and vibrational transitions on the electronic transitions gives a combination of overlapping lines. This appears as a continuous absorption band.

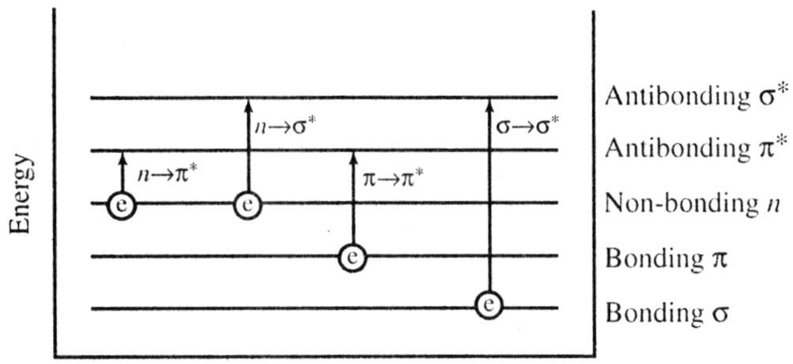

Fig. 5-2

Possible electronic transitions of π, σ, and n electrons are

5.2.1 $\sigma \longrightarrow \sigma^*$ transitions

An electron in a bonding σ orbital is excited to the corresponding antibonding orbital. The energy required is large. Absorption maxima due to $\sigma \longrightarrow \sigma^*$ transitions are not seen in typical UV–Vis. spectra (200–700 nm).

5.2.2 $n \longrightarrow \sigma^*$ transitions

Saturated compounds containing atoms with lone pairs (non-bonding electrons) are capable of $n \longrightarrow \sigma^*$ transitions. These transitions usually need less energy than $\sigma \longrightarrow \sigma^*$ transitions. They can be initiated by light whose wavelength is in the range 150–250 nm. The number of organic functional groups with $n \longrightarrow \sigma^*$ peaks in the UV region is small.

5.2.3 $n \longrightarrow \pi^*$ and $\pi \longrightarrow \pi^*$ transitions

Most absorption spectroscopy of organic compounds is based on transitions of n or p electrons to the π^* excited state. This is because the absorption peaks for these transitions fall in an experimentally convenient region of the spectrum (200–700 nm). These transitions need an unsaturated group in the molecule to provide the π electrons.

Molar absorbtivities from $n \longrightarrow \pi^*$ transitions are relatively low, and range from 10 to 100 L mol^{-1} cm^{-1}. $\pi \longrightarrow \pi^*$ transitions normally give molar absorbtivities between 1000 and 10,000 L mol^{-1} cm^{-1}. The solvent in which the absorbing species is dissolved also has an effect on the spectrum of the species. Peaks resulting from $n \longrightarrow \pi^*$ transitions are shifted to shorter wavelengths (blue shift) with increasing solvent polarity. This arises from increased solvation of the lone pair, which lowers the energy of the n orbital. Often the reverse (i. e. red shift) is seen for $\pi \longrightarrow \pi^*$ transitions. This is caused by attractive polarization forces between the solvent and the absorber, which lower the energy levels of both the excited and unexcited states. This effect is greater for the excited state, and so the energy difference between the excited and unexcited states is slightly reduced-resulting in a small red shift. This effect also influences $n \longrightarrow \pi^*$ transitions but is overshadowed by the blue shift resulting from solvation of lone pairs.

5.3 Charge-Transfer Absorption

Many inorganic species show charge-transfer absorption and are called charge-transfer complexes. For a complex to demonstrate charge-transfer behaviour, one of its components must have electron donating properties and another component must be able to accept electrons. Absorption of radiation then involves the transfer of an electron from the donor to an orbital associated with the acceptor.

Molar absorbtivities from charge-transfer absorption are large (greater that 10,000 L mol^{-1} cm^{-1}).

Why do solutes absorb photic (light) energy within characteristic wavelength bands?

Atom and molecules absorb light energy in such a manner that an alteration in atomic or molecular energy state occurs. In light absorption changes can occur in atomic or molecular energy states in one of three ways:

1. by the boosting of electrons to higher energy levels;
2. by increasing the rate at which atoms within a molecule vibrate relative to one another;
3. by increasing the rotational speed of molecules in the solution.

The physical phenomena underlying light absorption in various regions of the electromagnetic spectrum are:

Region	Wavelength	Effect on molecule
X-rays	1–100 nm	Subvalence electrons excited to higher energy
UV	100–400 nm	Valence electrons excited to higher energy
Visible	400–800 nm	Valence electrons
Infrared	800 nm–100 μm	Molecular vibaration
Microwave	100 μm–30 cm	Molecular rotation

Unit

Micrometer $= 10^{-6}$ micron
Nanometer $= 10^{-9}$ millimicron
Angstrom $=10^{-10}$

Spectrophotometric techniques are used to measure the concentration of solutes in solution. To do this measure the amount of light that is absorbed by the solutes in solution in a cuvette in the spectrophotometer. Spectrophotometry takes advantage of the dual nature of light. Namely, light has:

1. A particle nature which gives rise to the photoelectric effect (used in the spectrophotometer).
2. A wave nature which gives rise to the visible spectrum of light.

A spectrophotometer measures the intensity of a light beam after it is directed through and emerges from a solution. As an example, let's look at how a solution of copper sulfate

($CuSO_4$) absorbs light. The red part of the spectrum has been almost completely absorbed by $CuSO_4$ and blue light has been transmitted. Thus, $CuSO_4$ absorbs little blue light and therefore appears blue. In spectrophotometry, greater sensitivity is obtained by directing red light through the solution because $CuSO_4$ absorbs strongest at the red end of the visible spectrum.

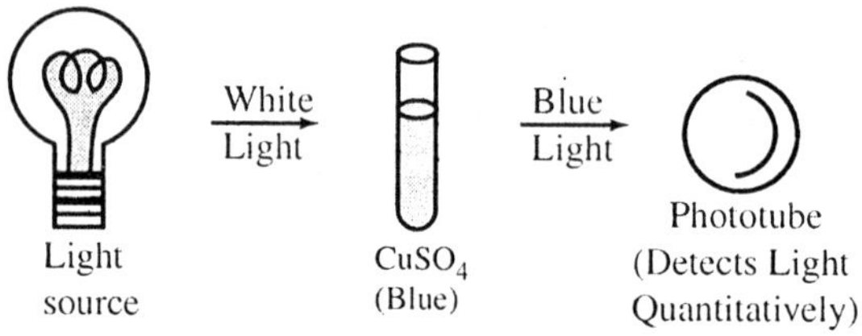

Fig. 5-3

5.4 Spectrum of Visible Light

In a spectrophotometer, a light source gives off white light which strikes a prism, separating the light into its component wavelengths: Now, isolate the red wavelengths and pass them through the $CuSO_4$ solution and measure the amount of red light absorbed. The spectrophotometer will actually measure the amount of light transmitted by the $CuSO_4$ solution. The important point to note here is that, colored compounds absorb light differently depending on the I of incident light.

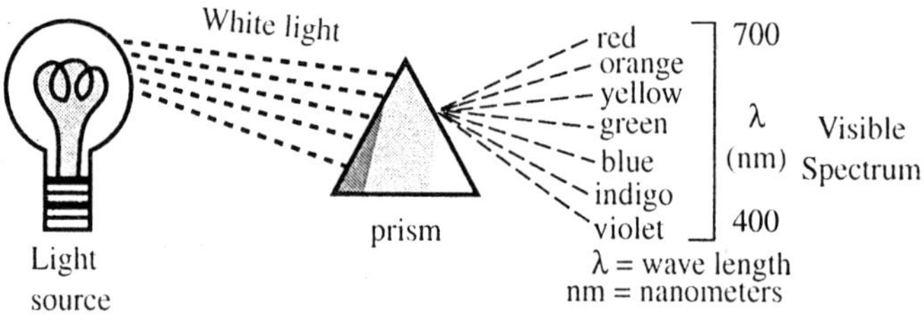

Fig. 5-4

5.5 Design of a Spectrophotometer

I_0 is the incident light and represents 100% of the light striking the cuvette. I is the transmitted light. This is the light which has not been absorbed by the solution in the cuvette and will strike the phototube. The photons of light which do strike the phototube will be converted into electrical energy. This current which has been produced is very small and must be amplified before it can be efficiently detected by the galvanometer. The deflection of the needle on the galvanometer is proportional to the amount of light which originally struck the phototube and is thus an accurate measurement of the amount of light which has passed through (been transmitted by) the sample.

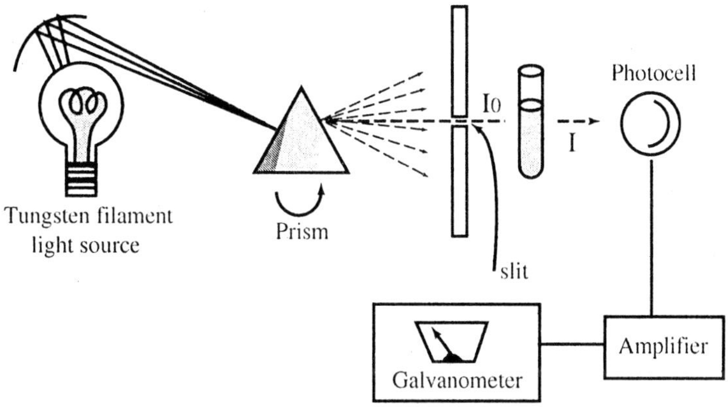

Fig. 5-5

5.6 The Blank

In order to effectively use a spectrophotometer first zero the machine, using "the blank." The blank contains everything except the compound of interest which absorbs light. Thus, by zeroing the machine using "the blank", any measured absorbance is due to the presence of the solute of interest.

5.7 Absorption Spectrum

Different compounds having dissimilar atomic and molecular interactions have characteristic absorption phenomena and absorption spectra which differ. The point (wavelength) at which any given solute exhibits maximum absorption of light (the peaks on the curves on the figure below) is defined as lamda max of that particular compound.

Colorimetry is based on spectrophotometry. Spectral measurements of reflectance, radiance factor and transmittance made at suitable intervals at visible wavelengths can be converted into colorimetric data using appropriate mathematical function that have been defined by the International Lighting Commission (CIE).

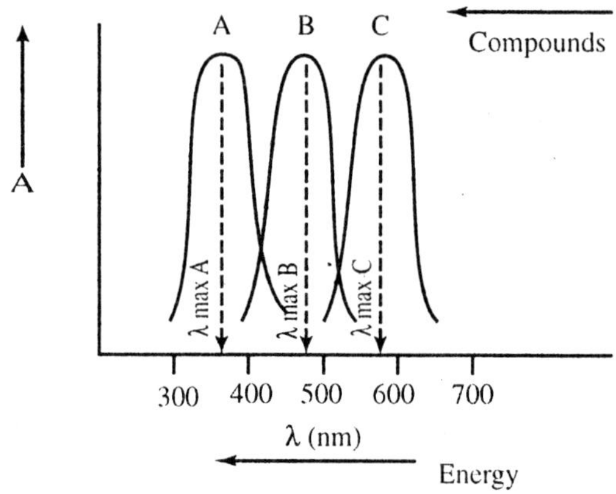

Fig. 5-6

Biochemical Calculations

1. Calculate the absorbance and the transmission at 260 nm and 340 nm of the following solutions in a 1 cm cuvette (a) 2.2×10^{-5} M NADH (b) 7×10^{-6} M NADH plus 4.2×10^{-5} M ATP. The a_m of NADH is 15,000 $M^{-1} \times cm^{-1}$ at 260 nm and 6220 $M^{-1} \times cm^{-1}$ at 340 nm. The a_m of ATP is 15,400 $M^{-1} \times cm^{-1}$ at 260 nm and zero at 340 nm.

At 260 nm

(a) $$A = \log \frac{I_0}{I} = a_m \times c \times \ell$$

$$= 15 \times 10^3 \times 2.2 \times 10^{-5} \times \ell$$

$$= 33 \times 10^{-2}$$

$$= 0.33$$

$$A = 0.33$$

ℓ = 1 c (1 cm cuvette)

I = transmission

$c = 2.2 \times 10^{-5}$

$$\log \frac{I_0}{I} = 0.33$$

Taking $I_0 = 100\%$ transmission

$$\log \frac{100}{I} = 0.33$$

$$\log 100 - \log(I) = 0.33$$

$$\log I = 2 - 0.33 = 1.67$$

$$I = \text{antilog } 1.67$$

$$= 0.468$$

$$I = 46.8\%$$

At 340 nm

$$\text{Absorbance } A = \log\left(\frac{I_0}{I}\right) = 6220 \times 2.2 \times 10^{-5} \times 1$$

$$= 0.13684 = 0.137$$

$$A = 0.137$$

$$\log\left(\frac{I_0}{I}\right) = 0.137$$

Taking $I_0 = 100$

$$\log\frac{100}{I} = 0.137$$

$$\log 100 - \log I = 0.137$$

$$\log I = 2 - 0.137 = 1.863$$

$$I = 0.73; \qquad I = 73\%$$

At 260 nm

$$A_1 = a_m \times CA \times \ell$$

$$= 15000 \times 7 \times 10^{-6} + 15400 \times 4.2 \times 10^{-5}$$

$$= 105 \times 10^{-3} + 646.8 \times 10^{-3}$$

$$= 751.8 \times 10^{-3} = 0.752$$

$$A = 0.752$$

$$\log\frac{100}{I} = 0.752 \qquad \therefore \quad \log I = \log 100 - 0.752$$

$$= 2 - 0.752$$

$$= 1.248$$

$$\therefore \qquad I = 0.177; \qquad I = 17.7\%$$

At 340 nm

$$A_2 = 6220 \times 7 \times 10^{-6} + 0 = 43540 \times 10^{-6} \qquad (a_m \text{ for ATP} = 0)$$

$$= 0.044$$

$$A = 0.044$$

$$\log \frac{100}{I} = 0.044 \qquad \therefore \quad \log I = \log 100 - 0.044$$
$$= 2 - 0.044$$
$$= 1.956$$
$$\therefore \quad I = 0.904; \quad I = 90.4\%$$

2. Find the concentrations of ADP and NADH in solutions with absorbances (in a 1 cm cuvette) of (a) 0.15 at 340 nm and 0.90 at 260 nm (b) zero at 340 nm and 0.75 at 260 nm (c) 0.22 at 340 nm and 0.531 at 260 nm. The a_m values of NADH at 340 nm and 260 nm are 6220 and 15000 $M^{-1} \times cm^{-1}$ respectively and the a_m values of ADP at these wave lengths are zero and 15,400 $M^{-1} \times cm^{-1}$ respectively.

(a) Let C_1 and C_2 denote the concentrations of NADH and ADP.

Then

$$0.15 = 6220 \times C_1 + 0 \qquad (1)$$

$$0.90 = 15 \times 15000 \times C_1 + 15,400 \times C_2 \qquad (2)$$

from (1) $C_1 = \dfrac{0.15}{6220} = 0.0000241 = 2.41 \times 10^{-5}$

Using this in (2)

$$C_2 = \frac{0.90 - 15 \times 10^3 \times \dfrac{0.15}{6220}}{15400}$$

$$= \frac{0.90 - 0.3617}{15,400} = 0.0000349$$

$$= 3.5 \times 10^{-5}$$

$C_1 = [NADH] = 2.41 \times 10^{-5}$ M
$C_2 = [ADP] = 3.5 \times 10^{-5}$ M

(b) Absorbance $= 0 = 6220 \times C_1 + 0 \qquad (1)$

$\therefore \quad C_1 = 0$

$$0.75 = 15 \times 10^3 \times C_1 + 15,400 \times C_2 \qquad (2)$$

By (1) $C_1 = 0$

$$\therefore \quad 0.75 = 15,400\ C_2$$

$$\text{or,} \quad C_2 = \frac{0.75}{15,400} = 0.0000487$$

$$= 4.87 \times 10^{-5}$$

$$C_1 = [\text{NADH}] = 0$$

$$C_2 = [\text{ADP}] = 4.87 \times 10^{-5}\ \text{M}$$

(c) $0.22 = 6220C_1 + 0$

$$\therefore \quad C_1 = \frac{0.22}{6220} = 0.00003536$$

$$= 0.0000354 = 3.54 \times 10^{-5}\ \text{M}$$

$$0.531 = 15 \times 10^3 \times C_1 + 15,400 \times C_2$$

$$\therefore \quad C_2 = \frac{0.531 - \dfrac{15 \times 10^3 \times 0.22}{6220}}{15,400}$$

$$= \frac{0.531 - 0.531}{15.400} = 0$$

$$\therefore \quad C_1 = 3.54 \times 10^{-5}\ \text{M and } C_2 = 0$$

3. A solution of D-Tryptophan (4 g/100 ml of 1 M HCl) had an observed rotation of $-0.41°$ in a 20 cm polarimeter tube. Calculate (a) the specific rotation $[\alpha]$ and (b) the molar rotation $[\alpha]_M$ of D-Tryptophan in 1 M HCl.

(a) $[\alpha] = \dfrac{\text{Å}}{l_{\text{dm}} \times C_{\text{g/ml}}}$

$= 20$ cm

$= 2$ dm

$l = 20 \text{ cm} = 2 \text{ dm}$

$C = 4 \text{ g/100 ml}$

$\doteq 0.04 \text{ g/ml}$

$$\therefore \quad [\alpha] = \frac{-0.41°}{2 \times .04} = -5.125 \qquad [\alpha] = -5.125°$$

(b) $[\alpha]_M = [\alpha]\text{mol. wt}$

$= -5.125 \times 155.2 \qquad [\alpha]_M = -795.4°$

$= -795.4$

4. Find the carotene and vitamin A content of 20 gms of saponified butter; the nonsaponifiable fraction was extracted into 30 ml of chloroform. The absorbance of the chloroform solution in a 1 cm cuvette was 0.53 at 328 nm and 0.48 at 458 nm. The extinction coefficients for carotene and vitamin A at the above two wavelengths are given below.

	$\mathbf{a_{1\%}^{1\ cm}}$ **in** $\mathbf{CHCl_3}$	
Compound	328 nm	458 nm
Carotene	340	2200
Vitamin A	1550	0

Since vitamin A has no absorption at 458 nm, the concentration of carotene can be calculated from the absorbance at 458 nm.

$$A = a_{1\%} \times C_{g/100\ ml} \times l$$

$$0.48 = 200 \times\ Cg/100\ ml \times l$$

$$\text{Carotene} = \frac{0.48}{2200} = \frac{4.8 \times 10^{-1}}{2.2 \times 10^{3}}$$

$$= 2.18 \times 10^{-4}\ g/100\ ml$$

The total carotene content of the chloroform extract (30 ml) is equal to

$$\frac{2.18 \times 10^{-4}\ g/100\ ml}{100/30} = 2.18 \times 3 \times 10^{-5}$$

$$= 6.54 \times 10^{-5}$$

$$= 0.0654 \times 10^{-3}\ g$$

$$= 0.0654\ mg$$

$$\text{The carotene content of butter is} = \frac{0.0654\ mg}{20\ gm}$$

$$= 3.27 \times 10^{-3}\ \text{mg carotene/g butter}$$

$$= 3.27\ \mu\ \text{g carotene/g butter}$$

To obtain the vitamin A content of the butter we use the absorbance of the chloroform solution at 328 nm after correction of the absorbance for the carotene present.

$$A_{328}(\text{carotene}) = 340 \times \text{C car} \times L$$

$$= 340 \times 2.18 \times 10^{-4} \times 1$$

$$= 741.2 \times 10^{-4} = 0.0741$$

$$\text{Total } A_{328} = 0.530 \quad (\text{given})$$

$$A_{328}\ (\text{carotene}) = 0.074$$

$$\therefore \quad 0.456 = 1550 \times C_{\text{vita.A}} \times 1$$

$$\therefore \quad C_{\text{vit.A}} = \frac{0.456}{1550} = 2.94 \times 10^{-4}\ \text{g/100 ml}$$

$$\text{Total } A_{328} = 0.530 \quad (\text{given})$$

$$A_{328}\ (\text{carotene}) = 0.074$$

$$\therefore \quad A_{328}\ (\text{vit. A}) = 0.456$$

$$\therefore \quad 0.456 = 1550 \times C_{\text{vita.A}} \times 1$$

$$\therefore \quad C_{\text{vit.A}} = \frac{0.456}{1550} = 2.94 \times 10^{-4}\ \text{g/100 ml.}$$

$$\text{The vitamin A content of the chloroform extract} = \frac{2.94 \times 10^{-4}\ \text{g/ml}}{100/30}$$

$$= 8.82 \times 10^{-5} = 0.0882 \times 10^{-3}\ \text{g}$$

$$= 0.0882\ \text{mg}$$

$$\therefore \quad \text{The vitamin A content of the butter} = \frac{0.0882\ \text{mg}}{20\ \text{mg}} = \frac{88.2\mu\text{g}}{20\ \text{gm}}$$

$$= 4.41\mu\ \text{g vit A/g butter.}$$

5. Calculate the concentration of two absorbing compounds X and Y if the absorbance of this solution in a 1 cm cuvette is 0.52 at 350 nm and 0.44 at 385 nm. Compound A has an a_m of 10,000 $M^{-1}\times cm^{-1}$ at 350 nm and 3000 $M^{-1}\times cm^{-1}$ at 385 nm. Compound B has an a_m of 5000 $M^{-1}\times cm^{-1}$ at 350 nm 12,000. $M^{-1}\times cm^{-1}$ at 385 nm.

Let C_1 and C_2 denote the concentrations of compounds X and Y.

	a_m 350 nm ($M^{-1}\times cm^{-1}$)	a_m 385 nm ($M^{-1}\times cm^{-1}$)
X	10,000	3000
Y	5000	12,000

$$0.52 = 10,000 \times C_1 \times 1 + 5000 \times C_2 \times 1 \quad (1)$$

$$0.44 = 5000 \times C_1 \times 1 + 12,000 C_2 \times 1 \quad (2)$$

solving for C_1 we get

$$(10 \times 10^3 \times 12 - 5 \times 10^3 \times 5)C_1 = 0.52 \times 12 - 0.44 \times 5$$

$$\therefore \quad C_1 = \frac{6.24 - 2.20}{(120 - 25)10^3} = \frac{4.04}{95} \times 10^{-3}$$

$$= 0.0425 \times 10^{-3} = 4.25 \times 10^{-5}$$

$$C_2 = \frac{0.52 - 10 \times 10^3 \times 4.25 \times 10^{-5}}{5000}$$

$$= \frac{0.52 - 0.425}{5000} = \frac{0.095}{5} \times 10^{-3}$$

$$= 0.019 \times 10^{-3} = 1.9 \times 10^{-5}$$

6. Calculate the protein concentrations of the (a) undiluted and (b) diluted solutions given below. The measurements were made using a 1 cm cuvette.

Solution	$A_{280\text{ nm}}$	$A_{260\text{ nm}}$	$A_{225\text{ nm}}$	$A_{215\text{nm}}$
Undiluted	0.45	0.03	—	—
Diluted (1:10)	—	—	0.25	0.48

(a) $[\text{Protein}]_{mg/ml} = 1.55 A_{280}^{1\text{ cm}} - 0.76 A_{260}^{1\text{ cm}}$ (*Formula*)

$$= 1.55 \times 0.45 - 0.76 \times 0.03$$

$$= 0.6945 - 0.0228$$

$$= 0.6717 \text{ mg/ml}$$

(b) $[\text{Protein}]_{\mu g/ml} = 144 \times \left(A^{1\text{ cm}}_{215\text{ nm}} - A^{1\text{ cm}}_{225\text{ cm}}\right)$ (*Formula*)

$$= 144 \times (0.48 - 0.25)$$

$$= 144 \times 0.23 = 33.12\ \mu g/ml.$$

7. A mixture of glucose-6-phosphate and glucose-1-phosphate was present in 1.5 ml of a solution. To this was added a 1.5 ml solution containing excess $NADP^+$, $MgCl_2$ and glucose-6-phosphate dehydrogenase. The absorbance in a 1 cm cuvette increased to 0.66 at 340 nm. When no further increase in absorbance was observed, an additional 1.5 ml of a phosphoglucomutase solution was added. The absorbance then decreased to 0.60. Calculate the (a) glucose-6-phosphate and (b) glucose-1-phosphate concentration in the original solution.

(a) $A = a_m \times C_{\text{NADPH}} \times l.$ Where $a_m \times 10^{-3} = 6.22$ for NADPH

$$0.65 = (6.22 \times 10^3) C_{\text{NADPH}} \times (l).$$

$$\therefore \quad C_{\text{NADPH}} = \frac{0.65}{6.22 \times 10^3} = 10.45 \times 10^{-5}\text{M}$$

$$C_{G-6-P_{\text{original}}} = 10.45 \times 10^{-5} \times (\text{dilution factor})$$

$$= 10.45 \times 10^{-5} \times 2$$

$$= 20.9 \times 10^{-5}$$

$$= 2.09 \times 10^{-4} M$$

$$= 2 \times 10^{-4} M$$

Now $A_{\text{orig. corrected}} = (0.66) \times \frac{2}{3} = 0.44$

$$\Delta A_{\text{corrected}} = 0.60 - 0.44 = 0.16$$

Now $\Delta A = a_m \times \Delta C_{\text{NADPH}} \times \ell$

$$0.16 = (6.22 \times 10^3) \times \Delta C_{\text{NADPH}} \times \ell$$

$$\therefore \quad \Delta C_{\text{NADPH}} = \frac{0.16}{6.22 \times 10^3}$$

$$= \frac{16 \times 10^{-2} \times 10^{-3}}{6.22} = 2.57 \times 10^{-5}$$

$$\therefore \quad C_{G-1-P(\text{original})} = 2.57 \times 10^{-5} \times (\text{dilution factor})$$

$$= 2.57 \times 10^{-5} \times 3$$

$$= 7.71 \times 10^{-5} \text{ M}$$

$$C_{G-1-P(\text{original})} = 7.71 \times 10^{-5} \text{ M}$$

8. 1.25 ml of a 2.5×10^{-4} M solution of NADPH were added to 1.25 ml of a solution containing oxidized glutathione and of the enzyme glutathione reductase. The final absorbance of the solution at 340 nm was 0.30 in a 1 cm cuvette. Calculate the concentration of oxidized glutathione in the original 1.25 ml. The complete reaction catalyzed by glutathione reductase is GSSG + NADPH + H^+ $\longrightarrow$ 2GSH + $NADP^+$.

We can calculate the concentration of NADPH from the absorbance at 340 nm. After correction for dilution the concentration of glutathione in the original solution can be calculated.

$$\text{Now} \quad A = a_m \times C \times l$$

$$0.30 = (6.22 \times 10^3) \times C \times 1$$

$$\therefore C_{\text{NADPH}} = \frac{0.30}{6.22 \times 10^3 \times 1} = 0.048 \times 10^{-3}$$

$$= 0.48 \times 10^{-4}$$

$$\therefore \quad C_{\text{glu-original}} = 0.048 \times 10^{-3} \times \text{ dilution factor}$$

$$= 0.48 \times 10^{-4} \times 2/1$$

$$= 0.96 \times 10^{-4}$$

$\therefore$ The concentration of oxidized glutathione in the original solution

$$= 2.5 \times 10^{-4} - 0.96 \times 10^{-4}$$

$$= 1.54 \times 10^{-4} \text{ M}$$

9. An equilibrium mixture of α and β-D galatose has an α_D^{25} of 80.2° Pure α-D galatose has an $[\alpha]_D^{25}$ of 150.7° and pure β-D galactose has an $[\alpha]_D^{25}$ of 52.8°. Calculate the proportion of α and β-D galatose in the equilibrium mixture.

The contribution of each anomer toward the total Å is directly proportional to the concentration of each anomer present.

Let, $X = \%\beta$ so that $100 - X = \%\alpha$

$$(52.8) \times X + (150.7)(100 - X) = 100 \times 80.2$$

$$97.9X = 100 \times (150.7 - 80.2)$$

$$X = \frac{70.5 \times 100}{97.9}$$

$$= 0.72 \times 100$$

$$= 72$$

$\therefore \quad 100 - X = 28$

$\therefore \quad \alpha = 28\%$ and $\beta = 72\%$.

10. A solution containing an equilibrium mixture of α and β forms of D-glucose has an observed rotation of 112° in a 10 cm polarimeter tube at 25°C. Calculate the concentration of D-glucose in the solution. The $[\alpha]_D^{25}$ for an equilibrium mixture of α and β-D-glucose is 52°.7.

$$\text{Å} = [\alpha] \times \ell \times C$$

$$\therefore \quad C = \frac{\text{Å}}{[\alpha] \times \ell}$$

$$= \frac{112}{52.7 \times 1}$$

$$= 2.125$$

$$\therefore \quad C = 2.125 \text{ g/ml}$$

11. Calculate the erythrocyte transketolase activity/ml of RBC using the ribose-5-phosphate as substrate. The rate of decrease in absorbance at 340 nm consequent upon NADH oxidation is 0.05 per minute where the volume of assay mixture is 3.11 ml. The dilution factor for glycerol-3-phosphate dehydrogenase-trios phosphate isomerase is 10.

One unit of the activity is expressed as the μ moles of NADH oxidised/hr/ml of red cells.

ΔA = decrease in absorbance in 1 min at 340 nm.

Molar extinction of NADH $= 6.22 \times 10^3$

$v \quad = \Delta A/\text{min} = 0.05/\text{min}$

$\Delta A = (a_m)\ (\Delta\ C)\ (l)$

or, $\Delta C = \dfrac{\Delta A}{(a_m)(1)} = \dfrac{0.05}{(6.22 \times 10^3)} = \dfrac{0.05}{6.22} \times 10^{-3}$ M

$= \dfrac{0.05}{6.22} \times 10^3 \mu$ moles/litre

$v = \dfrac{0.05}{6.22} \times 10^3 \mu$ moles/litre$^{-1} \times min^{-1}$

$= \dfrac{0.05}{6.22} \times 10^3 \times 10^{-3} \mu$ moles/ml$^{-1} \times$ min^{-1}

The activity $= \dfrac{0.05}{6.22} \times \dfrac{3.11 \times 10 \times 60}{0.10}$

$= 0.05 \times 3000/$ ml

$= 150$ units/ml of RBC.

12. Calculate the activity of acetyl choline esterase using DTNB where the reaction of volume is 3 lit/10 mg pr. The charge in optical density as followed is 0.6/min at 412 nm.

The molar extinction coefficient of SH group of DTNB is 14.3×10^3 at 412 nm.

For a 3.0 ml reaction volume it is $\dfrac{14.3 \times 10^3}{3} = 4.77 \times 10^3$

Activity of acetylcholine esterase (μ moles/min) $= \dfrac{0.6 \times 10^3}{4.77 \times 10^3}$

$= 0.126$

$= 0.126/\mu$ moles/min

Specific activity = 0.126/mg pr present in assay mixture

i.e. S.A. $= \dfrac{0.126}{10}$

S.A. = 0.0126

13. A 20 μg of mitochondrial protein solution is given to you. Calculate the first order velocity constant for the oxidation cytochrome c. The absorbance of reduced cytochrome c and completely oxidized cytochrome c at time t_1 (15 sec) and t_2 (75 sec) is 0.2 and 0.7 respectively.

A millimolar extinction coefficient of 19.1 is employed for reduced vs oxidized cytochrome c at 550 nm.

$$\text{The velocity constant } K = \frac{\log A_{t_2} - \log A_{t_1}}{t_2 - t_1} \times 2.3$$

$$= \frac{\log 0.7 - \log 0.2}{75 - 15} \times 2.3 - \frac{\log 3.5}{60} \times 2.3$$

$$= \frac{0.5440 \times 2.3}{60} \text{ sec}^{-1} = 0.0209 \text{ sec}^{-1}$$

$$= 2.09 \times 10^{-2} \text{ sec}^{-1}.$$

$$\text{Velocity constant } K = 2.09 \times 10^{-2} \text{ sec}^{-1}.$$

6

Isotopes in Biochemistry

6.1 Isotopes

Elements are defined by the number of protons, positively charged subatomic particles, in an atom's nucleus. The number of protons in an atom's nucleus is termed its *atomic number.* Isotopes of a given element carry different numbers of neutrons, or neutrally charged particles, in their nuclei. The sum of the number of neutrons and protons in an atom's nucleus defines its apporximate *atomic weight.* For example, all carbon atoms have six protons; isotopes of carbon can have 6,7,or 8 neutrons.

Table 1: Isotopes of Carbon

Element	Protons	Neutrons	Atomic Number	Approximate Atomic Weight (gm/mol)	Name
Carbon	6	6	6	12	Carbon 12, ^{12}C
Carbon	6	7	6	13	Carbon 13, ^{13}C
Carbon	6	8	6	14	Carbon 14, ^{14}C

Elements are defined by the number of protons (Z) in their nucleus. The mass number (A) of an element is equal to the sum of both protons and neutrons (N) in the nucleus, or

$$A = N + Z$$

A single element can have two or more mass numbers due to differences in the number of neutrons that can occur in the nucleus. These different forms of a single element are called isotopes. While protons have a positive charge, neutrons have no charge, so the number of neutrons does not affect the charge of a molecule. Some isotopes are stable, while others are radioactive and release particles and energy to decay into a more stable form.

Elements usually have a common isotope that is the form most often found in nature. Because carbon, oxygen, and hydrogen are the elements that make up all organic matter, biologists are often interested in the isotopes of these elements. Each has common and rare forms. For instance, 98.8% of carbon atoms contain 6 protons and 6 neutrons with a mass number of 12; the notation for this form is ^{12}C. 1.1% of carbon has 6 protons and 7 neutrons, noted as ^{13}C. Similarly, 99.98% of hydrogen is found as ^{1}H, but two stable isotopes and one radioactive isotope are known. 99.6% of oxygen in ^{16}O, in addition there are three stable isotopes and five radioactive isotopes. Nitrogen, an important plant nutrient, is also of interest to biologists. It is found as 99.6% ^{14}N and 0.4% ^{14}N.

A brief listing of the stable isotopes and their abundances for the elements most commonly used in global change research would include:

Hydrogen	Isotope	Abundance (%)
Hydrogen	^{1}H	99.985
	^{2}H	0.015
Carbon	^{12}C	98.89
	^{13}C	1.11
Nitrogen	^{14}N	99.63
	^{15}N	0.37
Oxygen	^{16}O	99.759
	^{17}O	0.037
	^{18}O	0.204
Sulfur	^{32}S	95.00
	^{33}S	0.76
	^{34}S	4.22
	^{36}S	0.014
Strontium	^{84}Sr	0.56
	^{86}Sr	9.86
	^{87}Sr	7.02
	^{88}Sr	82.56

Isotopes influence the physical and chemical properties of matter. For instance, $^{12}CO_2$ will behave differently than $^{13}CO_2$ during certain processes and chemical reactions. Light isotopes form weaker chemical bonds than heavy isotopes, so light isotopes are somewhat more chemically reactive. Molecules containing light isotopes also evaporate and diffuse more quickly than their heavier counterparts. Therefore an evaporating liquid will contain more of the light isotope in the gas phase, and more of the heavy isotope in the liquid phase.

Isotopic fractionation occurs when isotopes are partitioned differently between two phases or two substances. If we understand the processes underlying fractionation, we can use isotopes to determine how plants, animals, and whole ecosystems function. We can gain information from isotopes that naturally occur in the environment in studies of natural abundance, or we can artificially apply in high concentrations to use them as tracers.

6.2 Radioactivity

Radioactive isotopes (also called *radioisotopes*) have unstable nuclei. These isotopes disintegrate to form atoms with stable nuclei by the release of subatomic particles and gamma rays (akin to X-rays). The radioactive elements are referred to as parent atoms; the atoms they disintegrate to form are called daughter products.

Some isotopes release an *alpha particle* during nuclear disintegration; an alpha particle consists of two protons and two neutrons (equivalent to the nucleus of an atom of helium). Others release a *beta particle*, which is an electron, or negatively charged nuclear particle. Beta Particles originate in the nucleus, presumably by breakdown of a neutron into its proton-electron components. Gamma rays are released during both types of radioactive decay.

When an isotope emits an alpha particle, the resultant daughter product has an atomic number two units less than its parent's atomic number, and an atomic weight four units less than its parent's atomic weight. When an isotope emits a beta particle, it decays to a daughter with an atomic number one unit greater and an essentially unchanged atomic weitht. Some isotopes decay and immediately produce a stable daughter product. For example, one-step decays to stable daughters are completed by the radiogenic isotopes ^{14}C (decaying to ^{14}N by the beta process), and ^{87}Rb (decaying to ^{87}Sr by the beta process).Others decay and produce unstable daughters, which then become the parent products of their own daughters. Unstable isotopes producing unstable daughters form a *radioactive decay chain*. For example, the ^{235}U decay chain eventually produces ^{207}Pb, a stable daughter.

Using empirical data, it is possible to statistically forecast what percentage of a radioisotope's population will decay over a given period of time. This has enabled workers to define a *half-life* for each radioisotope, the period required for one-half of the original parent population to decay to its stable daughter product. Each radioisotope has its own characteristic half-life.

Suppose that at its inception, a sample contains 100 units of a parent radioisotope. After one half-life has passed, there will remain 50 units of the parent isotope, and 50 units of the daughter product will have been produced. After another half-life, 25 units of the parent isotope will remain, and 75 units of the daughter product will have been produced. After another half-life, 12.5 units of the parent isotope will remain, and 87.5 units of the daughter product will have been produced. Through time, the number of parents constantly decreases while the number of daughters constantly increases. Theoretically, although the number of parents will become insignificantly small, there should never come a time when all of the parent population has decayed to daughters.

Knowing the value of a specific isotope's half-life, it is possible to determine the age of a geologic or archaeological sample by evaluating the amount of parent and daughter isotopes in it. For example, given the half-life of U-235 is 7×10^8y, suppose you have a rock sample containing minerals haveing 1 unit of ^{235}U and 3 units of ^{207}Pb. The sample must have originally contained 4 units of parent material, and 25% of the parent material (U-235) remains. From calculation one can arrive that time equivalent to two half-lives have passed, or apporximately 1.4 billion years.

The decay of a radioactive substance follows an exponential relationship. This relationship can be written:

$$N = N_0 e^{-ct}$$

Where N is the number of parent atoms at time t, and N_0 is the number of parent atoms at $t = 0$ (the original amount in the sample). The *decay constant*, c, with dimensions of reciprocal time, is related to the half-life ($t_{\frac{1}{2}}$) by the following relationship:

$$c = \frac{0.693}{t_{\frac{1}{2}}}$$

Conversely:

$$t_{\frac{1}{2}} = \frac{\ln 2}{c}$$

6.3 Definitions

Becquerel (Bq) This is the SI unit of radioactivity. 1 Bq = one disintegration per second (1 d.p.s.).

Counts per minute (c.p.m.) The number of β particles detected per minute; c.p.m. = d.p.m. $\times$ efficiency of the counting device.

Curie (Ci) The Curie was the standard unit of radioactivity before the introduction of the Becquerel. Nevertheless, it is still widely used in laboratories through widespread familiarity. 1 Ci = 2.22 $\times$ 10^{12} d.p.m.

Disintegrations per minute (d.p.m.) The number of atoms disintegrating per minute. **Electron volt (eV). The energy acquired by an electron when accelerating along a potential gradient of 1 volt. 1 eV =** 1.6 $\times$ 10^{-12} **erg. 1 MeV =** 10^6 **eV.**

Gray: The SI unit of absorbed dose. 1 Gray = 1 joule per kg tissue = 100 rads.

Rad (radiation absorbed dose) The earlier (non-SI) unit of absorbed dose. 1 rad = 100 erg per g tissue.

Rbe (relative biological effectiveness) The ratio of body damage caused by a particular radiation dose to that caused by the same dose of x-rays (average specific ionization of 100 ion pairs per μm of water).

Rem (Rad equivalent man) Rem = dose in rad $\times$ RBE.

Sievert (SV) SI unit equivalent to 100 rem.

Specific activity (Sa) The radioactivity of an element per unit mass.

6.4 Conversion between Units

Conversion between Curies and d.p.m.

Curie (Ci) $= 2.22 \times 10^{12}$ d.p.m.

1 milliCurie (mCi) $= 2.22 \times 10^{9}$ d.p.m.

1 microCurie (μ Ci) $= 2.22 \times 10^{6}$ d.p.m.

1 nanoCurie (nCi) $= 2.22 \times 10^{3}$ d.p.m.

1 picoCurie (pCi) $= 2.22$ d.p.m.

Conversion between Curies and Becquerels

1 Ci $= 3.7 \times 10^{10}$ Bq $= 37$ GBq (gigaBq)

1 mCi $= 3.7 \times 10^{7}$ Bq $= 37$ MBq (megaBq)

1 mCi $= 3.7 \times 10^{4}$ Bq $= 37$ kBq (kiloBq)

1 GBq $= 2.7 \times 10^{-4}$ Bq $= 27.027$ mCi (milliCi)

1 MBq $= 2.7 \times 10^{-7}$ Ci $= 27.027$ μCi (microCi)

1 kBq $= 2.7 \times 10^{-10}$ Ci $= 27.027$ nCi (nanoCi)

Radio nuclides

Table 2: Decay characteristics of biological radionuclides

Radionuclide	Half-life	Type of emission
^{3}H	12.4 years	β
^{7}Be	55.3 days	Ec[a]
^{11}C	20.5 min	β^{+}
^{14}C	57600 years	β^{-}
^{22}Na	2.6 years	$\beta^{+}\gamma$
^{24}Na	14.8 h	$\beta^{-}\gamma$
^{28}Mg	21.4 h	β^{-}
^{31}Si	170 min	β^{-}
^{32}P	14.3 days	β^{-}
^{33}P	25 days	β^{-}
^{35}S	87.4 days	β^{-}
^{38}Cl	38.5 min	β^{-}k
^{42}K	12.4 h	$\beta^{-}\gamma$
^{45}Ca	165 days	$\beta^{-}\gamma$
^{47}Ca	4.54 days	β^{-}
^{51}Cr	28 days	β^{-}

Contd. Table 2

Radionuclide	Half-life	Type of emission
^{52}Mn	5.8 days	γk
^{54}Mn	310 days	$\beta^+ k$
^{55}Fe	2.94 years	γk
^{57}Co	270 days	γ
^{58}Co	72 days	$\beta^+ \gamma$
^{59}Fe	46.3 days	$\beta^- \gamma$
^{60}Co	5.3 years	$\beta^- \gamma$
^{63}Ni	100 years	β^-
^{64}Cu	12.8 h	$\beta^- \beta^+ \gamma k$
^{65}Zn	250 days	$\beta^+ k$
^{75}Se	121 days	$\beta^+ k$
^{76}As	26.8 h	γk
^{82}Br	36 h	$\beta^- \gamma$
^{86}Rb	18.7 days	$\beta^- \gamma$
^{89}Sr	51 days	$\beta^- \gamma$
^{90}Sr	28.5 years	β^-
^{99}Mo	68 h	β^-
^{109}Cd	462 days	$\beta^- \gamma$
^{110}Ag	249.8 days	Ec[a]
^{111}Ag	7.47 days	β^-
^{115}Cd	44.6 days	EC[a]
^{125}I	60.0 days	β^-
^{131}I	8.04 days	γEC[a]
^{133}Ba	10.8 year	$\beta^- \gamma$
^{134}Cs	2.06 year	EC[a]
^{135}I	9.7 h	β^-
^{137}Cs	30.17 years	β^-
^{185}W	75.1 days	β^-
^{197}Hg	64.4 h	β^-
^{198}Au	2.698 days	EC[a]
^{199}Au	3.13 days	β^-
^{203}Hg	46.6 days	β^-
^{36}Cl	310000 years	β^-

Table 3: Emission energies for important radionuclides

Radionuclide	Particle	Emission energy (MeV)	% of total disintegrations
^{3}H	β^-	0.018	100
^{14}C	β^-	0.155	100
^{32}P	β^-	1.71	100
^{35}S	β^-	0.167	100
^{125}I	[a]EC	—	—
	γ	0.035	—
^{131}I	Te KX-ray	0.027–0.032	—
	β^-	0.25	3
	β^-	0.33	9
	β^-	0.61	87
	β^-	0.81	1
	γ	0.08	2
	γ	0.28	5
	γ	0.36	80
	γ	0.64	9
	γ	0.72	3

Table 4: Decay table for ^{3}H

Years	Months											
	0	1	2	3	4	5	6	7	8	9	10	11
0	1.0000	0.995	0.991	0.986	0.982	0.977	0.973	0.968	0.963	0.959	0.955	0.950
1	0.946	0.941	0.937	0.933	0.928	0.924	0.920	0.916	0.911	0.907	0.903	0.899
2	0.894	0.890	0.886	0.882	0.878	0.874	0.870	0.866	0.862	0.858	0.854	0.850
3	0.846	0.842	0.838	0.834	0.831	0.827	0.823	0.818	0.815	0.811	0.808	0.804
4	0.800	0.796	0.793	0.789	0.786	0.782	0.778	0.775	0.771	0.767	0.764	0.760
5	0.757	0.753	0.750	0.746	0.743	0.739	0.736	0.733	0.729	0.726	0.722	0.719
6	0.716	0.712	0.709	0.706	0.703	0.699	0.696	0.393	0690	0.686	0.683	0.680
7	0.677	0.674	0.671	0.668	0.665	0.661	0.658	0.655	0.652	0.649	0.646	0.643
8	0.640	0.637	0.634	0.631	0.629	0.626	0.623	0.620	0.617	0.614	0.611	0.608
9	0.606	0.603	0.600	0.597	0.595	0.592	0.589	0.586	0.583	0.581	0.578	0.575
10	0.573	0.570	0.568	0.565	0.562	0.560	0.557	0.554	0.552	0.549	0.547	0.544
11	0.542	0.539	0.537	0.534	0.532	0.529	0.527	0.524	0.522	0.520	0.517	0.515
12	0.512	0.510	0.508	0.505	0.503	0.501	0.498					

Table 5: Decay table for ^{35}S

Days	0	1	2	3	4	5	6	7	8	9
0	1.000	0.992	0.984	0.976	0.969	0.961	0.954	0.946	0.939	0.931
10	0.924	0.916	0.909	0.902	0.895	0.888	0.881	0.874	0.867	0.860
20	0.853	0.847	0.840	0.833	0.827	0.820	0.814	0.807	0.801	0.795
30	0.788	0.782	0.776	0.770	0.764	0.758	0.752	0.746	0.740	0.734
40	0.728	0.722	0.717	0.711	0.705	0.700	0.694	0.789	0.683	0.678
50	0.673	0.667	0.662	0.657	0.652	0.646	0.641	0.636	0.631	0.626
60	0.621	0.616	0.612	0.607	0.602	0.597	0.592	0.588	0.583	0.579
70	0.574	0.569	0.565	0.560	0.556	0.552	0.547	0.543	0.539	0.534
80	0.530	0.526	0.522	0.518	0.514	0.510	0.506	0.502	0.498	

Table 6: Decay table for ^{32}P

Days		6	12	18	24	30	36	42
0	1.000	0.988	0.976	0.964	0.953	0.941	0.930	0.919
2	0.908	0.897	0.886	0.876	0.865	0.854	0.844	0.834
4	0.824	0.814	0.804	0.794	0.785	0.775	0.766	0.757
6	0.748	0.739	0.730	0.721	0.712	0.704	0.695	0.687
8	0.679	0.671	0.662	0.654	0.646	0.639	0.631	0.623
10	0.616	0.609	0.601	0.594	0.587	0.580	0.573	0.566
12	0.559	0.552	0.546	0.539	0.533	0.526	0.520	0.513
14	0.507	0.501	0.495					

(Column headings 6–42 are Hours.)

Table 7: Decay table for ^{125}I

Days	0	1	2	3	4	5	6	7	8	9
0	1.000	0.988	0.977	0.966	0.955	0.944	0.933	0.922	0.912	0.901
10	0.891	0.881	0.871	0.861	0.851	0.841	0.831	0.821	0.812	0.803
20	0.794	0.785	0.776	0.767	0.758	0.750	0.741	0.732	0.724	0.716
30	0.707	0.699	0.691	0.683	0.675	0.667	0.660	0.653	0.645	0.637
40	0.630	0.623	0.616	0.609	0.602	0.595	0.588	0.581	0.574	0.567
50	0.561	0.555	0.548	0.542	0.536	0.530	0.524	0.518	0.512	0.506
60	0.500									

Table 8: Decay table for ^{131}I

Days	Hours 0	3	6	9	12	15	18	21
0	1.000	0.989	0.979	0.969	0.958	0.947	0.937	0.927
1	0.917	0.908	0.898	0.888	0.879	0.870	0.860	0.851
2	0.842	0.833	0.824	0.815	0.806	0.797	0.789	0.781
3	0.772	0.764	0.756	0.748	0.740	0.732	0.724	0.716
4	0.708	0.701	0.693	0.685	0.678	0.671	0.664	0.657
5	0.650	0.643	0.636	0.629	0.622	0.616	0.609	0.602
6	0.596	0.590	0.583	0.577	0.571	0.565	0.559	0.553
7	0.547	0.541	0.535	0.529	0.524	0.519	0.513	0.507
8	0.502	0.496						

6.5 Detection of Radioactivity

Scintillation counting: Many scintillation cocktails are available commercially or can be readily prepared in the laboratory, each suited to particular types of sample (e.g. aqueous samples versus filters). Physical and chemical data on common fluors and solvents of scintillation cocktails are given in Tables 9 and 10 respectively.

Table 9: Scintillants: Physical and Chemical data

Compounds	Mol. wt.	Fluoroscent maxima(nm) 1	2	3	m.p. (°C)	b.p. (°C)
Anthracene	178.2	405	428	454	218	342
BBOT	430.6	438	–	–	200	–
Butyl-PBD	354.5	367	382	–	135	–
Naphthalene	128.2	325	336	351	80	218
PBD	298.4	364	377	–	168	–
PPO	221.3	365	380	–	73	360
Secondary scintillants						
Bis-MSB	310.4	423	–	–	182	–
DMMPOPOP	392.5	430	–	–	234	–
POPOP	364.4	420	441	–	245	–

Table 10: Solvents for scintillation counting: physical and chemical data

Compound	Mol. wt.	m.p. (°C)	b.p. (°)	f.p.[a]	Specific gravity
Acetone	58.1	−95	56	−20	0.79
Anisole (methyl phenyl ether)	108.1	−37	155	41	0.99
1,3-dimethoxybenzene	138.2	—	86	87	1.06
1,2-dimethoxyethane (ethylene glycol dimethyl ether)	90.1	−58	85	1	0.87
1,4-dioxan	88.1	11	101	12	1.03
Methanol (methyl alcohol)	32.0	−98	65	12	0.79
2-methoxyethanol (ethylene glycol monomethyl ether)	76.1	−86	125	46	0.96
Tetrahydrofuran	72.1	−108	66	−17	0.89
Toluene	92.1	−95	111	7	0.86
Xylene (mixed isomers)	106.2	−95	140	25	0.86

High energy β-particles traveling through water cause the polarization of molecules along their trajectory which then emit photons of light (350–600 nm) as their energy returns to the ground state (the Cerenkov effect). Data relevant to counting ^{32}P in aqueous buffer in a scintillation counter without added scintillation fluid (Cerenkov counting) are given below.

Per cent of ^{32}P energy spectrum above 0.5 MeV
(the threshold for Cerenkov counting) = 80%

Counting efficiency:

Glass vials = 50%

Plastic vials = 60%

A low energy window must be used when counting under these conditions. The ^3H channel is sometimes used but it is better to calibrate the counter specifically.

Biochemical Calculations

1. Calculate the decay constant λ of an isotope with half life of 2 years in terms of year^{-1}, day^{-1}, hr^{-1}, min^{-1} and sec^{-1}.

Decay constant $\lambda = \dfrac{0.693}{t_{1/2}} = \dfrac{0.693}{2} = 0.347$

$$= 0.347\text{yr}^{-1} = \frac{0.347}{365}\ \text{day}^{-1}$$

$$= 0.0009506\ \text{day}^{-1} = \boxed{9.5 \times 10^{-4}\ \text{day}^{-1}}$$

$$= \frac{0.347}{365 \times 24}\ \text{hr}^{-1}$$

$$= 0.0000396\ \text{hr}^{-1} = \boxed{3.96 \times 10^{-5}\ \text{hr}^{-1}}$$

$$= \frac{0.347}{365 \times 24 \times 60}\ \text{min}^{-1}$$

$$= 0.00000066\ \text{min}^{-1} = \boxed{6.6 \times 10^{-7}\ \text{min}^{-1}}$$

$$= \frac{6.6}{60} \times 10^{-7}\ \text{sec}^{-1}$$

$$= 0.11 \times 10^{-7}\ \text{sec}^{-1} = \boxed{1.1 \times 10^{-8}\ \text{sec}^{-1}}$$

2. Ca^{45} has half life of 163 days. Calculate (a) the fraction of the Ca^{45} atoms that decays per day and per minute and (b) the specific activity of pure Ca^{45} in terms of Ci/g, Ci/g atom and DPM/g. $t_{1/2} = 163$ days

(a) $$\lambda = \frac{0.693}{t_{1/2}} = \frac{0.693}{163}$$

$$= 0.0042515\ \text{day}^{-1}$$

i.e. $$\lambda = 4.25 \times 10^{-3}\ \text{day}^{-1}$$

$$= \frac{4.25 \times 10^{-3}}{24 \times 60}\ \text{min}^{-1}$$

$$= 0.00295 \times 10^{-3}\ \text{min}^{-1}$$

$$= 2.95 \times 10^{-6}\ \text{min}^{-1}$$

Now the fraction of the Ca^{45} atoms that decays per day $= 1/\lambda$

$$= \frac{1}{4.25 \times 10^{-3}} = 0.235290 \times 10^{3}$$

$$= 235.3$$

The fraction of the Ca^{45} atoms that decays per minute

$= 1/\lambda$ where λ is expressed in min^{-1}

$$= \frac{1}{2.95 \times 10^{-6}} = 0.33898305 \times 10^{6}$$

$$= 338983$$

(b) $1\text{gm } Ca^{45} = \dfrac{1\text{g}}{45 \text{ g/g atom}}$

$$= 0.0222 \text{ g.atom}$$

$$N = 0.0222 \text{ g.atom} \times 6.023 \times 10^{23} \text{ atom/g.atom}$$

$$= 0.1337 \times 10^{23}$$

$$= 1.337 \times 10^{22}$$

$$\text{DPM/g} = \lambda N = 2.95 \times 10^{-6} \times 1.34 \times 10^{22}$$

$$= 3.95 \times 10^{16}$$

Specific activity $= 3.95 \times 10^{16}$ DPM/g

$$= \frac{3.95 \times 10^{16}}{2.22 \times 10^{12}} = 1.779 \times 10^{4}$$

$$= 1.78 \times 10^{4} \text{ Ci/g} = 1.78 \times 10^{4} \times 45$$

$$= 80.1 \times 10^{4} = 8.01 \times 10^{5} \text{ Ci/g.atom}$$

$\therefore$ Specific activity $= 3.95 \times 10^{16}$ DPM/g
$= 1.78 \times 10^{4}$ Ci/g
$= 8.01 \times 10^{5}$ Ci/g.atom

3. Calculate the weight in grams of I^{131} in 1 mCi of carrier-free I^{131}. The half life of I^{131} is 8.1 days.

$1\text{Ci} = 2.22 \times 10^{12}$ DPM

$1\text{m Ci} = 2.22 \times 10^9$ DPM

$$\lambda = \frac{0.693}{8.1} \text{ day}^{-1}$$

$$= \frac{0.693}{8.1 \times 24 \times 60} \text{ min}^{-1}$$

$$= 0.0000594 \text{ min}^{-1}$$

$$= 5.94 \times 10^{-5} \text{ min}^{-1}$$

$$\text{DPM} = \lambda N$$

$$N = \frac{2.22 \times 10^{12}}{\lambda} = \frac{2.22 \times 10^{12}}{5.94 \times 10^{-5}}$$

i. e. $N = 0.373737 \times 10^{17} = 3.74 \times 10^{16}$ atoms/Ci

1 gm atom of I^{131} contains 6.023×10^{23} atom

$\therefore$ 3.74×10^{23} atoms weight $= \dfrac{3.74 \times 10^{16} \times 131}{6.023 \times 10^{23}}$

$$= 81.345 \times 10^{-7} = 8.135 \times 10^{-6} \text{ g/mCi}$$

4. Uniformly labelled serine C^{14} contains 40 mCi in 5.5 ml of solution. The specific activity is given as 160 mCi/n mole. Calculate (a) the concentration of serine in the solution and (b) the activity of the solution in terms of CPM/ml at a counting efficiency of 70%.

one mole is equivalent to 160 mCi.

$\therefore$ 4 mCi is equivalent to 4/160 n moles/mCi $= 0.025$

$$= 2.5 \times 10^{-2} \text{ n moles/mCi}$$

For a solution of 5.5 ml it is $= 2.5 \times 10^{-2}$ n moles

$$\therefore \quad \text{Concentration} = \frac{2.5 \times 10^{-2}}{5.5} = 0.4545 \times 10^{-2}$$

$$= 4.5 \times 10^{-3} \text{ M}$$

$$1\text{mCi} = 2.2 \times 10^9 \text{ DPM}$$

$$4\text{mCi} = 2.2 \times 10^9 \times 4 \text{ DPM}$$

$$\text{Total activity} = 0.70 \times 2.22 \times 10^9 \times 4 \text{ CPM}$$

$$\therefore \quad \text{The Activity} = \frac{0.70 \times 2.22 \times 10^9 \times 4}{5.5} = 1.13 \times 10^9 \text{ CPM/ml}$$

5. A small amount of K^{42} ($t_{1/2}$ = 12.4 hr) was injected into the blood stream of an animal. Blood samples were taken periodically and counted immediately. The specific activities of the samples are given below (a) From the data calculate the effective half life of the isotope in the blood stream. (b) Also find the zero time specific activity (CPM/ml).

Sample time (hrs)	Specific Activity (CPM/ml)
2	7120
4	5070
8	3610
12	1830

Sample time (hrs)	Specific Activity (CPM/ml)
16	770
20	230

(a) Taking any two sample times say 12 and 4

$$2.3 \log \frac{5070}{1830} = \frac{0.693}{t_{1/2 \text{ eff}}}(12 - 4)$$

$$\therefore \quad t_{1/2 \text{ eff}} = \frac{0.693 \times 8}{2.3 \times \log(2.7705)}$$

$$= \frac{5.544}{2.3 \times 0.4426}$$

$$= 5.446$$

$$t_{1/2 \text{ eff}} = 5.4 \text{ hrs.}$$

Now

$$\frac{1}{t_{1/2 \text{ eff}}} = \frac{1}{t_{1/2 \text{ r}}} + \frac{1}{t_{1/2 \text{ b}}} \qquad r = \text{radioactive}$$

$$\therefore \quad \frac{1}{t_{1/2\ \mathrm{b}}} = \frac{1}{t_{1/2\ \mathrm{eff}}} - \frac{1}{t_{1/2\ \mathrm{r}}} \qquad b = \text{biological}$$

$$= \frac{1}{5.4} - \frac{1}{12.4}$$

$$= 0.1852 - 0.0806 = 0.1046$$

$$\therefore \quad t_{1/2} = \frac{1}{0.1046} = 9.56 = 9.6$$

$$\therefore \quad t_{1/2\ \mathrm{b}} = 9.6 \text{ hrs}$$

(b) Let N_0 be zero time S.A.

$$2.3 \log \frac{N_0}{7120} = \frac{0.693}{5.4}(2-0)$$

$$\log N_0 - \log 7120 = \frac{0.693 \times 2}{5.4 \times 2.3}$$

$$= 0.1116$$

$$\log N_0 = \log 7120 + 0.1116 = 3.8525 + 0.116 = 3.9641$$

$$\therefore \quad N_0 = 9206 \text{ CPM/ml}$$

Part II

Biostatistics

I. Introduction

1. Statistics is a science of figures. Its scope comprises methods of collection of data, classification, summarizing, and interpretation, drawing inferences and testing the significance of data.
2. Biostatistics is that branch of statistics, which is applied to data obtained from biological and medical sciences. Since biostatistics deals with facts and measurements, it can also be called quantitative medicine or biometry.
3. Statistics makes it possible to draw inferences from an analytical study of the collected figures. Statistical methods are useful in measuring numerical changes in complex groups and interpreting collective phenomena.

Selected Statistical Terms

1. **Variables:** A characteristic which takes on different values in different places, persons or things and it is denoted by X, and if numbering as a sequence is required it is represented as $X_1, X_2, \ldots X_n$, n being the symbol for the number in the series. Data values taken by variable is called data. Characteristic of a person is a variable (like age) 20 years of age is a data.
2. **Constants:** Quantities which do not vary such as $\pi = 3.14$. In biostatistics some examples of constants are mean, standard deviation, standard error, correlation coefficient and proportion of a specified population.
3. **Observation:** An event and its measurements. Example: weight 60 kg.
4. **Data:** A set of values recorded over a number of observational units.
5. **Population:** It is the complete group of people or entire collection of measurements on which conclusions are drawn. Populations are often very large, it may be finite or infinite but it has to be completely defined. Example, all children in the age group of 3 to 5 years, all serving women doctors and so on. Such a population gives a qualitative data and can be counted if it is finite. On the other hand quantitative data is one that refers to data such as weight of children at birth, height of students in a class.
6. **Sample:** It is a part of a population under investigation, which is selected as a representative of the whole population. The characteristic of a large population can be inferred from the study of the corresponding character of representative samples selected from the population.

7. **Parameter:** It is the value or a constant of a variable, which describes the population like the mean variance, proportion, correlation coefficient etc.
8. **Parametric test:** It is the test in which population constants are used and data tend to follow one assumed distribution like normal, binomial, Poisson, etc.
9. **Nonparametric test:** Here no constant of population is used. We don't make any assumption but allocate arbitrary numbers which are marked to each category. Example: χ^2 test.

Note: Characteristics of population are called parameters while characteristics of a sample are called "statistics". It is an estimate of a population parameter.

Notations: For a population and sample values: Roman letters are used for statistics of samples and Greek letters are used for parameters of population.

Commonly used notations and symbols

Characteristics	Parameter	Statistics
Mean	μ	X
Standard deviation	σ	s
Variance	σ^2	s^2
Probability	P	p
Complement of Proportion	Q	q
Pearson Correlation Coefficient	R	r
Number of Cases	N	n

Other commonly used symbols

$=$ Equal to
$\neq$ Not equal to
$>$ Greater than
$<$ Less than
$\geq$ Greater than or equal to
$\leq$ Less than or equal to
O Observed value
E Expected value

K	Number of groups of classes
ρ	Spearman's Correlation coefficient or Rank Correlation Coefficient
df	Degrees of freedom
$\lvert x\rvert$	Absolute/Positive value of x
$p(A)$	Probability of event A
$p(A/B)$	Probability of event A given that event B has happened.
H_0	Null Hypothesis
H_1	Alternative Hypothesis
α	(Alpha) probability of Type I error
β	(i) (Beta) probability of Type II error
	(ii) Population value of the slope of the regression line
δ	(Delta) Population Mean difference between 2-paired measurements
λ	(Lambda) Used to denote terms in the log-linear model
μ	(Mu) Population Mean
π	(pi) Population proportion
ρ	(Rho) Population Correlation
σ	(Sigma-Lower Case) Population Standard Deviation
Σ	(Sigma-Upper Case) Symbol to denote a sum
CV	Coefficient of variation
e	Base of Natural logarithm
E	Expected Frequency
F	Symbol for F test and distribution
n	Sample size
r	Sample correlation
r^2	Squared Correlation
SD	Sample Standard Deviation
SE	Standard Error of the Mean
t	Symbol for the t ratio
χ	Symbol for Chi-Square test
$\bar{X}$	(X Bar) Sample Mean
z	Symbol for Z ratio

Nature of Measurement and Types of Variables

S.No.	Type of variable	Nature of measurement	Analysis of data for which useful	Calculations that can be used
1.	Nominal	Qualitative	Nonparametric	Simple statistics like percentage, Proportion
2.	Ordinal	Qualitative	Statistic	Simple statistics like percentage, Proportion
3.	Interval	Quantitative	Parametric	Median, Centiles, Rank Correlation Coefficient
4.	Ratio	Quantitative	Statistic	Mean, S.D., t-test, z-test, F-test

Nominal Variable takes values which are categories or names when the number of categories of a qualitative variable is two, like male/female. It is called binary, dichotomous or an attribute variable.

Ordinal Variable has an order in the categories.

Discrete Variables are quantities which can take only integral values e.g. no. of students in a class, no. of white blood cells in 2 mm^3 of blood. They are also called Discontinuous or **Ameristic** variables.

Continuous Variables are quantities, which can take any value in a given range, a value between any two possible values e.g. body weight, height etc., are continuous variables.

Note: Ratio, Interval and ordinal scale data may be either continuous or discrete. But nominal scale data are discrete.

II. Collection, Classification and Tabulation of Data

For any statistical investigation, the purpose and the expected outcome from the investigation must be clearly known.

Data collected for such purpose can be primary or secondary. The collected data is said to be primary, when the investigator collects the data directly by personal survey, by appointing enumerators to gather information or by preparing questionnaires and get the replies to these questionnaires even by post. Secondary data are data available from public records or registers. Experiments and surveys are mainly applied to obtain data required for specific purposes whereas records provide on the spot data for routine and continuous information.

The merit in collecting primary data is that the information gathered are more reliable than those from secondary data. On the other hand secondary data is more economical and it involves less time to collect the information required.

The data collected can be classified into two types, qualitative and quantitative.

Qualitative Data

- The fundamental rule to present a qualitative data is to enumerate the number of observations for each category of the variable and present the numbers and percentage in a table.
- The data presented in the table should be self explanatory.
- The presentation should be compact and the table should be readily comprehensible.
- Percentage should be rounded to one decimal place.
- The number of observations is referred to as frequency.
- Give the total frequency in the table to get the total number of observations. Qualitative data can also be presented diagrammatically. The types of such diagrams are *Bar and Chart.*

Bar diagram

In the Bar diagram, by choosing a proper scale and making the horizontal and vertical axes, categories of variables and their frequencies are shown on these two axes respectively. Bars are constructed to show the frequency or relative frequency for each class of the variable.

Generally the bars are equal in width. There is a space between the bars. The height of the bar indicates the frequency of each class. The scale should start at zero, only then the heights of the bar will be proportional to the frequencies. The bar diagrams are suitable for representing spatial series. When the variable is represented on the vertical axis and the frequency on the horizontal axis the bars will be horizontally placed.

Pie chart/Pie diagram

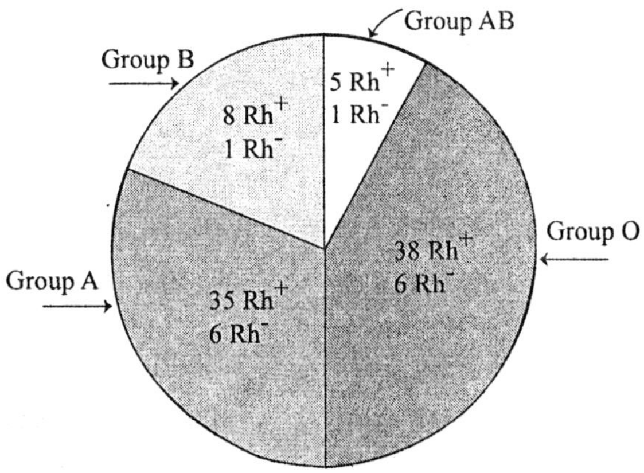

100 blood samples collected in an area showing blood group distribution with Rh^+ and Rh^- factors

Fig. 7.1

Here a circle is drawn whose total area will represent the total frequency. The area of the circle is divided into number of segments whose areas will be proportional to the observed frequencies in each class of the variables.

Note: In most cases, pictorial representation of a qualitative variable by means of bar diagram is preferred, as it is also easy to construct. It can also be used to display quantitative data.

Quantitative Data

Quantitative data can be presented by means of a histogram. It is a bar chart for a quantitative data. Before presenting a data by means of a histogram the following should be borne in mind.

- Whether the variable is discrete or continuous, the number of times it occurs must be counted and recorded as its frequency.
- When the variable is a continuous one, the variable value should be grouped into class intervals to completely cover the entire data.
- There should be no overlapping.
- The limits for these intervals are chosen to fit each observation in one of these classes.
- The width of the class intervals are kept equal.
- If the upper class limit of one class is less than the lower class limit of the next class in the table true class limits are formed to equalize and have complete continuity (see example).
- The class interval will be the difference between the true upper class limit and the true lower class limit.
- The class midpoint will be the value of the variable midway between these true class limits.

Example:

Number of flowers/twig	11–15	16–20	21–25	26–30	31–35	36–40	41–45
Number of flowers	3	4	11	12	9	7	4

Classes (with true class limits)	**Frequency**	**Midpoints**
10.5–15.5	03	13
15.5–20.5	04	18
20.5–25.5	11	23
25.5–30.5	12	28

contd. from previous page

Classes (with true class limits)	Frequency	Midpoints
30.5–35.5	09	33
35.5–40.5	07	38
40.5–45.5	04	43

Histogram

Histogram Table

Selling Price (Rs.)	Frequencny	Selling Price (Rs.)	Frequencny
50,000–74,999	1	75,000–99,999	9
100,000–124,999	22	125,000–149,999	47
150,000–174,999	31	175,000–199,999	23
225,000–249,999	2	250,000–274,999	3
275,000–299,999	2	300,000–324,999	1

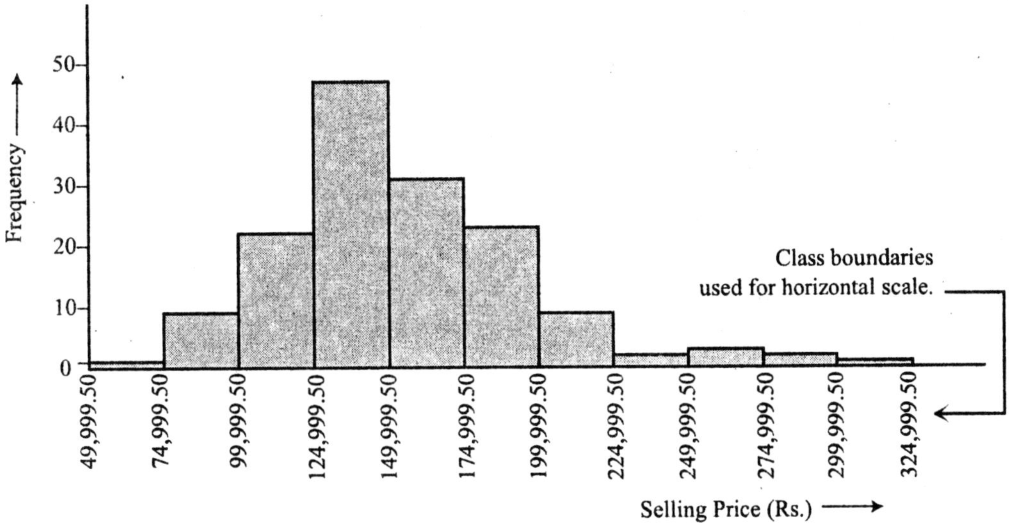

Fig. 7.2

In the case of histogram the area of the bar is more important. Since class intervals are equal, the area is proportional to the height. The total area of all bars will be proportional to the total frequency. The bars have no space in between them unlike the bars for qualitative presentation. In the bar diagram the order in which the bars are drawn is immaterial. Histogram gives a clear picture of the shape of the distribution.

Points to remember while constructing the histogram.

- There should be no overlapping of the class intervals.
- They should completely cover the data.

- Based on the number of variables under consideration the entire range should be divided into class intervals between 5 and 20 in numbers.
- Class intervals should be maintained to be equal.
- Open ended intervals should be avoided.
- The edge of the bars in the histograms must be drawn on the true class limits.
- Number of class intervals should be neither too many nor too few.

Frequency polygon

An alternative method of presenting a frequency distribution is by means of a *Frequency Polygon*.

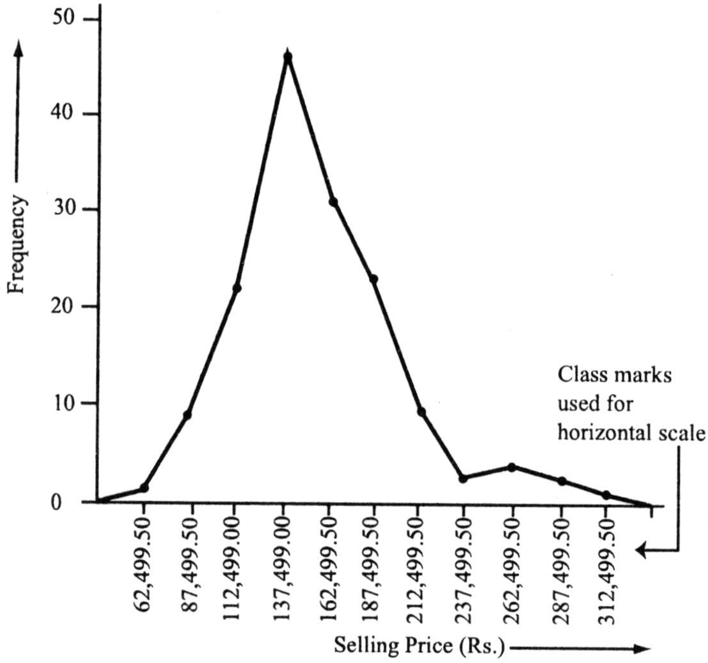

Fig. 7.3

A *frequency polygon* is a variation of the histogram with the following changes.

1. Bars are replaced by dots that are conected to form a line graph.
2. *Class marks* on the horizontal scale locate the dots directly above them.
3. The line graph is extended at the left and right so that it begins and ends with a frequency of 0.

The frequency polygon is constructed by joining the midpoints of each bar of the histogram by straight lines and by joining the top of the 1st bar to the horizontal axis at the empty class next to it.

The area under the frequency polygon will be equal to the area of the bars with histogram. Just like the histogram, the area enclosed by the frequency polygon is proportional to the total frequency.

Frequency curve

When the number of observations is very large, the data is represented by means of a frequency curve. The histogram in this case will also have similar shape. It will have more number of bars but the width of each bar will be narrower. Therefore the midpoints of each bar will become closer so that the frequency polygon will tend to become a curve, the frequency curve.
The frequency curve enables us to know more than the information furnished by the data. If the given data is only for a sample from a population, its frequency curve enables to gather an idea of the population to which the sample belongs. Frequency between any two units can also be found, as the area between the two limits is proportional to the total frequency between these limits.

Cumulative polygon/Ogive

Similar to representation of frequencies graphically, cumulative distribution can also be depicted pictorially. They are presented in cumulative frequency polygons (called Ogives). The cumulative frequency polygon is obtained as follows:

- The upper limit of the classes are plotted along the horizontal axis.
- The cumulative frequencies of various classes are marked along the vertical axis.
- The points corresponding to cumulative frequency at each upper limit of the frequencies are joined by a smooth curve.
- This curve is called the cumulative frequency curve or "Ogive".

Note:

1. The choice of the direction of cumulation is immaterial.
2. The graph of the curve plotted by starting from the lowest to the highest value of the variable is called the "Less than Ogive".
3. The curve drawn by starting from the highest class boundary of the variable is called "More than Ogive".

Uses

1. Using this mean, quartiles and percentiles or values of the variable can be found.
2. Cumulative frequency corresponding to a value of the variable can be found.
3. Number of observations that are expected to lie between two given values can also be found.

Histogram and Frequency Polygon Compared

Histogram	Frequency polygon
1. Used for graphical representation of frequency distribution	1. Used for graphical representation of frequency distribution
2. Essentially a Bar diagram	2. It is a line graph of the distribution
3. Trend of the distribution is not known	3. Part of the curve itself is easy to let us know the trend of the distribution
4. Gives clear and accurate picture of the relative proportions of frequency for various intervals.	4. Two or more curves can be drawn on the same axis to represent two or more distribution and hence more useful.
	5. Points out merely the graphical relationship between midpoints and frequency. The distribution frequencies within each class interval cannot be found.

III. Measures of Central Tendency

Having collected the data and presented the data in any form and having got a frequency distribution of the data alone cannot give us all the requirements of types of data summarization.

While analyzing and interpreting the classified data, it may be noted that middle of the range of collected data will communicate more information about the data. This value is an average or a measure of central tendency. These measures of central tendency describe a property of population.

The various parameters of measures of central tendency are

1. Mean (Arithmetic mean)
2. Median
3. Mode
4. Geometric mean

Arithmetic Mean

The arithmetic mean or simply mean is the most frequently used measure of central tendency in statistics. It is the arithmetic average of all observations.

The population mean is given by the formula,

$$\mu = \frac{\sum X_i}{N}$$

where, Σ (capital sigma) means summation and X_i takes on all values from X_i to X_N when N refers to the number of observations.

The sample mean denoted by $\bar{X}$ is the most unbiased and efficient estimate of the population mean and it is given by

$$\bar{X} = \sum_{i=1}^{n} \frac{X_i}{n} \quad \text{(or)} \quad \bar{X} = \frac{\sum X_i}{n}$$

where n denotes the sample size.

Note:

1. Mean has the same unit of measurement as the individual observations.
2. When the individual observations are small in size use the above method to calculate the mean.
3. When individual observations are large in size either use the above method or assume an arbitrary mean or working origin A. Then find the difference between each origin and A. Sum up these differences. Divide this sum by n, the no. of observations. This gives the mean of differences or deviations d given by $\bar{x}$, i.e.,

$$\bar{x} = \frac{[(X_1 - A) + (X_2 - A) + \cdots + (X_n - A)]}{n} = \frac{\sum(X - A)}{n}$$

$$= \frac{\sum X}{n} - \frac{nA}{n}$$

$$= \left(\frac{\sum X}{n}\right) - A$$

Required arithmetic mean $\bar{X} = \dfrac{\sum X}{n} = A + \bar{x}$

Note:

4. If the sample contains identical data for a number of values of the variable one can present the data by grouping them and form the frequency distribution table. In such case the sample mean is found by forming the sum of the products of 'f' and 'x' and dividing this sum by 'n'

$$\text{i.e.,}\quad \bar{X} = \frac{\sum fx}{n} = \frac{\sum fx}{\sum f}$$

where, x is the individual value

f is the frequency with which x occurs

$n = \sum f$ represents the number of observations

Note:

5. Given a grouped data i.e., a continuous series proceed as follows:

(i) Calculate mid values m of the class interval.

(ii) Form the products fm, where f is the frequency of each class.

(iii) Obtain the sum $\sum fm$.

(iv) Divide this sum by $n = \sum f$

(v) Calculate the arithmetic mean.

$$\bar{X} = \frac{\sum fm}{\sum f}$$

Note:

6. Series with open end classes. Here either the lower end of the 1st class interval or the upper end of the last class interval or both are not known. Depending upon the class-intervals, make an assumption for these class intervals and proceed.

Median

Another important and widely used measure of central tendency is the median. The median can be defined as that of the variable, which divides the group into two equal parts, one part comprising all values greater and the other part all values less than the median.

Calculation of median

1. A series of individual observations

- Arrange the data in ascending or descending order
- The median is the central observation, i.e., it is the size of the $\frac{(n+1)}{2}$ th term where, n is the number of observation

Note:

(a) If n is odd say 25, $\frac{(n+1)}{2} = \frac{26}{2} = 13$, the median is the value corresponding to 13th observation.

(b) If n is even say 24, $\frac{(n+1)}{2} = \frac{25}{2} = 12.5$, median will be the mean of the 12th and 13th observations.

2. Discrete series of observations

- Arrange the data in ascending or descending order.
- Obtain the cumulative frequencies.
- Median = size of the $\frac{(n+1)}{2}$th item where n is the total number of observations.
- Depending on whether n is odd or even the median will be the value corresponding to the cumulative frequency which equals $\frac{(n+1)}{2}$ or the next higher value to this.

3. Continuous series

- Let n be the number of observations.
- Find $\frac{n}{2}$, which will be the rank of the median to decide the class in which the median lies.
- Median $= L + \frac{\left(\frac{n}{2} - cf\right)}{f} \times i$

where,

L = lower limit median class,
cf = cumulative frequency of the class preceding the median class,
f = frequency of the median class,
i = the class interval

4. **Unequal class intervals** The formula in (3) holds good straight away. There is no need to make the class intervals equal.

5. **Open end classes**

As the values of the extreme ends are not involved, the formula holds good for calculation of median in this case also.

Mode

It is the value that occurs most frequently.

1. **Individual observations:** By inspection we conclude mode as the value that occurs maximum number of times.
2. **Discrete series:** As mode is affected by frequencies of neighbouring items too, grouping is essential to estimate the point of maximum concentration.
3. **Continuous series:**
   - Locate the modal class which will be that class with maximum frequency.
   - Calculate the mode by the formula,

$$\text{Mode} = L + \frac{\Delta_1}{\Delta_1 + \Delta_2} \times i$$

where,

L = lower limit of the modal class
Δ_1 = difference between the frequency of modal class and the preceding class (with sign ignored)
Δ_2 = difference between the frequency of modal class and the succeeding class (with sign ignored)
i = interval of the modal class

Note:

- This formula is applicable when the class intervals are uniform.
- It is not applicable in bimodal distribution.
- When class intervals are unequal make them equal and change the frequencies accordingly before applying the formula.

4. **Graphical location of Mode**
   - Draw the histogram of the data.
   - Modal class is the tallest rectangle.

- Draw a straight line from the top right corner of the tallest rectangle to the top right corner of the immediate left rectangle.
- Draw a straight line from the top left corner of this tallest rectangle to the top left corner of the rectangle to its immediate right.
- The x value corresponding to the meeting point of these 2 line is the mode, which is located by drawing a perpendicular line to the x-axis from this intersecting point.

In a perfectly symmetrical distribution, all the three central measures will be equal. In a positively skewed distribution the values of mean, median and mode will be in descending order. In a negatively skewed distribution this order will be mode, median and mean.

In general an empirical relationship exists between these three, which is expressed by Karl Pearson.

i.e., Mean − Mode = 3 (Mean − Median)

or, Mode = 3 Median − 2 Mean

Geometric Mean

Among the measures of central tendencies, the less used ones are the geometric mean and the harmonic mean. The geometric mean is the n-th root of the product of n items in a series.

i.e. $\text{GM} = \sqrt[n]{x_1 x_2 \dots x_n}$

n = no. of items

x = value of the variable

Note: When the number of items is more we take the logarithms.

$$\begin{aligned}\text{i.e.,}\quad \log \text{GM} &= \frac{1}{n}[\log(x_1 \dots x_n)]\\ &= \frac{1}{n}[\log x_1 + \log x_2 + \dots + \log x_n] = \frac{\sum \log x}{n}\\ &= \text{the arithmetic mean of the logarithms of the values of the variables}\end{aligned}$$

$$\text{Therefore, GM} = \text{antilog}\left(\frac{\sum \log x}{n}\right)$$

1. **Individual observations**

- Obtain the logarithms for each x.
- Sum up the logarithms ($\sum \log x$).

- Divide this sum by n, the number of observations $\left(\frac{\sum \log x}{n}\right)$.
- Find the antilogarithm of the above value to get, GM = antilog $\left(\frac{\sum \log x}{n}\right)$.

2. **Discrete series**

- Find the logarithms for each x.
- Multiply these logarithmic values by the respective frequencies .
- Find the sum of these products $\sum f(\log x)$.
- Divide this sum by $\sum f$.
- Find the antilogarithm of this value to get GM.

i.e., GM = antilog $\frac{\sum f(\log x)}{\sum f}$,

where, $\sum f$ is the total frequency, x is the value of the variables.

3. **Continuous series**

Here instead of individual values x, take the mid points of the various class intervals.

- Find the mid points m for each class interval.
- Find the logarithms of m, $(\log m)$
- Multiply these logarithmic values by the respective frequencies $f(\log m)$.
- Find the sum of these products $\sum f(\log m)$.
- Divide the sum by $\sum f$.
- Find the antilogarithm of this value to get GM.

Geometric mean GM = i.e., GM = antilog $\frac{\sum f(\log m)}{\sum f}$

where, $\sum f$ is the total frequency, m is the mid value of the class intervals.

Harmonic Mean

This is mainly used to find an average rate. It is defined as the reciprocal of the arithmetic mean of the reciprocals of the variables. Thus, the harmonic mean (HM) of the n variables $x_1, x_2, \cdots, x_n$

$$\text{HM} = \frac{n}{\frac{1}{x_1} + \frac{1}{x_2} + \ldots + \frac{1}{x_n}}$$

$$= \frac{n}{\left(\sum \frac{1}{x_i}\right)}$$

$$\text{or} \quad \frac{1}{\text{HM}} = \frac{\sum \frac{1}{x}}{n}$$

1. **Individual observations**

- Find the reciprocals of the values of the variables $\left(\frac{1}{x}\right)$
- Find their mean $\frac{\sum \frac{1}{x}}{n}$
- Find its reciprocal

$$\text{HM} = \frac{n}{\left(\sum \frac{1}{x}\right)}$$

2. **Discrete series**

- Divide each frequency by the respective value of the variable $\left(\frac{f}{x}\right)$.
- Find their sum $\sum \left(\frac{f}{x}\right)$.
- Divide this sum by $\sum f$.
- Take the reciprocal of this value to get HM

$$\text{HM} = \frac{\left(\sum f\right)}{\left(\sum \frac{f}{x}\right)}$$

3. **Continuous series**

- Find the mid points m of class intervals .
- Divide each frequency by the corresponding mid value m.
- Find the sum of these values $\left(\sum \frac{f}{m}\right)$.
- Divide this sum by $\sum f$.
- Find the reciprocal of this value to get HM

$$\text{HM} = \frac{\sum f}{\sum \frac{f}{m}}$$

Note: AM > GM > HM except when the number of variables and values of the variables are identical.

Comparison of Mean, Median and Mode

Mean	Median	Mode
Characteristics:	**Characteristics:**	**Characteristics:**
1. Mean is always definite.	1. Easily calculated and not affected by extreme values.	1. Simple and easily understandable.
2. All items of a series must be included for calculation based on all observations.	2. For quantitative data an ideal measure of central tendency	2. Not affected by values of extreme observations.
3. It is capable of further algebraic treatment.	3. Easily defined, easy to understand.	3. All values in the series are not considered.
4. Simple and easy to calculate. It does not require large number of observation.	4. Can be used for unequal class distributions	4. Simple and precise.

contd. from previous page

Mean	Median	Mode
Characteristics:	**Characteristics:**	**Characteristics:**
5. Rigidly defined	5. Can be determined graphically.	5. Can be located in open end distribution.
6. Good basis for comparison.	6. Used for Ordinal data or numerical data when the distribution is skewed.	6. Can be found graphically.
7. Most reliable.	7. Can be determined not only for interval and ratio scale but also for data on the ordinal scale.	7. Can be used on nominal, ordinal, interval and ratio scale mainly for bimodal distribution.
8. Not affected by fluctuation of sample.		
9. Cannot be determined graphically.		
10. Used for numerical data and symmetric distribution.		
11. Applicable to ratio or interval scale.		
Demerits:	**Demerits:**	**Demerits:**
1. Affected by extreme items.	1. May not be representative of a series in many cases.	1. It is not based on all observation; therefore it is not rigidly defined.
2. Cannot be determined even if one value is not known.	2. Not suitable for further mathematical treatment.	2. Gives only rough idea of central tendency.
3. Assumption has to be made in open-end classes to calculate Arithmetic mean.	3. Calculations of mean in a continuous series requires interpolation.	3. Not capable of further mathematical treatment.
	4. Likely to be affected by fluctuation of sample.	4. There may be more than one modal value.
	5. Unsuitable for large and small items in a series.	5. Most unstable.
		6. Not clearly defined.

Worked out Examples

1. *Calculate the arithmetic mean from the following data.*

x: 5, 10, 15, 20, 25, 30, 35, 40, 45, 50

f: 20, 43, 75, 76, 72, 45, 39, 9, 8, 6

x	f	fx
5	20	100
10	43	430
15	75	1125
20	76	1520
25	72	1800
30	45	1350
35	39	1365
40	9	360
45	8	360
50	6	300
	$\sum f = 393$	$\sum fx = 8710$

$$\text{Mean} = \frac{\sum fx}{\sum f} = \frac{8710}{393} = 22.16$$

2. *Find the Arithmetic mean given the following data.*

Age at last birthday:	16–20	21–25	26–30	31–35	36–40	41–45	46–50
Number of Persons:	12	25	31	42	36	24	30

Age in Years	Number of persons f	Mid x m	fm
16–20	12	18	216
21–25	25	23	575
26–30	31	28	868
31–35	42	33	1386
36–40	36	38	1368
41–45	24	43	1032
46–50	30	48	1440
	$\sum f = 200$		$\sum fm = 6885$

$$\text{Mean} = \frac{\sum fm}{\sum f} = \frac{6885}{200} = 34.43$$

3. *From the following table of monthly wages (in Rs) of 12 people, find out the Median.*

300, 350, 250, 420, 400, 280, 270, 480, 360, 325, 340, 290.

Arrange the wages values in ascending order.

250, 270, 280, 290, 300, 325, 340, 350, 360, 400, 420, 480.

Number of items = 12 (An even number)

$$\therefore \quad \frac{N}{2} = \frac{12}{2} = 6$$

There are two middle terms 6th and 7th.

$\therefore$ The median is the average of these two terms and it is $= \dfrac{325 + 340}{2}$

$$= \frac{665}{2}$$

$$= 332.5$$

Hence the estimated median wages is Rs. 332.5

4. *The following table gives the age of married women at first childbirth. Find the median.*

Age at birth of first child	16	17	18	19	20	21	22	23	24	25	26	27
Number of married woman	28	184	205	256	380	452	240	175	90	85	59	46

Age at birth of 1st child (x)	Number of married women (f)	fx	Cumulative frequency (cf)
16	28	448	28
17	184	3128	212
18	205	3690	417
19	256	4864	673
20	380	7600	1053
21	452	9492	1505
22	240	5280	1745
23	175	4025	1920

contd. from previous page

Age at birth of 1st child (x)	Number of married women (f)	fx	Cumulative frequency (cf)
24	90	2160	2010
25	85	2125	2095
26	59	1534	2154
27	46	782	2200
	$\sum f = 2200$	$\sum fx = 45128$	

$$\text{Mean} = \frac{\sum fx}{\sum f} = \frac{45128}{2200} = 20.51; \quad N = \sum f = 2200$$

$$\text{Median} = \text{Size of } \frac{(N+1)}{2} \text{ observation} = \text{Size of } 1100.5 \text{ th observation}$$

And this lies in the interval 1053 to 1505. Therefore its value is 21.

5. *Find the median from the following table.*

Soil pH	5	5.5	6	6.5	7	7.5	8	8.5	9	9.5
Number of plants germinated	4	7	7	12	14	20	24	30	16	15

Soil pH (x)	Number of plants (f)	Cumulative Frequency (cf)
5.0	4	4
5.5	7	11
6.0	8	19
6.5	12	31
7.0	14	45
7.5	20	65
8.0	24	89
8.5	30	119
9.0	16	135
9.5	15	150

$$\text{Median} = \text{Size of the } (N+1)/2\text{th term}$$
$$= \text{Size of the } (150+1)/2\text{th term}$$
$$= \text{Size of the } 75.5\text{th term } (N = \text{Total frequency} = 150)$$
$$= 8$$

6. *Calculate the median for the following data.*

Marks:	25–30	30–35	35–40	40–45	45–50	50–55	55–60	60–65	65–70	70–75
Number of Students:	6	4	15	30	27	18	25	12	8	5

Marks (x)	Number of Students (f)	Cumulative frequency (cf)
25–30	6	6
30–35	4	10
35–40	15	25
40–45	30	55
45–50	27	82
50–55	18	100
55–60	25	125
60–65	12	137
65–70	8	145
70–75	5	150

$N = 150 \quad \therefore \ \frac{N}{2} = 75$; The median class is 45–50.

Here $\frac{N}{2} = 75$, $l = 45$, $C = 5$.

$$\text{Median } M = l + \frac{\left(\frac{N}{2} - m\right)}{f} \times C = 45 + \frac{75 - 55}{27} \times 5$$

$$= 45 + 20 \times \frac{5}{27} = 45 + 3.704 = 48.7$$

7. *Compute the mode for the following distribution.*

Age in Months	3	4	5	6	7	8	9	10
Number of Children	10	12	10	20	15	18	20	10

The frequency table gives

Number of Children	Frequency
10	3
12	1
15	1
18	1
20	2

Since 10 occurs maximum number of times i.e. 3, the modal value is 10.

8. *Calculate the mode from the following table.*

Monthly Income:	400–450	450–500	500–550	550–600
Number of Workers:	10	15	122	207

Monthly Income:	600–650	650–700	700–750	750–800
Number of Workers:	328	415	294	256

x	f
400–450	10
450–500	15
500–550	122
550–600	207
600–650	328
650–700	**415**
700–750	294
750–800	256

The maximum frequency is 415. Therefore 650–700 is the modal class.

$$\text{Mode } M = L + \left(\frac{\Delta_1}{\Delta_1 + \Delta_2}\right) \times i$$

L is the lower limit of the modal class = 650

Δ_1 = Diff. between the frequencies of the modal class and the previous class

$= 415 - 328 = 87$

Δ_2 = Diff. between the frequencies of the modal class and the next class

$= 415 - 294 = 121$

$i = 50$ (class interval)

$$\therefore \quad M = 650 + \frac{87}{87 + 121} \times 50 = 650 + \frac{87 \times 50}{208}$$

$$= 650 + 20.913 = 670.9$$

9. *Calculate the mode from the following.*

Number of seeds /plant	50–100	100–150	150–200	200–250	250–300	300–350	350–400
Number of plants	10	15	28	29	36	22	9

x	f
50–100	10
100–150	15
150-200	28
200–250	29
250–300	**36**
300-350	22
350–400	9

The maximum frequency i.e. 36 gives the modal class as 250–300.

$$\text{Mode } M = L + \frac{\Delta_1}{\Delta_1 + \Delta_2} \times i = 250 + \frac{(36-29)}{(36-29)+(36-22)} \times 50$$

$$= 250 + \frac{7}{7+14} \times 50 = 250 + \frac{1}{3} \times 50 = 250 + 16.67$$

$$= 266.7$$

10. *Find the mean, median and mode from the following frequency distribution, which gives the age at first walking of infants in a locality.*

Age in months	8–8.9	9–9.9	10–10.9	11–11.9	12–12.9
Number of infants	1	9	20	29	60
Age in months	13–13.9	14–14.9	15–15.9	16–16.9	17–17.9
Number of infants	32	30	14	8	1

Age in months x	Number of Infants m	Mid value of x (m)	(fm)	Cumulative frequency (cf)	
8–8.9	1	8.45	8.45	1	
9–9.9	9	9.45	85.05	10	
10–10.9	20	10.45	200.90	30	
11–11.9	29	11.45	332.05	59	
			.		102
12–12.9	60	12.45	747.00	119	
13–13.9	32	13.45	430.4	151	
14–14.9	30	14.45	433.5	181	
15–15.9	14	15.45	216.3	195	
16–16.9	8	16.45	131.6	203	
17–17.9	1	17.45	17.45	204	
	$\sum f = 204$		$\sum fm = 2602.70$		

$$\text{Mean} = \frac{\sum fm}{\sum f} = \frac{2602.7}{204} = 12.76$$

Upper Class Limit	8.95	9.95	10.95	11.95	12.95	13.95	14.95	15.95	16.95	17.95
Cumulative Frequency	1	10	30	59	119	151	181	195	203	204

$$\frac{N}{2} = \frac{204}{2} = 102.$$

This corresponds to median. The corresponding end values are 11.95 and 12.95, which represents the median class end values.

$$\text{Median} = L + \frac{(\frac{N}{2} - m)}{f} \times c = 11.95 + \frac{(102 - 59)}{60} \times 1$$

$$= 11.95 + \frac{43}{60} = 11.95 + 0.7166$$

$$= 12.67$$

Mode: 12–12.9 is the modal class as it has the maximum frequency. The lower boundary of the class is 11.95. $f_1 = 29$, $f_2 = 32$, $c = 1$

$$\text{Mode} = l + \frac{f_2}{f_1 + f_2} \times c = 11.95 + \frac{32}{29 + 32} \times 1$$

$$= 11.95 + 0.525 = 12.475 = 12.48$$

IV. Other Measures of Location and Measures of Dispersion

The already discussed measures of central tendency tend to locate the middle of a distribution. Other points in the distribution may also have to be located in some cases. The median divides the distribution into 2 halves with 50% above and 50% below it. The median is called as the 50th percentile. It is now worthwhile to find out quartiles, centiles and all these can be termed as quartiles.

Quartiles can be used to form smaller groups from a numerical distribution. These quartiles determine a cut off point. The quartiles divide the data into 4 groups, Q_1, the lower quartile will have 25% of the observations below it and 75% above it, whereas Q_3 or upper quartile will have 75% of the observations below it and 25% above it. The nine deciles divide the distribution into 10 equal parts with the 1st decile or the 10th percentile dividing the distribution into 10% and 90%, while the 5th decile and 50th percentile will correspond to the median.

Locating the quartiles

If n is the number of observations in a set of individual observation arranged in ascending order of magnitude, the 1st quartile Q_1 will correspond to the $(I+1)$th observation or the average of I and $(I+1)$th observations according as $\frac{n}{4} = I + F$ or I, I and F being the integral and decimal part of $\frac{n}{4}$. For Q_3, the third quartile it will be corresponding to the $(I+1)$th observation or the average of I and $(I+1)$th observation where I is found in $\frac{3n}{4} = I + F$, I and F having the same meaning as above.

For frequency distribution quartiles are given by the formula $Q = l + \left(\frac{n}{4} - m\right) \times \frac{c}{f}$.

l = Lower boundary of the class c or the cumulative frequency $\frac{n}{4}$

m = Cumulative frequency up to l

f = Frequency of Q_1

c = Class interval

n = The total frequency

Measures of Dispersion

In order to get a clear picture of a distribution, measures of central tendency alone are not enough since this gives only an idea of the central value. Hence, a measure of variability or dispersion is also needed. They enable us to study the character of the distribution. The various measures of dispersion are.

1. Range
2. Mean deviation/Average deviation
3. Variance
4. Standard deviation
5. Quartile deviation
6. Coefficient of variation

Dispersion is the degree of the variation of the variable about a central value. Measures of dispersion of a population are parameters of the population and the sample measure of dispersion that estimate them are statistics.

The Range

It is defined as the difference between the highest and lowest value of the variable in a distribution.

Thus, if the measurements are arranged in an increasing order of magnitude as x_1, x_2, x_n then the range $= x_n - x_1$.

Note:

- The range is the simplest but most crude method of dispersion.
- Since only the extreme values of the distribution and no intermediate value is considered it is based on extreme values.
- In a given population range may change drastically from one sample to another.
- There is no range in an open end distribution as there is no upper limit or lower limit for such a distribution.
- Range is applicable to ordinal, interval and ratio scale data.

Calculation of range

(a) Interval scale data Range is calculated as the difference between the upper limit of highest class and lower limits of the smallest classes.

(b) Data grouped into frequency distribution.

- Calculate the difference between the upper limit of the highest class and lower limit of the lowest class.
- Calculate the difference by subtracting the midpoint of the lowest class from the midpoint of the highest class.

Mean Deviation

It is the average of deviations from the arithmetic mean. In other words, take the absolute values of the differences of observations from the mean, i.e., take these differences ignoring the sign and then sum up these values and divide the sum by the number of observations.

$$\text{Mean deviation} = \frac{\sum |X - \bar{X}|}{n}$$

Calculation of mean deviation

1. Set of individual observations.

- Calculate the mean $\bar{X}$.
- Find the difference between the variable values and $\bar{X}$ i.e., deviations of the values from the mean, ignoring the sign and denote these deviations by $|d|$.
- Obtain the sum of these deviations $\sum |d|$.

- Mean deviation is obtained by dividing this sum by the number of values of the variable. i.e.,

$$\text{Mean Deviation} = \frac{\sum |d|}{n}$$

where n is the number of values of the variable.

2. Discrete series

- Calculate the mean $\bar{X}$.
- Obtain the deviation of the values of the variable from the mean ignoring the sign $|d\bar{X}|$.
- Multiply the deviations with corresponding frequencies $f|d\bar{X}|$.
- Form the sum of these products $\sum f|d\bar{X}|$
- Divide the sum by the total number of observations $\sum f$ to get the mean deviation as

$$\text{Mean Deviation} = \frac{\sum f|d\bar{X}|}{\sum f}$$

3. Continuous series

- Obtain the mid values (m) of various class
- Then proceed as in Case 2

$$\text{Mean Deviation} = \frac{\sum f|d\bar{X}|}{\sum f}$$

where $|d\bar{X}|$ = Numerical value of the deviations of mid values from the mean.

Note:

- This measure has the same units as the data.
- The sum of all deviations from the mean i.e. $\sum(X - \bar{X})$ will always be equal to zero.
- Hence such a summation will not be useful as a measure of dispersion.

Variance

Variance is also called mean square or mean square deviation. For a population, it is denoted by σ^2 and as the best estimate of the population variance we have the sample variance s^2

$$\sigma^2 = \frac{\sum(X_i - \mu)^2}{N}$$

and

$$s^2 = \frac{\sum(X_i - \bar{X})^2}{n - 1}$$

Note:

- If in s^2 n is used instead of $n-1$, it results in a biased estimate of σ^2.
- s^2 being a square is always positive. It can never be negative.
- Because of the important properties with regard to probability and hypothesis testing, variance is preferred to mean deviation.
- The variance has square units.

The expression for s^2 can be simplified to be represented in the following form as a working formula

$$s^2 = \frac{\sum X^2 - \frac{(\sum X)^2}{n}}{n-1}.$$

For data with the continuous series in frequency tables

$$s^2 = \frac{\sum fX^2 - \frac{(\sum fX)^2}{n}}{n-1}$$

where f is the frequency corresponding to the variable observation X.

$n = \sum f$ is the total frequency.

Standard Deviation

It is the most useful and most widely used measure of dispersion. Simply stated, it is a measure of dispersion of different variables around the central value. It is the root mean square deviation from the arithmetic mean. From the above definition of variance it follows that the standard deviation is the positive square root of the variance.

For a population S.D

$$\sigma = \sqrt{\frac{\sum X^2 - \frac{(\sum X)^2}{N}}{N}}$$

and for a sample S.D

$$s = \sqrt{\frac{\sum X^2 - \frac{(\sum X)^2}{n}}{n-1}}$$

Note:

- Standard Deviation has the same unit as the original measurements.
- $n-1$ is used instead of n in the denominator of the formula because by using $\bar{X}$ which is mean of the given values, in the deviations the number of degrees of freedom is effectively reduced from n to $n-1$.

Calculation of Standard Deviation

1. Given a set of individual observations.
2. Given a discrete series.
3. Given a continuous series.

1. Individual observation

(a) Calculate the mean $\bar{X}$

- Obtain the difference of each observation from the mean $X - \bar{X}$.
- Square the difference $(X - \bar{X})^2$.
- Obtain the sum of these squares $\sum(X - \bar{X})^2$.
- Divide this sum by $n - 1$, where n is the number of observations.
- Calculate the square root of this value to get the Standard Deviation as

$$s = \sqrt{\frac{\sum(X - \bar{X})^2}{n - 1}}$$

(b) Simple method

This method is applied when the mean involves a function and therefore the calculations become laborious

- Instead of the mean $\bar{X}$ an assumed mean M is selected.
- Find the values $X - M = d$ for each variable X.
- Square each d.
- Calculate $\sum d^2$.
- Calculate $(\sum d)^2$.
- SD is given by:

$$s = \sqrt{\frac{\sum d^2 - \frac{(\sum d)^2}{n}}{n - 1}}$$

n being the number of observations.

2. Discrete series

(a) Procedure for direct method:

- Obtain the mean $\bar{X}$.
- Calculate the deviations from the mean $X - \bar{X}$.
- Find the squares of deviations $(X - \bar{X})^2$.
- Find the product of these squares with the respective frequency as $f(X - \bar{X})^2$.
- Obtain the sum $\sum f(X - \bar{X})^2$.
- Divide the sum by $n - 1$ where n is the number of observations

$$\frac{\sum f(X - \bar{X})^2}{n - 1}$$

- Finally take the square root of this sum to get standard deviation. Thus Standard Deviation

$$s = \sqrt{\frac{\sum f(X - \bar{X})^2}{n - 1}}$$

(b) Short method

As in case 1(b) when the mean involves decimal, starting with an assumed mean M, calculate $X - M$ and as above calculate the Standard Deviation as

$$s = \sqrt{\frac{\sum f(X - M)^2}{n - 1}}$$

where, $n = \sum f$.

3. Continuous series

(a) In the case of continuous series

- Calculate the mid values of class intervals (X).
- Obtain the arithmetic mean $\bar{X}$.
- Get the deviations $X - \bar{X}$.
- Square these deviations $(X - \bar{X})^2$.
- Multiply each square by the respective frequency $f, f(X - \bar{X})^2$.
- Find the sum of these products $\sum f(X - \bar{X})^2$.
- Divide the sum by $\sum f - 1$, where $\sum f$ is the total frequency.

- Calculate the square root of this value.

$$\text{S.D} = \sqrt{\frac{\sum f(X-\bar{X})^2}{(\sum f - 1)}}$$

or, calculate $\sum fX^2$ and $(\sum fX)^2$ and use these values in

$$s = \sqrt{\frac{\sum fX^2 - \frac{(\sum fX)^2}{n}}{n-1}}$$

where $n = \sum f$.

(b) Short Method

- Calculate the mid values of class intervals X.
- Start with an assumed mean M.
- Compute $d = \dfrac{X-M}{C}$.
- Compute d^2.
- Obtain fd^2.
- Obtain fd.
- Find the sums $\sum fd$ and $\sum fd^2$.
- Calculate $(\sum fd)^2$.
- Use these value in the formula.

$$\text{Standard Deviation, S.D} = \left(\sqrt{\frac{\sum fd^2 - \frac{(\sum fd)^2}{n}}{n-1}} \times c\right)$$

where,

M = assumed mean

X = mid values

C = Class width

d = $\frac{X-M}{C}$

n = $\sum f$ = Total frequency

Merits and Demerits of Standard Deviation

Merits		Demerits	
1.	It is rigidly defined.	1.	It gives more weightage to extreme values.
2.	It is based on all observations.	2.	Compared to other measures of dispersion it is difficult to calculate.
3.	It is less affected by fluctuations of sampling.		
4.	It is widely used in correlation.		

Uses

- Standard Deviation summarizes the deviations of a large distribution from the mean.
- It indicates whether the observed difference of an individual item from the mean is by chance or due to some special reason i.e., significant.
- It helps in calculation of standard error which determines whether the difference between mean of two similar samples is by chance or real .
- It also helps in finding suitable sample size.

Comparison of Mean Deviation and Standard Deviation

Mean Deviation		Standard Deviation	
1.	Deviations are calculated from mean, median and mode.	1.	Deviations are mainly calculated only from mean.
2.	Algebraic signs are ignored while calculating mean deviation.	2.	Algebraic signs are taken into account.
3.	It is simple to calculate.	3.	It is difficult to calculate.
4.	It is not mathematically sound as algebraic signs are ignored.	4.	As algebraic signs are retained it is mathematically sound.

A sampling distribution is a distribution of sample statistics. $\sigma_{\bar{x}}$ is actually the standard deviation of the mean. A standard deviation of a statistic is referred as a standard error. Therefore $\sigma_{\bar{x}}$ can be taken as the standard error of the mean. $\sigma_{\bar{x}} = \frac{\sigma^2}{n}$ is called the variance of the mean.

Note:

- Standard error value is expressed up to two significant figures.
- Standard deviation is also expressed to two significant figures.
- Variance is expressed with twice the number of decimal places as the standard deviation.

Coefficient of variation or the coefficient of variability is defined as

$$CV = \frac{s}{\bar{X}}$$

$$\text{or} \quad CV = \frac{s}{\bar{X}} \times 100\%$$

It is expressed as a percentage.

Note:

- Since s and X have same units CV is a number i.e. it is a relative measure.
- The coefficient of variation of a sample is an estimate of the coefficient of variation of the population out of which the sample is taken i.e. it is an estimate of $\frac{\sigma}{\mu}$.
- Coefficient of variation can be calculated only for ratio scale data.

Comparison between Standard deviation and Standard error of the mean

	Standard deviation		Standard error of the mean
1.	It is the measure of dispersion of a population or a sample	1.	It is the standard deviation of the sample distribution of the mean
2.	It belongs to descriptive statistics	2.	It belongs to inferential statistics
3.	In a normal distribution only gives a range within which 95% of observations lie: Mean ± 1.96 (SD).	3.	Measures the precision to which sample mean reflects the population mean.
		4.	Gives a 95% confidence interval for the mean
		5.	It does not describe the spread of a sample/population

Quartile Deviation

The Quartile Deviation is another method of measuring dispersion and it is given by $\frac{Q_3-Q_1}{2}$. It is also known as the semi-inter quartile range. It is most useful when extreme values have to be neglected and also for distribution with open class as the ends.

It is always associated with median. Hence, it is used in skewed distribution. In normal distribution, the quartile deviation on either side of the median will cover 50% of the distribution and when 4 quartile deviations are taken, the entire distribution will be covered in normal distribution.

$$Q = 0.6745 \times \text{ S.D}$$

Coefficient of Quartile deviation

It is the ratio of quartile deviation to its mean multiplied by 100.

$$\text{Coefficient of Quartile deviation} = \frac{\text{Quartile Deviation}}{\text{Mean}} \times 100$$

Skewness and Kurtosis

1. If the mean and the median are equal the distribution of observations is symmetric.
2. If the mean is larger than the median, the distribution is skewed to the right.
3. If the mean is smaller than the median, the distribution is skewed to the left.

Skewness is defined as the lack of symmetry in the shape of a frequency distribution.

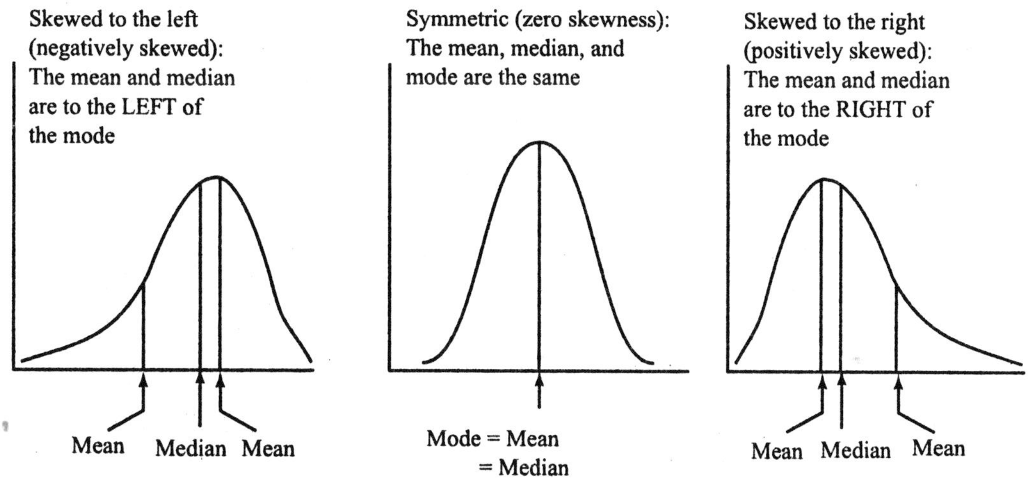

Fig. 7.4

For positive skewness the tail of the frequency curve is longer towards the right of the ordinate at the mode and for negative skewness the tail of this curve is longer towards the left of the ordinate at the mode. When skewness vanishes the curve will be symmetrical about the ordinate at the mode.

The mean, median and the mode will be in an ascending or descending order of magnitude according as the skewness is negative, positive. Hence a measure of skewness based on the averages is Mean − Mode. Of course this measure is positive, negative or zero according as the curve is positively skewed, negatively skewed or symmetric about the ordinate at the mode.

Pearson's coefficient of skewness is defined as the ratio of this measure of skewness to the standard deviation of the distribution.

Thus it is equal to $Sk_p = \frac{\text{Mean} - \text{Mode}}{\text{SD}} = \frac{3(\text{Mean} - \text{Median})}{\text{SD}}$ where S.D is the Standard Deviation.

The Quartile measure of skewness

If Q_1, M, Q_3 denote the 1st quartile, the median and the 3rd quartile of a symmetric distribution respectively, $M - Q_1 = Q_3 - M$

i.e. $Q_1 + Q_2 - 2M = 0$

If the distribution is skewed $Q_1 + Q_2 - 2M \neq 0$.

The skewness will be positive if $Q_1 + Q_2 - 2M > 0$ and negative if $Q_1 + Q_2 - 2M < 0$.

To obtain a relative measure of skewness we take the ratio of $Q_1 + Q_2 - 2M$ to twice the quartile deviation, so that the quartile measure of skewness can now be given as

$$\frac{Q_1 + Q_2 - 2M}{2\ (\text{Quartile deviation})}$$

Therefore the quartile coefficient of skewness $= \frac{Q_1 + Q_2 - 2M}{Q_3 - Q_1}$

Note:

- This measure of quartile deviation always lies between −1 and +1.
- Extreme values of observations are not taken into account.
- Hence it is unsuitable when extreme values are to be considered.
- Since the quartile coefficient of skewness is a ratio it is independent of unit of the observation and therefore the skewness of 2 different distributions can be compared.

Moments of a Frequency Distribution

For a given population distribution the first moment about the arithmetic mean is the mean of the algebraic deviations of the values of the variable from the arithmetic mean.

If μ_1 is the 1st moment about the arithmetic mean,

$\mu_1 = \frac{\sum f(X-\bar{X})}{N}$

$\bar{X}$ = arithmetic mean

f = frequency corresponding to X

N = total frequency

Since $\sum f(X - \bar{X}) = 0$ the first moment about the mean is zero for any distribution .

The second moment about the mean is $\mu_2 = \frac{\sum f(X - \bar{X})^2}{N}$

It is to be noted that μ_2 for any frequency distribution is equal to the variance of the distribution similarly the third and fourth moments about the mean are given by

$$\mu_3 = \frac{\sum f(X-\bar{X})^3}{N} \quad \text{and} \quad \mu_4 = \frac{\sum f(X-\bar{X})^4}{N}.$$

1. μ_1 the first moment of a frequency distribution with zero as the origin gives a measure of central tendency of the distribution.
2. Since the second moment is the square of the SD, it gives a measure of dispersion of the distribution
3. μ_3 gives a measure of skewness of the distribution.
4. μ_4 serves to measure kurtosis of the frequency distribution.

The third moment about the mean becomes zero only for symmetrical distribution. Hence this can be utilized to obtain a measure of skewness. The coefficient of skewness based on third moment is $\frac{\mu_3^{\frac{1}{3}}}{\sigma}$.

This takes on positive, negative and zero values according as the skewness is positive, negative, or when the curve is symmetric .

This coefficient of skewness is sometimes denoted by α_3

$$\text{i.e.} \quad \alpha_3 = \frac{\mu_3^{\frac{1}{3}}}{\sigma}$$

$$\alpha_3^6 = \frac{\mu_3^2}{\sigma^6} = \frac{\mu_3^2}{(\sigma^2)^3}$$

$$= \frac{\mu_3^2}{\mu_2^3} \qquad \text{since } \mu_2 = \sigma^2$$

$$= \beta_1 (\text{say})$$

$$\text{Thus } \beta_1 = \frac{\mu_3^2}{\mu_2^3} = \alpha_3^6$$

Note: $\sqrt{\beta_1}$, gives a measure of skewness of the frequency distribution. It has the same sign as μ_3.

Kurtosis

Kurtosis of a population is often characterized in terms of 'peakedness' or tailedness. It gives the extent to which a unimodal frequency curve is peaked. "A measure of kurtosis indicates the degree to which a curve of a frequency distribution is peaked or flat-topped".

If the peak of the frequency near the mode is flat the curve is said to be *Platykurtic.* Such a distribution might be the composite of two normal populations with the same variance by different means.

When the peak of the curve is normal the curve is *Mesokurtic*.

If the peak is sharper than the normal curve it is said to be *Leptokurtic*. Such a distribution might be the composite of 2 normal populations with the same mean but different variance. Symmetric frequency distributions:

(i) Distribution is Mesokurtic (normal)

(ii) Distribution is Platykurtic

(iii) Distribution is Leptokurtic

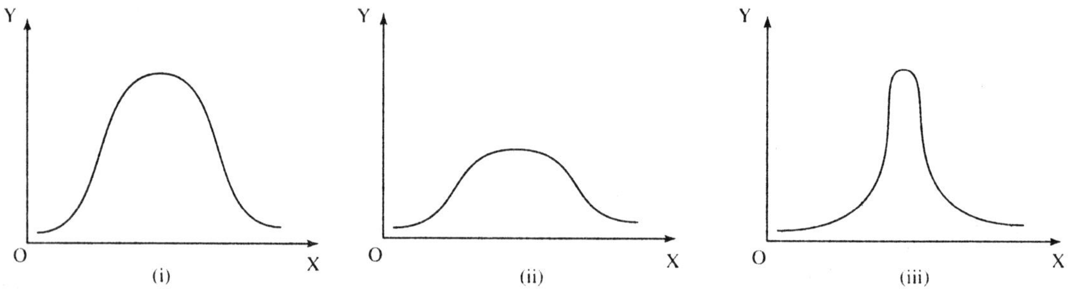

A measure of kurtosis based on moments is the ratio of the fourth moment to the square of the second moment of the frequency distribution. This measure is denoted by

$$\beta_2 = \frac{\mu_4}{\mu_2^2}$$

A second measure of kurtosis is the ratio of the percentile range to the quartile deviation of the distribution. It equals

$$\frac{P_{90} - P_{10}}{\left(\frac{Q_3 - Q_1}{2}\right)}$$

Note:

1. Unless otherwise specified the moments are always taken about the mean.
2. The second measure of kurtosis doesn't consider the extreme items. Hence it is not a satisfactory measure of kurtosis.
3. This measure of kurtosis is greater than 3.8 for leptokurtic curves; less than 3.8 for platykurtic curves; equals 3.8 for mesokurtic curves.
4. The standard value of β_2 is taken as 3.
5. The greater the value of β_2, the more peaked is the distribution.

Worked out Examples

1. *Calculate the mean deviation and the coefficient of mean deviation from the following*

Wt in grams (Vegetables)	100	120	130	140	150	160	200

Wt. in grams (X)	$dX(X-\bar{X})$	$\|dX\|$
100	−42.86	42.86
120	−22.86	22.86
130	−12.86	12.86
140	−2.86	2.86
150	7.14	7.14
160	17.14	17.14
200	57.14	57.14
$\sum X = 1000$		$\sum \|dX\| = 162.86$

$$\text{Mean } \bar{X} = \frac{\sum X}{n} = \frac{1000}{7} = 142.86.$$

$$\text{Mean deviation from the mean} = \frac{S|dX|}{n} = \frac{162.86}{7} = 23.27.$$

$$\text{Coefficient of mean deviation} = \frac{\text{Mean deviation}}{\text{Mean}} = \frac{23.27}{142.86} = 0.163$$

2. *Calculate the mean, the variance. standard deviation and the coefficient of variation from the following data on plant yield of a green gram variety.*

Plant yield in grams	18.1	19.2	20.1	20.4	22.8	22.9	23.7	24.5	25.2	26.2

Wt in grams (X)	$(X-\bar{X})$	$(X-\bar{X})^2$
18.1	−4.21	17.72
19.2	−3.11	9.67
20.1	−2.21	4.88
20.4	−1.91	3.65
22.8	0.49	0.24

contd. from previous page

Wt in grams (X)	$(X - \bar{X})$	$(X - \bar{X})^2$
22.9	0.59	0.35
23.7	1.39	1.93
24.5	2.19	4.79
25.2	2.89	8.35
26.2	3.89	15.13
$\sum X = 223.1$		$\sum(X - \bar{X})^2 = 66.71$

$$\text{Mean } \bar{X} = \sum Xn = \frac{223.1}{10} = 22.31$$

$$\text{Variance } s^2 = \frac{\sum(X - \bar{X})^2}{n-1} = 66.719 = 7.41$$

$$\text{Standard deviation} = \sqrt{7.41} = 2.72$$

$$\begin{aligned}\text{Coefficient of variation} &= \frac{s}{\bar{X}} \times 100 = \frac{2.72}{22.31} \times 100 \\ &= 0.1219 \times 100 \\ &= 12.19\end{aligned}$$

3. *The runs scored by 2 cricket players in 10 matches are given below. Determine the more consistent player of the two.*

Player A	74	75	78	77	81	79	72	71
Player B	87	80	84	88	89	80	82	86

Player A			Player B		
Score (X)	$d = X - 78$	d^2	Score (X_1)	$d_1 = X_1 - 86$	d_1^2
74	−4	16	87	1	1
75	−3	9	80	−6	36
78	0	0	84	−2	4
77	−1	1	88	2	4
81	3	9	89	3	9
79	1	1	80	−6	36
72	−6	36	82	−4	16
71	−7	49	86	0	0
$\sum d = -17, \sum d^2 = 121$			$\sum d_1 = -12, \sum d_1^2 = 106$		

78 and 86 are taken as the assumed means for the set of data for player A and player B respectively.

Player A: $\bar{X} = A + \frac{\sum d}{n}$

$$= 78 + \frac{(-17)}{8} = 78 - 2.13$$

$$= 75.87$$

$$s_2 = \frac{\sum d^2 - \frac{(\sum d)^2}{n}}{n-1}$$

$$= \frac{121 - \frac{(17)^2}{8}}{8-1} = \frac{121 - \frac{289}{8}}{7} = \frac{121 - 41.26}{7}$$

$$= 11.39$$

Variance = 11.39

Standard deviation = $\sqrt{11.39} = 3.37$

Coefficient of variation $= \frac{s}{\bar{X}} \times 100$

$$= \frac{\sqrt{11.39}}{75.87} \times 100$$

$$= \frac{3.37}{75.87} \times 100 = 4.44\%$$

Player B: $\bar{X}_1 = 86 + \frac{\sum X_1}{n_1}$

$$= 86 - \frac{(12)}{8} = 86 - 1.5$$

$$= 84.5$$

$$s_1^2 = \frac{\sum d_1^2 - \frac{(\sum d_1)^2}{n_1}}{n_1 - 1} = \frac{106 - \frac{(12)^2}{8}}{8-1}$$

$$= \frac{106 - 18}{7} = \frac{88}{7}$$

$$= 12.56$$

Variance = 12.56

Standard deviation = $\sqrt{12.56} = 3.54$

$$\text{Coefficient of variation CV} = \frac{s_1}{\bar{X}_1} \times 100$$

$$= \frac{3.54}{84.5} \times 100$$

$$= 4.19$$

CV of player $A = 4.44$
CV of player $B = 4.19$
As the CV for player B less than that of player A, B is a more consistent player than A.

4. *The following data gives the frequency distribution of 200 persons for blood sugar levels. Calculate quartile deviation and coefficient of quartile deviation.*

mg% glucose	30	40	50	60	70	80
No. of persons	15	25	48	42	52	18

Glucose mg% X	No. of persons f	Cumulative frequency cf
30	15	15
40	25	40
50	48	88
60	42	130
70	52	182
80	18	200
	$\sum f = 200$	

$$Q_1 = \text{size of } \left(n + \frac{1}{4}\right) \text{th item} = \text{size of } \left(\frac{200+1}{4}\right) \text{th item}$$

$$= 50.25\text{th item}$$

$$= 50\text{th item}$$

and it is 50.

Therefore, $Q_1 = 50$

Similarly Q_3 = size of $3\left(n + \frac{1}{4}\right)$ th item = 150.75 th item i.e., 150th item which is 70, therefore, $Q_3 = 70$.

$$\text{Quartile deviation} = \frac{Q_3 - Q_1}{2} = \frac{70 - 50}{2} = 10.$$

$$\text{Coefficient of quartile deviation} = \frac{Q_3 - Q_1}{Q_3 + Q_1} = \frac{70 - 50}{70 + 50}$$

$$= \frac{10}{120} = \frac{1}{12} = 0.083$$

5. *Calculate quartile deviation and coefficient of quartile deviation for the following data.*

Length of tadpoles (mm)	0–6	6–12	12–18	18–24	24–30
Frequency	4	8	14	15	19

Length (mm) X	Frequency f	Cumulative frequency cf
0–6	4	4
6–12	8	12
12–18	14	26
18–24	15	41
24–30	19	60

$$Q_1 = \text{size of } \left(\frac{n+1}{4}\right)\text{th item} = \text{size of } \left(60 + \frac{1}{4}\right)\text{th item} = 15\text{th item}.$$

Therefore, Q_1 lies in the group 12–18.

$$Q_1 = L + \frac{(\frac{n}{4} - cf)}{f} \times i$$

(L = lower boundary of the class in which Q_1 lies; cf = cumulative frequency of the previous class; f = frequency in the quartile class)

$$= 12 + \frac{15 - 12}{14} \times 6 = 12 + \frac{18}{14} = 12 + 1.29$$

$$= 13.29$$

Similarly, Q_3 = size of $3(n + \frac{1}{4})$th item

= size of 3×15th item

= size of 45th item

Q_3 is in the group 24–30

$$Q_3 = 24 + \frac{(\frac{3n}{4} - cf)}{f} \times i = 24 + \frac{45 - 41}{19} \times 6 = 24 + \frac{24}{19} = 24 + 1.26$$

$$= 25.26$$

$$\therefore \quad \text{Quartile deviation} = \frac{Q_3 - Q_1}{2} = \frac{25.26 - 13.29}{2} = \frac{11.97}{2} = 5.985$$

$$\text{Coefficient of QD} = \frac{Q_3 - Q_1}{Q_3 + Q_1} = \frac{25.26 - 13.29}{25.26 + 13.29} = \frac{11.97}{38.55} = 0.31$$

6. *Calculate the standard deviation of the marks obtained by 15 students in Statistics Exam.*

62, 85, 73, 81, 74, 58, 66, 72, 84, 65, 83, 80, 86, 71, 75

$$\bar{X} = \sum \frac{X}{n} = \frac{1115}{15} = 74.33$$

X	$X - \bar{X}$	$(X - \bar{X})^2$
62	12.33	152.0289
85	10.67	113.8489
73	−1.33	1.7689
81	6.67	44.4889
74	−0.33	0.1089
58	−16.33	266.6689
66	−8.33	69.3889
72	−2.33	5.4289
84	9.67	93.5089
65	−9.33	87.0489
83	8.67	75.1689
80	5.67	32.1489
86	11.67	136.1889
71	−3.33	11.0889
75	0.67	0.4489
$\sum X = 1115$		$\sum(X - \bar{X})^2 = 1089.335$

$$\text{Standard Deviation} = \frac{\sum(X - \bar{X})^2}{n - 1}$$

$$= \frac{1089.335}{1115 - 1} = 0.97785$$

$$= 0.98$$

7. *Calculate the mean, median and standard deviation of the following frequency distribution.*

Marks Scored	40–44	45–49	50–54	55–59	60–64	65–69	70–74	75–79
Number of Students	5	14	26	24	11	10	8	2

Class Interval	Mid Values (m)	Class Limits	f	cf	fm	$d = m - \bar{X}$	d^2	fd^2
40–44	42	39.5–44.5	5	5	210	-14.7	216.09	1080.45
45–49	47	44.5–49.5	14	19	658	−9.7	94.09	1317.26
50–54	52	49.5–54.5	26	45	1352	−4.7	22.09	574.34
55–59	57	54.5–59.5	24	69	1368	0.3	0.09	2.16
60–64	62	59.5–64.5	11	80	682	5.3	28.09	308.99
65–69	67	64.5–69.5	10	90	670	10.3	106.09	1060.90
70–74	72	69.5–74.5	8	98	576	15.3	234.09	1874.72
75–79	77	74.5–79.5	2	100	154	20.3	412.09	824.18
			$\sum f = 100$		$\sum fm = 5670$			$\sum fd^2 = 7043.00$

$$\text{Mean} = \bar{X} = \frac{\sum fm}{\sum f} = \frac{5670}{100} = 56.7$$

$$\text{Median} = L + \frac{\left(\frac{100}{2} - 45\right)}{24} \times 5 = 54.5 + 5 \times \frac{5}{24} = 54.5 + 1.042 = 55.54$$

$$\text{Standard Deviation s} = \sqrt{\frac{\sum fd^2}{\sum f - 1}} = \sqrt{\frac{7043}{99}} = \sqrt{71.14} = 8.43$$

8. *Calculate the mean, median, standard deviation and variance of the following data*

Height in cm	90–100	100–110	110–120	120–130	130–140	140–150
Number of Students	17	24	41	70	53	45

Class Interval	Mid values m	Number of student f	$m - \frac{A}{I} = m - \frac{125}{10}$ d	cf	d^2	fd	fd^2
90–100	95	17	−3	17	9	−51	153
100–110	105	24	−2	41	4	−48	96
110–120	115	41	−1	82	1	−41	41
120–130	125	70	0	152	0	0	0
130–140	135	53	1	205	1	53	53
140–150	145	45	2	250	4	90	180
		$\sum f = 250$				$\sum fd = 143 - 140 = 3$	$\sum fd^2 = 523$

Assumed mean $A = 125$

$$\text{Mean} = 125 + \frac{\sum fd}{\sum f} \times i \quad = 125 + \frac{3}{250} \times 10 = 125.12$$

$$\text{Median} \quad = L + \frac{\left(\frac{\sum f}{2} - C\right)}{f} \times i = 120 + \frac{125 - 82}{70} \times 10$$

$$= 120 + \frac{43}{7} = 120 + 6.14$$

$$= 126.14$$

$$\text{Standard Deviation} \quad = \sqrt{\frac{\sum fd^2 - \frac{(\sum fd)^2}{\sum f}}{\sum f - 1}} \times i = \sqrt{\frac{523 - \frac{3^2}{250}}{249}} \times 10$$

$$= \sqrt{\frac{523 - 0.036}{249}} \times 10 = \sqrt{\frac{522.964}{249}} \times 10$$

$$= 1.449 \times 10$$

$$= 14.49$$

$$\text{Variance (SD)}^2 = \frac{522.964}{249} \times 10^2 = 210.03.$$

9. *Calculate the mean, standard deviation and standard error of the mean for the following distribution.*

Class interval (Body weight in grams)	100–104	105–109	110–114	115–119	120–124	125–129	130–134
Number of Animals	7	19	42	26	13	8	5

Class Interval	Mid values m	Frequency f	cf	$d = \frac{X-A}{i}$ $A = 117, i = 5$	fd	d^2	fd^2
100–104	102	7	7	−3	−21	9	63
105–109	107	19	26	−2	−38	4	76
110–114	112	42	68	−1	−42	1	42
115–119	117	26	94	0	0	0	0

Contd. from previous page

Class Interval	Mid values m	Frequency f	cf	$d = \frac{X-A}{i}$ $A = 117, i = 5$	fd	d^2	fd^2
120–124	122	13	107	1	13	1	13
125–129	127	8	115	2	16	4	32
130–134	132	5	120	3	15	9	45
		$\sum f = 120$		$\sum fd = -101 + 44 = -57$			$\sum fd^2 = 271$

$A = 117$

$$\text{Mean} = 117 + \frac{\sum fd}{\sum f} \times i = 117 + \frac{-57}{120} = 117 - 2.375$$

$$= 114.625 = 114.63$$

$$\text{Standard Deviation} = \sqrt{\frac{\sum fd^2 - \frac{(\sum fd)^2}{\sum f}}{\sum f - 1}} \times i = \sqrt{\frac{271 - \frac{(-57)^2}{120}}{120 - 1}} \times 5$$

$$= \sqrt{\frac{271 - 27.075}{119}} \times 5 = \sqrt{\frac{243.925}{119}} \times 5$$

$$= 1.43 \times 5 = 7.15$$

$$\text{Standard error of the mean} = \frac{\text{SD}}{\sqrt{\sum f}} = \frac{7.15}{\sqrt{120}} = \frac{7.15}{10.954}$$

$$= 0.6527 = 0.65$$

10. *Calculate Pearson's Coefficient of Skewness.*

Spike Length (in cm):	12–13	13–14	14–15	15–16	16–17	17–18	18–19	19–20
Number of Plants:	30	35	52	60	80	100	75	68

Spike Length (x)	Number of plants (f)	Mid values (m)	fm	m^2	fm^2
12–13	30	12.5	375	156.25	4687.5
13–14	35	13.5	472.5	182.25	6378.75
14–15	52	14.5	754	210.25	10933.00
15–16	60	15.5	930	240.25	14415.00

contd. from previous page

Spike Length (x)	Number of plants (f)	Mid values (m)	fm	m^2	fm^2
16–17	80	16.5	1320	272.25	21780.00
17–18	100	17.5	1750	306.25	30625.00
18–19	75	18.5	1387.5	342.25	25668.75
19–20	68	19.5	1326	380.25	25857.00
	$\sum f = 500$	$\sum fm = 8315$	$\sum fm^2 = 140345.00$		

$$\text{Mean} = \bar{X} = \frac{\sum fm}{\sum f} = \frac{8315}{500} = 16.63$$

$$\text{Mode} = L + \frac{\Delta_1}{\Delta_1 + \Delta_2} \times I = 17 + \frac{20}{20+25} \times 1$$

$$= 17 + \frac{20}{45} = 17.44$$

$[\Delta_1 = 100 - 80, \Delta_2 = 100 - 75]$

$$\text{Standard Deviation S.D} = \sqrt{\frac{\sum fm^2 - \frac{(\sum fm)^2}{\sum f}}{\sum f - 1}} \times i = \sqrt{\frac{140345 - \frac{(8315)^2}{500}}{499}}$$

$$= \sqrt{\frac{140345 - 138278.45}{499}} = \sqrt{\frac{2066.55}{499}}$$

$$= 2.035$$

$$\text{Coefficient of Skewness } S_{kp} = \frac{\text{Mean} - \text{Mode}}{\text{SD}}$$

$$= \frac{16.63 - 17.14}{2.035} = -0.398$$

V. Probability and Probability Distribution

The word probability is used in the mathematical sense whenever it is desired to study the likelihood or certainty of an outcome of an experiment. For example when a coin is tossed the possible outcomes are getting the head or tail face of the coin. It is known that since these are the only two outcomes, the probability of getting H is $\frac{1}{2} = 0.5$ and that of getting T is also $\frac{1}{2} = 0.5$. These outcomes are called *events*.

In probability, an experiment is any planned process of data collection. It will consist of a number of independent trials i.e. it is repeated a number of times. It can be assumed that there will be one or more outcomes in each trial. Note that all these are performed under the same condition.

The relative frequency of an event may be defined as the proportion of the total observations of outcome that the event represents. Out of a total number of n trials, if the frequency of occurrence of an event is f, then the relative frequency of this event is $\frac{f}{n}$. It is to be noted that f can range from 0 to n so that the relative frequency can have values from 0 to 1.

It can now be said that the probability of an event is the likelihood of the event, which can be expressed by relative frequency observed from a large number of experiments under similar conditions.

If A is an outcome or event of an experiment, the probability of this event is denoted as $p(A)$. The event that is opposite of the event of interest is called a Complementary Event. The probability of a complementary event is opposite to the probability of the happening of the event. The probability of a complementary event is given by 1 − the probability of the event.

It is to be noted that:

- Probability is measured on a numerical scale in the range of 0 to 1.
- The probability of an event is always greater than or equal to zero.
- The sum of the probabilities of all outcomes is always equal to 1.
- A probability of 0 means the event is impossible and a probability of 1 implies that the event is a certainty.

Mutually Exclusive Events

Addition Rule

If the occurrence of one event excludes the occurrence of the other event or events these events are said to be mutually exclusive events. If two events A and B are mutually exclusive the probability of either A or B is the sum of the probabilities of A and B.

$$p(A \text{ or } B) = p(A) + p(B)$$

where A and B are mutually exclusive events.

Thus the probability of getting a head or tail when a coin is tossed is

$$\begin{aligned} p(H \text{ or } T) &= p(H) + p(T) \\ &= 0.5 + 0.5 = 1 \end{aligned}$$

It is to be noted that "Getting a tail" and "Getting a head" are mutually exclusive events.

The above property can be extended to more than 2 events if they are all mutually exclusive. The outcomes of rolling a die will be 1, 2, 3, 4, 5, or 6. The probability of getting 1 is $\frac{1}{6}$, that of getting 2 is $\frac{1}{6}$ and so on. The probability of getting 1 or 3 or 5 by throwing a die $= \frac{1}{6} + \frac{1}{6} + \frac{1}{6} = \frac{1}{2}$ (all these events being mutually exclusive).

$$P(A \text{ or } B \text{ or } C) = p(A) + p(B) + p(C)$$

Non-Mutually Exclusive Events

If two events are not mutually exclusive, the addition rule has to be modified. If A and B are not mutually exclusive events, i.e. if the occurrence of A does not totally negate the occurrence of B then

$$P(A \text{ or } B) = p(A) + p(B) - p(A \text{ and } B)$$

Extending this formula to three non-mutually exclusive events A, B, C

$p(A \text{ or } B \text{ or } C)$

$$= p(A) + p(B) + p(C) - p(A \text{ and } B) - p(B \text{ and } C) - p(C \text{ and } A) + p(A \text{ and } B \text{ and } C)$$

Since the last term is included in each of the six other terms, effectively it is added 3 times and subtracted three times. Hence it is added finally.

Example: To find the probability of selecting a queen or heart from a pack of 52 cards. Each category has 13 cards and in each category there is only one queen.

$$\therefore \quad p(\text{Queen}) = \frac{4}{52}.$$

$$\text{Similarly } p(\text{Hearts}) = \frac{13}{52}.$$

$\therefore$ The above two events queen and hearts have the common intersecting of element namely queen of hearts.

Therefore p (queen or hearts) $= p$ (Queen) $+ p$ (hearts) $- p$ (Queen and hearts)

$$= \frac{4}{52} + \frac{13}{52} - \frac{1}{52} = \frac{16}{52} = \frac{4}{13}$$

$$= 0.308$$

Multiplication rule

This rule is applied to find the probability of occurrence of 2 or more events when they are independent events.

If A and B are two independent events, probability of occurrence of A and B i.e.

$$p(A \text{ and } B) = p(A) \cdot p(B)$$

If there are three independent events A, B, C $P(A, B \text{ and } C)$ $P(A) \cdot P(B) \cdot P(C)$

Probability Distributions

Probability distributions are theoretical, frequency distributions of certain populations devised mathematically. That is probability distributions are not based on actual observations or experiments. But are mathematically deduced on the basis of some assumptions. These are based on the theory of probability.
The three widely used probability distributions are

1. Binomial Distribution
2. Poisson Distribution and
3. Normal Distribution.

Binomial distribution

One of the most widely used probability distribution, Binomial distribution is also known as Bernoulli distribution. It is a discrete probability distribution. It is derived from Bernoulli process wherein an experiment can have only two outcomes, which are mutually exclusive.

Suppose an event can have only either of two mutually exclusive outcomes A and B. Denote the $p(A)$ by p, then $p(B) = 1 - p$. Note also that the probability of either remains the same every time the event occurs. Each trial involving such a binomial probability is known as Bernoulli trial.

Assumption in Bernoulli process

1. Each trial has only two possible outcomes which are mutually exclusive.
2. The probability of one outcome p remains the same for each trial. Similarly the probability of the other outcome q also should be constant for all trials.
3. All trials must be independent of each other.

The binomial distribution gives the probability of occurrence of specific outcome in a given number of independent trials.

In a population of male and female students in a class let the proportion of male and female students be $p = 0.7$, $q = 0.3$ and let us take a random sample of 2 students. The probability of 1st student being a male is p and the second is also a male is p. As in this case there can be only two mutually exclusive outcomes male and female, the probability that both selected are males is $p \cdot p$ i.e. p^2, similarly the probability of both students are females is $q \cdot q$ i.e. q^2. The probability of one female and one male in the sample will be $pq + pq = 2pq$. In the present case $p^2 + 2pq + q^2$

$$= 0.7^2 + 2 \times 0.7 \times 0.3 + 0.3^2$$
$$= 0.49 + 0.42 + 0.09 = 1$$

Suppose we choose 3 samples

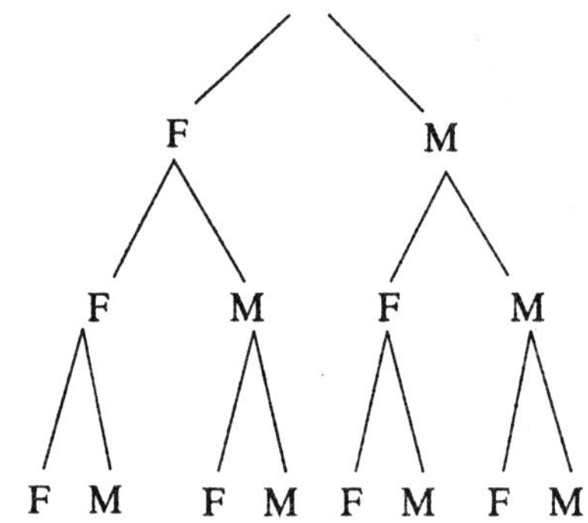

Now the outcomes will be 8 in number and they are MMM, MMF, MFM, MFF, FMM, FMF, FFM and FFF.

The probability of all 3 students being male = 0.7^3; that of all 3 being female $=0.3^3$. The probability of 2 male and one female is $3 \times 0.7 \times 0.7 \times 0.3$ and the probability of 2 female and one male is $3 \times 0.3 \times 0.3 \times 0.7$. Hence the required probability is

$$\begin{aligned} 0.7^3 + 3 \times 0.7^2 \times 0.3 + 3 \times 0.7 \times 0.3^2 + 0.3^3 &= 0.343 + 0.147 \times 3 + 3 \times 0.063 + 0.027 \\ &= 0.343 + 0.441 + 0.189 + 0.027 \\ &= 1.000 \end{aligned}$$

This also shows that there is 100% probability that 3 students selected will be one of the above combinations of sexes only.

In both these selection of samples we see the binomial coefficients in the expansion of $(p+q)^n$

$$(p+q)^2 = p^2 + 2(pq) + q^2$$

$$(p+q)^3 = p^3 + 3(p^2q) + 3(pq^2) + q^3$$

In general, if a random sample of size n is selected from a binomial population, the probability of r individuals being in one category $= p(r) = \binom{n}{r} p^r q^{n-r}$. This is the same as the probability of the $(n-r)$ individuals being in the other category.

> **Note:** $p^r q^{n-r}$ refers to the probability of sample with r items each having a probability of p and $n-r$ items each having a probability of q.

$\binom{n}{r}$ is the binomial coefficient, which also signifies number of combinations of n things, taken r at a time and has the value

$$\binom{n}{r} = \frac{n!}{r!(n-r)!}$$

A few observations about the binomial expansion:

1. There are $n+1$ terms in the expansion of $(p+q)^n$.
2. For any power n, the sum of the powers of p and q in any term is n.
3. The first term is p^n, the second term $p^{n-1}q$, next $p^{n-2}q^2$ and so on, the last term being q^n.
4. When $p+q = 1$, the sum of all the terms in the expansion will always be $1\{(p+q)^n = 1^n = 1\}$.
5. The coefficients in the expansion of $(p+q)^n$ can be exhibited and obtained from Pascal's triangle given below.

n	$(p+q)^n$
1	$p+q$
2	$p^2+2pq+q^2$
3	$p^3+3p^2q+3pq^2+q^3$
4	$p^4+4p^3q+6p^2q^2+4pq^3+q^4$ and so on

1
1 1
1 2 1
1 3 3 1
1 4 6 4 1
1 5 10 10 5 1
1 6 15 20 15 6 1

6. From the Pascal's Triangle we also note that the sum of all the coefficients in the expansion of $(p+q)^n$ is 2^n.
7. Any binomial coefficient is the sum of two coefficients in the line just above it.

Thus the probabilities of frequencies in random samples from binomial population can be calculated.

We need to know only n and p to define a binomial distribution and these are called the parameters of the binomial distribution.

Poisson Distribution

Like the binomial, Poisson distribution is also a discrete distribution used, when the outcome is the number of times an event occurs in a given number of trials. This distribution can be used to study the probability of rare events, i.e. when the number of trials is very large and the probability of occurrence of a specified outcome is very small/rare. Therefore this can be considered as the limiting distribution of binomial distribution when the number of trials n is very large and the probability p of success of every trial is small.

Consider a random variable representing the number of times an event occurs in a given time/space interval. Then the probability of exactly 'r' occurrences is given by

$$P(r) = e^{-\lambda} \cdot \frac{\lambda^r}{r!}$$

where $r = 1, 2, \ldots n$ successes. λ is the value of both the mean and the variance of the Poisson distribution and e is the base of the natural logarithm and equals 2.718. The term λ is called the parameter of the Poisson distribution. Hence any given Poisson distribution can be completely characterized by this single parameter λ.

Now from the formula

$$\begin{aligned} p(0) &= e^{-\lambda} \\ p(1) &= \lambda e^{-\lambda} \\ p(2) &= \frac{\lambda^2}{2} e^{-\lambda} \\ p(3) &= \frac{\lambda^3}{3 \times 2} \cdot e^{-\lambda} \text{ and so on.} \end{aligned} \tag{1}$$

Also

$$\begin{aligned} \frac{p(r)}{p(r-1)} &= e^{-\lambda} \cdot \frac{\lambda^r}{r!} \div e^{-\lambda} \cdot \frac{\lambda^{r-1}}{(r-1)!} \\ &= \frac{\lambda}{r} \end{aligned}$$

Thus

$$p(r) = \left(\frac{\lambda}{r}\right) p(r-1) \tag{2}$$

Hence for a given λ, the set of equations (1) together with (2) give a simple calculation formula.

It may be stressed that Poisson distribution is appropriate when the probability of a single event is very small. And this is similar to binomial distribution when n is large and p is small.

Note:

- This is one parameter distribution in which the mean equals variance.
- The property of a Poisson process is that the occurrence of one event is independent of all other events.
- Another property is that the probability of occurrence of any event remains constant over specified interval of time.

Normal Distribution

Normal distribution, otherwise known as Guassian distribution is one of the most important theoretical distribution in Statistics. The normal distribution always refers to the distribution of a variable in a population. Hence the mean and the standard deviation of such distribution are denoted by μ and σ respectively.

The normal distribution is unimodal, symmetric and bell shaped. Hence the values of the mean, median and mode are equal. A normal distribution is given by the formula

$$Y = \frac{1}{\sigma\sqrt{2\pi}} \exp\left[-\frac{(x-\mu)^2}{2\sigma^2}\right]$$

where e is the base of the natural logarithm and π is a constant and has value 3.1416. There are two mathematical constants e and π in this expression, also two parameters σ and μ. Hence a normal distribution is completely determined by its mean and standard deviation, i.e. μ and σ. In other words, any two normal distributions with the same mean and standard deviation will be identical. Thus for any given σ, for different values of μ, different normal curves are possible. It is also to be noted that not all bell shaped, unimodal symmetric distributions are normal distributions.

The normal distribution has certain distinguished area properties. 95% of the observations in a normal distribution lie within mean ± 1.96 (S.D). As it is a symmetric distribution, half of the curve is to the left and the other half to the right of the mean.

The normal distribution, unlike Binomial and Poisson distributions, is continuous. It is used to determine the probability of characteristics measured on a continuous numeric scale. Since the normal distribution is a probability distribution, the area under the curve is equal to 1. A normal curve with $\mu = 0$ and $\sigma = 1$ is said to be a standardized normal curve. Thus for a standardized normal distribution

$$Y = \frac{1}{\sqrt{2\pi}} \cdot e^{-\frac{x^2}{2}}$$

In other words any normal distribution can be transformed in to the standard normal distribution by using the transformation

$$z = \frac{X - \mu}{\sigma}$$

z is called z-score, a normal deviate, a standard score or a critical ratio.

Worked out Examples

1. *A bag contains 5 black, 4 white and 6 red balls. What is the probability of getting a black or white ball in a single draw?*

Total number of balls = 5 + 4 + 6 = 15.

Probability of getting one black ball $= \frac{5}{15}$

Similarly, probability of getting one white ball $= \frac{4}{15}$

$\therefore$ Probability of getting a black or a white ball $= \left(\frac{5}{15}\right) + \left(\frac{4}{15}\right) = \frac{9}{15} = \frac{3}{5}$

2. *A candidate applies for 3 posts for which the number of applicants is 5, 6 and 9. What is the probability that he is selected for at least one post?*

Probability of getting the 1st post = $P(A)$ say.

Let $P(B)$ and $P(C)$ denote the probabilities of getting the second and the third posts. Then,

$$P(A) = \frac{1}{5} = p_1 \qquad \therefore \quad q_1 = \frac{4}{5} \qquad (\text{as } p_1 + q_1 = 1)$$

$$P(B) = \frac{1}{6} = p_2 \qquad \therefore \quad q_2 = \frac{5}{6}$$

$$P(C) = \frac{1}{9} = p_3 \qquad \therefore \quad q_3 = \frac{8}{9}$$

$\therefore$ The probability of the candidate not getting any post $= q_1 \cdot q_2 \cdot q_3$

$$= \frac{4}{5} \times \frac{5}{6} \times \frac{8}{9} = \frac{16}{27}$$

$\therefore$ The probability of getting at least one post $= 1 - q_1 \cdot q_2 \cdot q_3$

$$= 1 - \frac{16}{27} = \frac{11}{27}$$

3. *In a pack of playing cards find*

(i) The probability of selecting a Jack of spade.

(ii) The probability of selecting a heart.

(iii) The probability of selecting a heart or queen.

(iv) The probability of selecting a heart or diamond king.

(i) There are 4 jacks in a pack of cards and there are 13 cards in each category.

The probability of selecting a Jack $= \dfrac{4}{52}$.

And that of selecting a spade $= \dfrac{13}{52}$.

$\therefore$ The probability of selecting Jack of spade $= \left(\dfrac{4}{52}\right) \cdot \left(\dfrac{13}{52}\right) = \dfrac{1}{52} = 0.0192$

(ii) The probability of selecting a heart $= \dfrac{13}{52} = \dfrac{1}{4} = 0.25$

(iii) The probability of selecting a heart $= \dfrac{13}{52}$

The probability of selecting a queen $= \dfrac{4}{52}$

$\therefore$ The probability of selecting a heart or queen $= \dfrac{13}{52} + \dfrac{4}{52} - p$ (heart and queen)

$$= \frac{1}{4} + \frac{1}{13} - \frac{1}{52} \quad \text{(From (i))}$$

$$= \frac{16}{52} = \frac{4}{13} = 0.308$$

(iv) The probability of selecting a king of heart $= \dfrac{4}{52} \cdot \dfrac{13}{52} = \dfrac{1}{52}$

The probability of selecting a king of diamond $= \dfrac{1}{52}$

$\therefore$ The probability of selecting a heart or diamond king $= \frac{1}{52} + \frac{1}{52}$

$$= \frac{2}{52} = \frac{1}{26}$$

$$= 0.038$$

4. *A basket contains 25 mangoes of variety A, 20 mangoes of variety B and 30 mangoes of variety C. Three mangoes are picked at random one after another without replacement. Find the probability that all mangoes belong to variety C.*

There are 30 mangoes of variety C out of a total of $25 + 20 + 30 = 75$ mangoes.

$\therefore$ The probability of selecting one mango of variety $C = \frac{30}{75}$

After taking out one mango of variety C there will be 29 mangoes of variety C in the remaining 74 mangoes.

The probability of selecting one mango of variety C from this group $= \frac{29}{74}$

Similarly the probability of taking the third mango of variety from the remaining 28 mangoes C out of a total of 73 mangoes $= \frac{28}{73}$.

$\therefore$ The probability of selecting three mangoes of variety C one after another

$$= \left(\frac{30}{75}\right) \cdot \left(\frac{29}{74}\right) \cdot \left(\frac{28}{73}\right)$$

$$= 0.06$$

5. *The probability that a boy will get admission to Postgraduate course is 0.7 and the probability that a girl will get admission to the Postgraduate course is 0.8. What is the probability that (i) Both get admission (ii) At least one of them gets admission?*

Let p_1 and p_2 represent the probability that a boy gets admission and the probability that a girl gets admission. Then $p_1 = 0.7$; $p_2 = 0.8$

The probability that both will get admission $= p_1 \cdot p_2 = 0.7 \times 0.8 = 0.56$.

The probability that the boy will not get admission $= q_1 = 1 - p_1 = 1 - 0.7 = 0.3$.

Similarly the probability that the girl will not get admission $= q_2 = 1 - p_2 = 1 - 0.8 = 0.2$.

The probability that neither of them will get admission $= q_1 \cdot q_2 = 0.3 \times 0.2 = 0.06$.

$\therefore$ The probability that at least one of them gets admission $= 1 - q_1 \cdot q_2 = 1 - 0.06 = 0.94$.

6. *The incidence of occupational disease in a factory is such that the workers have a 25% chance of suffering from it. Find out the probability that out of 8 workers 5 or more will contact the disease.*

The probability of 5 or more out of 8 workers getting the disease $= p(5)+p(6)+p(7)+p(8)$ where $p(5)$ represents the probability of 5 workers contacting the disease.

To calculate p(5)

Here $n = 8$, $r = 5$, $p = 20\% = \frac{1}{5}; q = 1 - p = \frac{4}{5}$

$$\therefore \quad p(5) = {}^8C_5 \left(\frac{1}{5}\right)^5 \cdot \left(\frac{4}{5}\right)^{8-5}$$

$$= {}^8C_5 \cdot \left(\frac{1}{5}\right)^5 \cdot \left(\frac{4}{5}\right)^3$$

$$= {}^8C_3 \frac{4^3}{5^8} = \left(\frac{8 \times 7 \times 6}{1 \times 2 \times 3}\right) \times \frac{4^3}{5^8}$$

$$= 8 \cdot 7 \cdot \frac{4^3}{5^8}$$

Similarly, $p(6) = {}^8C_6 \cdot \left(\frac{1}{5}\right)^6 \cdot \left(\frac{4}{5}\right)^2 = {}^8C_2 \cdot \left(\frac{1}{5}\right)^6 \cdot \left(\frac{4}{5}\right)^2 = \left(\frac{8 \cdot 7}{1 \cdot 2}\right) \times \frac{4^2}{5^8}$

$$p(7) = {}^8C_7 \cdot \left(\frac{1}{5}\right)^7 \cdot \left(\frac{4}{5}\right)^1 = {}^8C_1 \cdot \left(\frac{1}{5}\right)^7 \cdot \left(\frac{4}{5}\right) = \frac{8 \times 4}{5^8}$$

$$p(8) = {}^8C_8 \cdot \left(\frac{1}{5}\right)^8 \cdot \left(\frac{4}{5}\right)^0 = \frac{1 \times 1}{5^8}$$

$\therefore$ The required probability $= p(5) + p(6) + p(7) + p(8)$

$$= 8 \times 7 \times \frac{4^3}{5^8} + \frac{8 \times 7}{1 \times 2} \times \frac{4^2}{5^8} + 8 \times \frac{4}{5^8} + \frac{1}{5^8}$$

$$= \frac{(3584 + 448 + 32 + 1)}{5^8} = \frac{4065}{5^8}$$

$$= 0.0104$$

7. *The probability of Biochemistry student in a class getting distinction is 0.3. Determine the probability that out of 6 students 2 will get distinction.*

Here $p = 0.3; n = 6; r = 2; q = 1 - 0.3 = 0.7$

$\therefore$ The probability of 2 out of 6 students getting distinction $= {}^6C_2\,(0.3)^2(0.7)^4$

$$= \frac{6 \times 5}{1 \times 2}(0.3)^2(0.7)^4$$

$$= 0.324$$

8. *What is the probability of getting (i) two heads (ii) more than 2 heads when a coin is tossed 4 times?*

Probability of a head (single trial) $= p = \frac{1}{2}, q = 1 - \frac{1}{2} = \frac{1}{2}$

(i) $n = 4, r = 2$

Probability of getting 2 heads in 4 trials $= {}^4C_2 \left(\frac{1}{2}\right)^2 \left(\frac{1}{2}\right)^{4-2}$

$$= \left(\frac{4 \times 3}{1 \times 2}\right) \times \frac{1}{2^2} \times \frac{1}{2^2} = \frac{6}{2^4} = \frac{3}{8} = 0.375$$

(ii) Probability of getting more than 2 heads in 4 trials

= Probability of getting 3 successes + Probability of getting 4 successes

$$= {}^4C_3 \left(\frac{1}{2}\right)^3 \left(\frac{1}{2}\right)^1 + {}^4C_4 \left(\frac{1}{2}\right)^4 \left(\frac{1}{2}\right)^0$$

$$= 4 \times \frac{1}{2^4} + \frac{1 \times 1}{2^4} = \frac{5}{16} = 0.316$$

9. *Find the probability of 3 female births out of 5 births.*

Probability of getting a female child $= \frac{1}{2} = p$

Probability of getting a male child $= \frac{1}{2} = q$

Here, $n = 5; r = 3$

$\therefore\ p(3) = p(r) = {}^5C_3 \left(\frac{1}{2}\right)^3 \left(\frac{1}{2}\right)^{5-3}$

$$= \frac{5 \times 4 \times 3}{1 \times 2 \times 3} \times \frac{1}{2^3} \times \frac{1}{2^2} = \frac{10}{32} = \frac{5}{16} = 0.316$$

10. *The probability that a student gets first mark in physics is $\frac{13}{40}$. The probability of his getting first mark in both physics and chemistry is $\frac{1}{8}$. If the probability of his getting first mark in either physics or chemistry is $\frac{79}{120}$, find the probability of his getting first mark in chemistry alone.*

Let $p(A)$ = Probability of getting first mark in Physics; $p(B)$ = Probability of getting first mark in Chemistry.

$$p(A \text{ or } B) = p(A) + p(B) - p(A \text{ and } B)$$

i.e. $$\frac{79}{120} = \frac{13}{40} + p(B) - \frac{1}{8}$$

$$\therefore \quad p(B) = \left(\frac{79}{120}\right) + \frac{1}{8} - \frac{13}{40}$$

$$= \frac{(79 + 15 - 39)}{120} = \frac{55}{120} = \frac{11}{24}$$

The probability of getting first mark in Chemistry alone $= \dfrac{11}{24}$.

VI. Sampling Techniques

The term "Population" is used in statistics to represent a large collection of measurements having a common trait, about which one wishes to draw conclusions. The population under study may be very large or comparatively small. In the latter case, we may as well obtain all the measurements in the population itself. In the other case, we select a subset of the population called a Sample, which will be representative of the larger population. The size of the sample must be sufficient so as to be representative of the population.

- A sample should have the similar characteristics of the original population from which it has been selected.
- Selected samples should have similar nature. There should be no difference when compared with the population.
- The number of observations selected in the sample should be more to make the results more suitable.

Reasons for Sampling

- Samples can be studied more quickly than population. And speed will be the need of many clinical observations.
- A study of a sample is less expensive than studying an entire population. Follow up becomes lengthy for a design of large studies.
- In many situations, a study of entire population is impossible.
- Sample results are more accurate than the results based on population. Because we can spend more time and resources for collection of less volume of data and train people. To improve accuracy, since less number of procedures will be required for samples we can go for more expensive procedures.
- If samples are aptly selected, probability methods can be utilized to estimate the error in the resulting statistics.

- ❖ Sampling will reduce heterogeneity, as a population is likely to be less homogenous. We can always choose a sample with a specified characteristic for such investigation.

Thus the most judicious way is to choose the right size of sample for study prior to begining the research. The size of the sample depends upon a number of factors.

- ❖ In a population, with only perfectly homogenous units, a small sample is enough as all have the same characteristics. To get reliable results, large size sample should be selected in a heterogeneous population.
- ❖ As the population becomes larger, the sample size should also be larger.
- ❖ Nature of the study determines the size of the sample. For an intensive and continuous study we require only a small size sample.
- ❖ Again for reliability of results while classifying large number of classes we require a large size sample.
- ❖ Constraint on the sample size is effected by the availability of trained personnel, finance and time, besides other consideration.
- ❖ Sampling techniques also influence the size of the sample. In random sampling only in large sample, we can achieve greater accuracy whereas in stratified sample even a small size sample can give more accurate results.

Method of sampling

(a) Random sampling method

(b) Non-probability sampling method

Random sampling method

A random sample is a sample where each item in the population has an equal chance of being selected. "It indicates not a haphazard choice but a careful selection to ensure that every item has an equal chance of inclusion". Random sampling is a scientific method of getting a sample from a population. This method is also known as the "Unrestricted random sampling device". It completely depends upon the element of chance.

Different forms of random samplings are:

1. Simple random sampling
2. Systematic random sampling
3. Stratified Random Sampling
4. Multistage random sampling
5. The Cluster random sampling

Simple random sampling

For a simple random sampling from a given population, one must have a list of all the units in the population. This method is ideally suited when the population is small and homogenous. For such a simple random sample from a population, prepare the list of all units of the populations, referred to as the Sampling frame. Assign a specific number to each unit. The process of selection may be by Lottery method or from the table of random numbers.

Suppose one has to select 10 samples out of 200 available units, first form the sampling frame by assigning the numbers from 1 to 200 to these units at random. Then note the serial numbers of all the 200 units on 200 identical cards, shuffle them thoroughly and select a number.

After noting down the number, replace the card drawn, reshuffle and draw the second card, note down the number now selected. Each time a number is selected replace it shuffle and then draw another card. This process is continued until all the 10 required numbers are got. If in this process, any of the number is repeated reject that selection and proceed further. The 10 cards thus selected will show the numbers to be considered from the population and these 10 units will represent the random sample. This is sampling with replacement.

Alternatively, using the random number table, start at an arbitrary point in this table of random numbers and read the number down the column. Since 3 digit numbers between 001 and 200 are needed, read the first 3 digits of a column until 10 diferent 3 digit number are got. Reject the numbers greater than 200 and repetitions. The numbers chosen ultimately in this manner represent and identify the particular unit of the population to be considered for the sample.

Systematic random sampling

In this method, every rth item in the population is selected for sample. r is the number obtained by dividing the number of units in the sampling frame by the number of items in the sample needed. Thus 200 units divided by 10 gives 20. Hence every 20th unit is taken for sample. But the first number for the sampling frame should be chosen at random as the number between 1 and 20. If the number thus selected is, say 11, then the samples will be the units with identification number 11, 31, 51 and so on. The numbers are determined by adding 20 to each preceding number in this sequence to arrive at the required number of units is 10 samples.

Systematic sampling cannot be used if cyclic repetition is inherent in the sampling frame.

Stratified random sampling

The population in this case is divided in to a number of subgroups (Strata) based on some characteristics and the random sample is then selected from each of these strata. The results of the samples from all these strata are combined to give the results for the total sample. It should be noted that the sample proportion in each stratum must be determined in advance. Under such circumstances stratified sampling is preferred to simple random sampling.

When a sampling survey is done to compare two or more groups, better efficiency is achieved by taking the sample size for each group. If the groups in the population are not of the same size. we resort to stratified sampling by taking equal sample size form each stratum. This is called non-proportional stratified sampling.

Characteristics used to divide the population into subgroups must be related to measurement of interest. In medicine some such characteristics are age, sex, etc.,

Such criterion for selection of subgroups will make stratified sampling as the most efficient one requiring the smallest sample size.

Cluster sampling

It is also known as sampling in stages. The population is first divided into clusters and a subset of each cluster is randomly selected and all in each subset is studied. Choice of clusters is based mainly on natural groups of geographical areas such as towns, villages, factories, blocks etc,. Thus the sampling will be done in two stages, the practical work being very much reduced. Cluster sampling is slightly less efficient as it requires large sample size. It gives a higher standard error but the collection of data in this method is simpler and involves less time and cost than the other sampling techniques.

Multistage sampling

In the Multistage sampling a full list of the population to be surveyed is not required. Suppose a sample of workers of factories in a state is required. Avoid the task of taking complete list of all workers in all factories. Obtain the list of such factories in the particular state and take a simple random sample of these factories. Then take a simple random sample from the list of workers of these factories only. This drastically reduces the number of units to be studied. The difficulties involved in such multistage sampling must be overcome by consulting a statistician.

Non-probability sampling

In many situations of medical research when random or systematic sampling is rather impossible we can use non-probability sampling methods such as convenience samples and quota samples. Here the unit chosen is unknown. In non-probability sampling there will be bias of the person doing the study and therefore such samples may not fulfill the requirement of randomness, necessary for estimating sampling errors. In the quota sampling the objective is to adhere to certain quotas of individuals in well-defined groups. The quotas should be so arranged that the final samples reflect the population in relation to the quota criterion such as say age and sex (Boys in age group 15–20 years). Even though in this regard quota sampling may be similar to stratified sampling bias of the person undertaking the study is very likely to dominate.

In general, a sample in observational studies always implies that the sample has been randomly selected as per one of the methods.

VII. Correlation

According to Croxton and Cowdon, "When the relationship is of a quantitative nature, the approximate statistical tool for discovering and measuring the relationship and expressing in brief formula is known as Correlation". Karl Pearson introduced a method of assessing correlation by means of correlation coefficient.

Correlation is a measure of relationship between two variables. It is to be noted that this relationship is to be between two variables, which are measured on the same scale with both being numerical, both ordinal or both nominal.

Correlation coefficient, otherwise referred to as the Pearson's correlation coefficient is given by

$$r = \frac{\sum(X-\bar{X})(Y-\bar{Y})}{\sqrt{\{\sum(X-\bar{X})^2\}\{\sum(Y-\bar{Y})^2\}}} \quad (1)$$

which can also be given as

$$r = \frac{\sum XY - \frac{(\sum X \sum Y)}{n}}{\sqrt{\{\sum X^2 - \frac{(\sum X)^2}{n}\}\{\sum Y^2 - \frac{(\sum Y)^2}{n}\}}} \quad (2)$$

Note:

1. Correlation coefficient is generally given to 2 places of decimals.
2. The denominator in equation (1) is positive. Hence, depending on whether numerator is positive, negative or zero, the correlation coefficient will also be positive, negative or zero.

contd.

(a) A positive correlation means an increase in the value of one variable is followed by an increase in the value of the other.

(b) A negative correlation implies that an increase in the value of one variable leads to, decrease in the value of the other.

(c) A correlation of 0 means no relationship exists between the variables.

3. As the numerical or absolute value of the numerator in equation will always be less than or equal to the denominator, the value of r, correlation coefficient lies between -1 and $+1$; $+1$ gives a perfect positive correlation while -1 gives a perfect negative correlation.

4. The value of the correlation coefficient is independent of the particular unit of measurement of the variable.

5. Correlation coefficient is only a measure of the extent of the association between the two variables.

6. As the value of correlation coefficient is very much influenced by extreme values, it cannot be used to give a good relationship between two variables when their distributions are skewed.

7. In such case, change of scale of measurement is used to contain the effect of outlying values before computing correlation coefficient. Colton stated that "correlation from 0 to 0.25 (or -0.25) indicates little or no relationship; those from 0.25 to 0.50 (or -0.25 to -0.50) indicate a fair degree of relationship; those between 0.50 and 0.75 (or -0.50 to -0.75) a moderate to good relationship and those greater than 0.75 (or -0.75) a very good to excellent relationship".

The square of the correlation, r^2 is called the coefficient of determination. It represents the amount of variability in one of the variables by correlating that variable with the second variable. r^2 can also be considered to be a measure of the strength of the linear relationship.

The standard error of the correlation can be calculated as

$$S_r = \sqrt{\frac{1-r^2}{n-2}}$$

Rank correlation

If the data is obtained from a bivariable population which is not at all normal, or if the individual values of the variables are not available one cannot adopt the usual correlation procedures. In such cases, if the ranks or order for each variable is available the correlation between the ranks is determined and the coefficient called the rank correlation coefficient ρ is obtained. This is called as Spearman's rank correlation coefficient and is given by

$$r_s = 1 - 6\sum_{i=1}^{n} \frac{d_i^2}{n(n^2-1)}$$

where d_i represents the difference between the ranks of X and Y i.e. d_i = rank of X_i- rank of Y_i, r_s is an estimate of the population rank correlation coefficient ρ_s.

Note: (i) r_s varies from -1 to $+1$ and has no units (ii) This value will be different from the value of r calculated from original data. This is calculated only with the rank of the data.

Assumptions in correlation

1. The sample is randomly selected.
2. x and y have bivariate normal distribution.
3. The mean values of the y distribution is linearly related to x.

Test of significance for correlation coefficient

The correlation coefficient r calculated from a sample is an estimate of the correlation coefficient of a population that is sampled. This correlation coefficient of population is denoted by ρ. In the test of significance of the deviation of the estimate of correlation coefficient from zero, one may apply the "t" test with $n-2$ degrees of freedom using

$$t = \frac{r}{s_r}$$

s_r being the standard errors $t = r\sqrt{\frac{n-2}{1-r^2}}$. The t distribution can be used to test the hypothesis that the true value of the population parameter ρ is equal to zero.

The significance of an observed correlation can be tested with reference to the t values from the table. If the calculated t value exceeds the t value from the table for $p = 0.05$ or 0.01 one may conclude that the sample indicates a real correlation between the two variables and that the correlation is significant.

Fisher's Z-transformation

A significance test can also be used when the correlation coefficient ρ has a value other than zero. Under these circumstances a procedure called Fisher's Z-transformation is used to test any null hypothesis (other than $\rho = 0$) and, to form confidence interval. When the distribution of r is not normal the t test cannot be applied. To test the statistical significance of r when the population parameter $\rho = \rho_0$, where $\rho_0 \neq 0$ transform the correlation and then use standard normal (Z) distribution.

Fisher's Z-transformation is given by

$$Z = \frac{1}{2}\ln\left(\frac{1+r}{1-r}\right)$$

where ln represents natural logarithm.

Note:

1. For values of r between 0 and 1 the corresponding values of z will be between 0 and ∞ and when r has values between 0 and -1 the corresponding z values will be between 0 and $-\infty$.
2. One can utilize the table of values of z transformation for different values of r so that the computations need not be done using the formula to obtain z from a given value of r.
3. This transformation follows a normal distribution for a moderately sized sample and the following expression, for z test can be used.

$$z = \frac{z(r) - z(\rho)}{\sqrt{\frac{1}{n-3}}}$$

where $z(r)$ is the transform of r and $z(\rho)$ is the transform of the hypothesized coefficient ρ.

Confidence interval for the correlation coefficient

Using Fischer's transformation one can form confidence intervals.

The confidence interval is z transform of r $\pm$ (confidence coefficient $\times$ Standard error of z). After the limits are determined they must be transformed back to values corresponding to correlation coefficient. The confidence interval will not be symmetric about the sample value as the distribution of the correlation coefficient is also not symmetric.

VIII. Regression

The statistical analysis of the difference between two groups can be made in relation to quantitative variables and qualitative variables. Here one variable may be qualitative or both variables compared may be qualitative.

The third possible combination is that both the variables of the groups are quantitative. If some relationship exists between two variables, the statistical analysis of such data is called Bivariate Analysis. Under these circumstances, to know the magnitude of such relationship one may like to predict the value of one variable from the value of the other.

Analysis of the relationship between two such variables may arise due to

1. Examining whether the increase or decrease in one variable is related to that of the other, that is to determine the existence of such an association
2. Examining whether the magnitude of one variable changes when the magnitude of the second variable changes.

To be more specific, one variable depends on the other; that is one variable is independent and the other is dependent on it. Such a dependent relationship is called Regression. When only two variables are used the term is Simple Regression.

According to M. M. Blair, "Regression Analysis is a mathematical measure of the average relationship between two or more variables in terms of the original units of the data".

Regression analysis is also known as Linear regression; Simple linear regression or least square regression. As regression measures only on a straight line, it is termed as Linear regression. 'Simple Regression' means that one independent variable is used to predict an outcome. As against this in multiple regression, more than one independent variable is included in the equation. Obtaining mathematically, the statistical estimators from the regression equation refers to least square method of regression. It must be noted that 'regression' refers to linear regression based on least square method.

From fundamental geometry, the equation to a straight line can be written in its simplest form as $Y = a + bX$ where a, b are constants for the particular line with a measuring the intercept made by the line with the Y axis and b the 'Slope' of the line.

The simple linear regression equation for a population is given by

$$Y = a + \beta X$$

where α, and β are population parameters.

Unless all the data for the entire population is available it is not possible to determine the values of α and β with maximum accuracy. Hence the best estimate of these values are obtained from a sample of n pairs of values of X and Y.

The sample regression equation is $Y = a + bX$.

Here 'a' represents the intercept of this line on the y-axis and 'b' is the slope of this line. This slope measures the amount of change in Y for a unit change in the value of X. Also if this slope is positive, both X and Y increase or decrease together. When b is negative, Y decreases as X increases and vice versa. β in the population regression equation is called the regression coefficient. β and the y intercept α are the population parameters in regression.

When all the actual values of X, Y do not lie on the regression line, the regression model contains an error term ϵ. Thus the regression equation is given by

$$Y = a + \beta X + \epsilon$$

From the sample regression equation $Y = a + bX$, a and b are determined to be given by the formulas

$$b = \frac{\sum[(X - \bar{X})(Y - \bar{Y})]}{\sum(X - \bar{X})^2}$$

or,

$$b = \frac{\sum XY - \frac{(\sum X)(\sum Y)}{n}}{\sum X^2 - \frac{(\sum X)^2}{n}} \qquad (1)$$

$$a = \bar{Y} - b\bar{X} \qquad (2)$$

We have, $Y = \alpha + \beta X$

Therefore $\sum Y = (\sum \alpha) + \sum \beta X$

i.e. $\sum Y = n\alpha + \beta \sum X$

i.e. $\sum \frac{Y}{n} = \alpha + \beta \frac{\sum X}{n}$

i.e. $\bar{Y} = \alpha + \beta\bar{X}$. This shows that $(\bar{X}, \bar{Y})$ is a point on the regression line, $\bar{X}, \bar{Y}$ being the means for the two variables X and Y.

Note:

1. The denominator in the expression for b is positive. But the numerator can have positive, negative or zero values and the values of b can range through $-\infty$ to $+\infty$
2. If we determine a and b the regression line is uniquely determined.
3. Plotting this line requires only two points on this line. Two points $(\bar{X}, \bar{Y})$ and (0, a) are already known.
4. We should always draw the regression line using predicted points and not approximately.
5. It is safe to predict Y values for X values within the observed range of X. For values of X beyond this range either above or below the function may not be the same and we may not even have a linear relationship between X and Y.

Calculation of regression equation

Method (i) Using the raw data and equations 1 and 2 for determining a and b, one can calculate the regression equation. But this involves a number of calculations.

Method (ii) Here we make use of the relationship between r and b. When the correlation is having a positive value, the slope of the regression line through the observed values is also positive. Similarly the two values are together negative. When the correlation is zero, the slope is zero and the regression line is parallel to the x-axis. Hence there is a close relationship between the correlation and regression coefficients. Thus if the correlation coefficient (r) is already calculated the regression coefficient is derived as

$$b = r\frac{(SD)_Y}{(SD)_X}$$

where SD_Y represents Standard Deviation of Y series.

Assumptions in regression

1. In order to use the regression equation for each value of X, it is assumed that there exists in the population a normal distribution for the Y variable.
2. Irrespective of the value of X, the standard deviation of Y should be the same, i.e. there should be homogeneity of variance. There should be equal variance of Y for the entire range of X.
3. The mean values of Y's corresponding to various X values should lie on a straight line.
4. The values of Y should be independent of one another
5. The observations should be random samples of the population under study.

Standard error of estimate

The regression equation that is obtained is based on a sample of observations and is only an estimate of the true population regression equation. By considering different samples, it will be seen that the calculated regression equations will all be different from one another for each of these samples in terms of the values of a and b. An estimate of this variation is called the Standard error of the estimate and is denoted by $S_{Y,X}$ and is given by the formula

$$S_{Y,X} = \frac{\sum(Y - \hat{Y})^2}{n-2} \qquad (1)$$

where Y represents the observed value obtained and $\hat{Y}$ is the predicted value calculated from the regression equation for a given X value. $n-2$ is called the Residual Degree Of Freedom.

The standard error of b, the regression coefficient is given by

$$SE(b) = \frac{\sigma}{\sqrt{\sum(X-\bar{X})^2}}$$

$$= \frac{\sigma}{\sqrt{(n-1)S_X}} \qquad (2)$$

where σ is the standard deviation of the Y variable corresponding to each value of X. It is to be noted that σ has to be estimated from the data.

$$S_{Y,X} \text{ represents an estimate of } \sigma. \qquad (3)$$

In the light of this statement and equation 1 the standard error of b is estimated as

$$SE(b) = \frac{S_{Y,X}}{\sqrt{(n-1)S_X}}$$

Worked out Examples

1. *Obtain the Correlation Coefficient given the following data.*

X	1	2	3	4	5	6	7	8	9	10
Y	13	14	15	17	16	20	18	19	22	21

X	Y	XY	X^2	Y^2
1	13	13	1	169
2	14	28	4	196
3	15	45	9	225
4	17	68	16	289
5	16	80	25	256

X	Y	XY	X^2	Y^2
6	20	120	36	400
7	18	126	49	324
8	19	152	64	361
9	22	198	81	484
10	21	210	100	441
$\sum X = 55$	$\sum Y = 175$	$\sum XY = 1040$	$\sum X^2 = 385$	$\sum Y^2 = 3145$

$n = 10$

$$\text{Correlation Coefficient } r = \frac{n\sum XY - (\sum X)(\sum Y)}{\sqrt{\{n\sum X^2 - (\sum X)^2\}\{n\sum Y^2 - (\sum Y)^2\}}}$$

$$= \frac{10400 - 55 \times 175}{\sqrt{(3850 - 55^2)(31450 - 175^2)}}$$

$$= \frac{10400 - 9625}{\sqrt{(3850 - 3025)(31450 - 30625)}}$$

$$= \frac{775}{\sqrt{825 \times 825}} = 0.94$$

2. *Two competitors in a beauty contest are ranked by two judges as follows*

X	1	6	5	10	3	2	4	9	7	8
Y	6	4	9	8	1	2	3	10	5	7

On the basis of this ranking examine if the judges have common tastes in beauty.

$X(R_1)$	$Y(R_2)$	$D = R_1 - R_2$	D^2
1	6	−5	25
6	4	2	4
5	9	−4	16
10	8	2	4
3	1	2	4
2	2	0	0
4	3	1	1
9	10	−1	1
7	5	2	4
8	7	1	1
			$\sum D^2 = 61$

$n = 10$

$$\text{Rank Correlation } r_k = 1 - \frac{6\sum D^2}{n^3 - n} = 1 - \frac{6 \times 61}{1000 - 10}$$

$$= 1 - \frac{366}{990} = 1 - 0.37 = 0.63$$

3. *Find the rank correlation for the following distribution.*

Marks in Biostatistics	49	52	70	62	37	40	58	75	46	45
Marks in Chemistry	62	31	78	65	38	52	60	44	56	70

X (Marks in Biostatistics)	Y (Marks in Chemistry)	R_1	R_2	$R_1 - R_2 = D$	D^2
49	62	6	4	2	4
52	31	5	10	−5	25
70	78	2	1	1	1
62	65	3	3	0	0
37	38	10	9	1	1
40	52	9	7	2	4
58	60	4	5	−1	1
75	40	1	8	−7	49
46	56	7	6	1	1
45	70	8	2	6	36
					$\sum D^2 = 122$

$n = 10$

$$r_k = 1 - \frac{6\sum D^2}{n^3 - n} = 1 - \frac{6 \times 122}{1000 - 10}$$

$$= 1 - \frac{732}{990} = 0.26$$

4. *The following data gives the yield of maize grain (in kg) per plot of size 50 sq.m for following classes of nitrogen applications.*

Amount of Nitrogen (kg/Plot)	20	30	50	90	100
Field of Maize (Grain kg)	8.45	12.75	10.9	14.8	15.75

Calculate the correlation co-efficient

X	Y	XY	X^2	Y^2
20	8.45	169	400	71.4025
30	12.7	381	900	161.29
50	10.9	545	2500	118.81
90	14.8	1332	8100	219.04
100	15.75	1575	10000	248.0625
$\sum X = 290$	$\sum Y = 62.6$	$\sum XY = 4002$	$\sum X^2 = 21900$	$\sum Y^2 = 818.605$

The correlation coefficient

$$r = \frac{n\sum XY - (\sum X)(\sum Y)}{\sqrt{\{n\sum X^2 - (\sum X)^2\}\{n\sum Y^2 - (\sum Y)^2\}}}$$

$$= \frac{5 \times 4002 - 290 \times 62.6}{\sqrt{(5 \times 21900 - 290^2)(5 \times 818.605 - 62.6^2)}}$$

$$= \frac{20010 - 18154}{\sqrt{(109500 - 84100)(4093.025 - 3918.76)}}$$

$$= \frac{1856}{\sqrt{25400 \times 174.265}} = \frac{1856}{\sqrt{4426331}} = \frac{1856}{2103.885}$$

$$= 0.88$$

5. *Compute the coefficient of correlation for the following data*

Students	Height in cm	wt in kg
1	170	72
2	150	62
3	175	76
4	165	68
5	180	81
6	185	84

Let us consider an assumed mean, say 170 for ht and 76 for weight.

ht. in cm X	$dx = X - 170$	Wt.in kg Y	$dy = Y - 76$	dx^2	dy^2	$dxdy$
170	0	72	−4	0	16	0
150	−20	62	−14	400	196	280
175	5	76	0	25	0	0
165	−5	68	−8	25	64	40
180	10	81	5	100	25	50
185	15	84	8	225	64	120
	$\sum dx = 5$		$\sum dy = -13$	$\sum dx^2 = 775$	$\sum dy^2 = 365$	$\sum dxdy = 490$

$n = 6$

$$r = \frac{6\sum dxdy - \sum(dx)(dy)}{\sqrt{\{6\sum dx^2 - (\sum dx)^2\}\{6\sum dy^2 - (\sum dy)^2\}}}$$

$$= \frac{490 \times 6 - 5 \times (-13)}{\sqrt{(6 \times 775 - 5^2)(6 \times 365 - 13^2)}}$$

$$= \frac{2940 + 65}{\sqrt{(4650 - 25)(2190 - 169)}}$$

$$= \frac{3005}{\sqrt{4625 \times 2021}} = \frac{3005}{68.007 \times 44.956}$$

$$= 0.98$$

6. *The following gives the data from 25 observations on the pair of values* (x, y) $\sum x = 125$, $\sum y = 100$, $\sum x^2 = 650$, $\sum y^2 = 484$, $\sum xy = 533$. *Calculate the correlation coefficient.*

$$\text{Correlation coefficient } r = \frac{n\sum xy - (\sum x)(\sum y)}{\sqrt{\{n\sum x^2 - (\sum x)^2\}\{n\sum y^2 - (\sum y)^2\}}}$$

$$= \frac{25 \times 533 - 125 \times 100}{\sqrt{(25 \times 650 - 125^2)(25 \times 484 - 100^2)}}$$

$$= \frac{533 - 500}{\sqrt{(650 - 625)(484 - (100^2/25))}}$$

$$= \frac{33}{\sqrt{25 \times 84}} = \frac{33}{\sqrt{100 \times 21}}$$

$$= \frac{33}{10\sqrt{21}} = \frac{3.3}{4.58}$$

$$= 0.72$$

7. *The following data represents the percentage of marks in Mathematics and Physics of 10 students. Find the rank correlation.*

Students	1	2	3	4	5	6	7	8	9	10
Marks in Maths	84	70	75	70	92	70	75	75	85	58
Marks in Physics	75	60	48	68	90	60	52	48	70	60

When there is a tie t the modified formula for rank correlation is given by

$$r = 1 - \frac{\left\{6\sum d^2 + \frac{\sum(t^2-t)}{k}\right\}}{n(n^2-1)}$$

t = no. of tied items, k = total no. of tied items in both the variables.

Student	Marks in Maths	Rank R_1	Marks in Physics	R_2	$R_1 - R_2 = d$	d^2
1	84	3	75	2	1	1
2	70	8	60	6	2	4
3	75	5	48	9.5	−4.5	20.25
4	70	8	68	4	4	16
5	92	1	90	1	0	0
6	70	8	60	6	2	4
7	75	5	52	8	−3	9
8	75	5	48	9.5	−4.5	20.25
9	85	2	70	3	−1	1
10	58	10	60	6	4	16
					$\sum d = 0$	$\sum d^2 = 91.5$

There is a tie for 4th place among 3 students and a tie for 5th place among 3 students in maths Hence we assign average ranks to them.

$$75 \longrightarrow 3, 7, 8 \;\; \frac{4+5+6}{3} = \frac{15}{3} = 5$$

$$70 \longrightarrow 2, 4, 6 \;\; \frac{7+8+9}{3} = \frac{24}{3} = 8$$

Therefore the ranks for these set of students are 5 and 8 respectively.

Also for physics, there is a tie for the 5th place among 3 students (60 marks) and a tie between 2 students (48 marks)

Their ranks are given by $\frac{5+6+7}{3} = 6, \; \frac{9+10}{2} = 9.5$

Total number of tied values $=(3+3)+(3+2)=11=k$

(The summation is for each set of tied value t representing the number of tied items)

$$\frac{\sum t^3 - t}{11} = \frac{3^3-3}{11} + \frac{3^3-3}{11} + \frac{3^3-3}{11} + \frac{2^3-2}{11}$$

$$= \frac{24}{11} + \frac{24}{11} + \frac{24}{11} + \frac{6}{11} = \frac{78}{11}$$

$$r = 1 - \frac{6\left[\sum d^2 + \frac{\sum t^3 - t}{11}\right]}{n(n^2-1)} = 1 - \frac{6(91.5 + \frac{78}{11})}{990}$$

$$= 1 - \frac{6(91.5+7.09)}{990} = 1 - \frac{2}{330} \times 98.59$$

$$= 1 - 0.5975 = 0.4025$$

$$= 0.4$$

8.

Student	1	2	3	4	5	6	7	8	9	10
Entrance Exam Marks	79	59	81	62	70	80	89	92	60	75
Qualifying Marks	72	69	89	52	61	78	90	65	76	87

From the above data, find the rank correlation

Entrance test marks	Rank R_1	Exam Marks	Rank R_2	$R_1 - R_2 = d$	d^2
79	5	72	6	−1	1
59	10	69	7	3	9
81	3	89	2	1	1
62	8	52	10	−2	4
70	7	61	9	−2	4
80	4	78	4	0	0
89	2	90	1	1	1
92	1	65	8	−7	49
60	9	76	5	4	16
75	6	87	3	3	9
				$\sum d = 0$	$\sum d^2 = 94$

$$r = 1 - \frac{6\sum d^2}{n(n^2-1)} = 1 - \frac{6 \times 94}{10(100-1)}$$

$$= 1 - \frac{6 \times 94}{10 \times 99} = 1 - 0.569$$

$$= 0.431$$

9. *The following results were obtained from measurement of body weight (x) in gm and brain ACHE activity (y) of 25 Tilapia fish* $\sum x = 97.4$, $\sum y = 124.5$, $\sum x^2 = 459.80$, $\sum y^2 = 643.81$, $\sum xy = 444.68$. *Find out appropriate equation and establish brain ACHE activity of a fish having a body weight 5 gms.*

Here, $n = 25$

$$\bar{X} = \frac{\sum x}{n} = \frac{97.4}{25} = 3.896; \quad \bar{Y} = \frac{\sum y}{n} = \frac{124.5}{25} = 4.98$$

$$\text{Regression coefficient (by } x) = \frac{rSDy}{SDx}$$

i.e., $$b = \frac{\sum xy - \frac{\sum x \sum y}{n}}{\sum x^2 - \frac{(\sum x)^2}{n}}$$

$$= \frac{444.68 - (97.4 \times 124.5/25)}{459.80 - (97.4)^2/25}$$

$$= \frac{-40.372}{80.3296}$$

$$= -0.5$$

The regression equation is $(y \text{ on } x) y - \bar{y} = b(x - \bar{x})$

$$y - 4.98 = -0.5(x - 3.896)$$

$$y = -0.5x + 1.948 + 4.98$$

$$y = -0.5x + 6.928$$

when, $x = 5$ g,

$$y = -0.5 \times 5 + 6.928$$

$$= -2.5 + 6.928$$

$$= 4.428$$

10. *Calculate the correlation coefficient and regression coefficient between two measurements of quality of water in a lake. Find also their level of significance.*

Salinity%	3	4	6	7	8	10	12	14
Dissolved Oxygen (mg/l)	6	3	7	11	5	8	10	12

x	y	x^2	y^2	xy
3	6	9	36	18
4	3	16	9	12
6	7	36	49	42
7	11	49	121	77
8	5	64	25	40
10	8	100	64	80
12	10	144	100	120
14	12	196	144	168
$\sum x = 64$	$\sum y = 62$	$\sum x^2 = 614$	$\sum y^2 = 548$	$\sum xy = 557$

$n = 8$

$$\sum d_x^2 = \sum x^2 - \left(\frac{(\sum x)^2}{n}\right) = 614 - \frac{(64)^2}{8} = 614 - 512 = 102$$

$$\sum d_y^2 = \sum y^2 - \left(\frac{(\sum y)^2}{n}\right) = 548 - \frac{(62)^2}{8} = 548 - 480.5 = 67.5$$

$$\sum d_x d_y = \sum xy - \left(\frac{(\sum x \sum y)}{n}\right) = 557 - \frac{(64 \times 62)}{8} = 557 - 496 = 61$$

(i) Correlation coefficient: $r = \dfrac{\sum d_x d_y}{\sqrt{\sum d_x^2 \sum d_y^2}}$

$$= \frac{61}{\sqrt{102 \times 67.5}}$$

$$= \frac{61}{82.98} = 0.74$$

Significance test: $t = r\dfrac{\sqrt{n-2}}{\sqrt{1-r^2}} = 0.74\sqrt{\dfrac{8-2}{1-(0.74)^2}} = \dfrac{0.74 \times \sqrt{6}}{\sqrt{1-0.5476}}$

$$= \frac{0.74 \times 2.45}{\sqrt{0.4524}} = \frac{1.813}{0.67} = 2.71$$

Regression coefficient: $b = \dfrac{\sum d_x d_y}{\sum d_x^2} = \dfrac{61}{102} = 0.598$

Significance test: $t = \dfrac{b}{SE \text{ of } b}$

$$\text{SE of } b = \frac{\sqrt{1/n - 2\{\sum d_y^2 - \sum d_x d_y)^2 / \sum d_x^2\}}}{\sqrt{\sum d_x^2}}$$

$$= \sqrt{\frac{\frac{1}{8-2}\left\{67.5 - \frac{61^2}{102}\right\}}{102}} = \sqrt{\frac{67.5 - 36.48}{6(102)}}$$

$$= \frac{\sqrt{5.17}}{\sqrt{102}} = \frac{2.27}{10.09} = 0.225$$

$$t = \frac{b}{SE \text{ of } b} = \frac{0.598}{0.225} = 2.66$$

$$\bar{x} = \frac{\sum x}{n} = \frac{64}{8} = 8$$

$$\bar{y} = \frac{\sum y}{n} = \frac{62}{8} = 7.75$$

Regression equation y on x is $y = a + bx$

Now, $\bar{y} = a + b\bar{x} : a = \bar{y} - b\bar{x}$

$$= 7.75 - 0.598 \times 8$$

$$= 7.75 - 4.784$$

$$= 2.966$$

Therefore the equation is $y = 2.966 + 0.598x$

Correlation: $r = 0.74$ $t = 2.71$

tabulated value of t ($d.f = n - 2 = 6$) at $P = 0.05$ is 2.45

Regression: $b = 0.598$; $t = 2.66$

1. The correlation coefficient is significant as the calculated value of t is higher than the tabulated value at 6 d.f for $p = 0.05$.
2. The regression coefficient is also significant as the calculated value exceeds the tabulated values for $p = 0.05$.

11. *The following data regarding the ages of husbands* (x) *and ages of wives* (y) *are given.* $S(x) = 203$; $S(x^2) = 5963$; $S(y) = 178$; $S(y^2) = 4648$; $S(xy) = 5254$; $n = 7$

Find the correlation coefficient between the ages of husbands and wives and write down the two regression lines.

The correlation coefficient $r = \dfrac{\sum xy - \frac{(\sum x)(\sum y)}{n}}{\left(\sum x^2 - \frac{(\sum x)^2}{n}\right)\left(\sum y^2 - \frac{(\sum y)^2}{n}\right)}$

$$= \frac{5254 - 178 \times \frac{203}{7}}{\sqrt{\left(5963 - \frac{203^2}{7}\right)\left(4648 - \frac{178^2}{7}\right)}}$$

$$= \frac{5254 - 5162}{\sqrt{(5963 - 5887)(4648 - 4526.29)}}$$

$$= \frac{92}{\sqrt{9249.96}} = \frac{92}{96.18}$$

$$= 0.96$$

$$b = \frac{\sum xy - \dfrac{\sum x \sum y}{n}}{\sum x^2 - \dfrac{(\sum x)^2}{n}} = \frac{92}{76} = 1.2$$

Now $\bar{x} = \sum x/n = 203/7 = 29$

and $\bar{y} = \sum y/n = 178/7 = 25.43$

The regression equation of y on x is $y = a + bx$; $\bar{y} = a + b\bar{x}$

$a = \bar{y} - b\bar{x}$

$= 25.43 - 1.2 \times 29$

$= 25.43 - 34.8$

$= -9.37$

The regression equation of y on x is $y = -9.37 + 1.2x$

The regression equation of x on y is given by $x = a + by$, where

$$b = \frac{\sum xy - \frac{(\sum x)(\sum y)}{n}}{\sum y^2 - \frac{(\sum y)^2}{n}}$$

$$= \frac{92}{121.71} = 0.76$$

Now $a = \bar{x} - b\bar{y} = 29 - 0.76 \times 25.43$

$= 29 - 19.33$

$= 9.67$

Therefore the regression equation of x on y is $x = 9.67 + 0.76y$.

12. *Calculate the probable systolic blood pressure of a person aged 65 years from the following data.*

Age (x) %	56	42	60	50	54	49	39	45
Blood pressure (y)	160	130	125	135	145	115	140	120

First obtain the regression equation: Let $A_1 = 50$ and $A_2 = 135$ be assumed means for x and y

Age x	$dx = x - A_1$ $= x - 50$	dx^2	Blood pressure y	$dy = y - A_2$ $= y - 135$	dy^2	$dxdy$
56	6	36	160	25	625	150
42	−8	64	130	−5	25	40
60	10	100	125	−10	100	−100
50	0	0	135	0	0	0
54	4	16	145	10	100	40
49	−1	1	115	−20	400	20
39	−11	121	140	5	25	−55
45	−5	25	120	−15	225	75
$\sum x = 395$	$\sum dx = -5$	$\sum dx^2 = 363$	$\sum y = 1070$	$\sum dy = -10$	$\sum dy^2 = 1500$	$\sum dxdy = 170$

$$\bar{x} = \frac{395}{8} = 49.375 \qquad \bar{y} = \frac{1070}{8} = 133.75$$

Regression equation of y on x is given by $y - \bar{y} = r\dfrac{SDy}{SDx}(x - \bar{x})$

$$\frac{rSDy}{SDx} = \frac{\sum dxdy - \dfrac{\sum dx \sum dy}{n}}{\sum dx^2 - \dfrac{(\sum dx)^2}{n}}$$

$$= \frac{170 - \dfrac{(-5)(-10)}{8}}{363 - \dfrac{(-5)^2}{8}} = \frac{170 - \dfrac{50}{8}}{363 - \dfrac{(25)}{8}}$$

$$= \frac{170 \times 8 - 50}{363 \times 8 - 25} = \frac{1360 - 50}{2904 - 25}$$

$$= \frac{1310}{2879} = 0.46$$

Hence the regression equation of y on x is

$$y - 133.75 = 0.46(x - 49.38)$$
$$y = 0.46x + 133.75 - 22.72$$
$$\text{i.e.} \quad y = 0.46x + 111.03$$

The probable systolic blood pressure (y) of a person of age (x) 65 is

$$y = 0.46 \times 65 + 111.03$$
$$= 29.9 + 111.03 = 140.93$$

13. *In an experiment on effect of fertilizer application in fish ponds on fish yield the following data were obtained. Calculate the regression equation of y on x for the relation between the two variables.*

Fertilizer applied (kg/100 m^3)	5	8	10	15	18	22	25	28
Fish yield(kg/1000 m^2)	58	64	68	78	85	92	95	98

Fertilizer (x)	fish yield (y)	dx	dx^2	dy	dy^2	$dxdy$
5	58	−13	169	−20	400	260
8	64	−10	100	−14	196	140
10	68	−8	64	−10	100	80
15	78	−3	9	0	0	0
18	85	0	0	7	49	0
22	92	4	16	14	196	56
25	95	7	49	17	289	119
28	98	10	100	20	400	200
$\sum x = 131$	$\sum y = 638$	$\sum dx = -13$	$\sum dx^2 = 507$	$\sum dy = 14$	$\sum dy^2 = 1630$	$\sum dxdy = 855$

$$\bar{x} = \frac{\sum x}{8} = \frac{131}{8} = 16.375; \ \bar{y} = \frac{638}{8} = 79.75$$

The Regression coefficient

$$b_{y,x} = \frac{\sum dxdy - \dfrac{\sum dx \sum dy}{n}}{\sum dx^2 - \dfrac{(\sum dx)^2}{n}} = \frac{855 - \dfrac{(-13)(14)}{8}}{507 - \frac{(-13)^2}{8}}$$

$$= \frac{8 \times 855 + 13 \times 14}{507 \times 8 - 169}$$

$$= \frac{6840 + 182}{4056 - 169} = \frac{7022}{3877} = 1.81$$

The regression equation of y on x is given by

$$y - \bar{y} = b_{y,x}(x - \bar{x})$$

$$y - 79.75 = 1.81(x - 16.38)$$

$$y = 1.81x + 79.75 - 29.65$$

$$y = 1.81x + 50.1$$

14. *Find out the correlation coefficient and regression coefficient for the following data on the number of clusters and the number of pods in a pulse variety.*

No. of Clusters (x)	12	11	13	15	14	20	17	16	19	18
No. of pods (y)	55	50	75	30	35	45	70	80	65	70

x	y	$dx = x - 15$	dx^2	$dy = y - 55$	dy^2	$dxdy$
12	55	−3	9	0	0	0
11	50	−4	16	−5	25	20
13	75	−2	4	20	400	−40
15	30	0	0	−25	625	0
14	35	−1	1	−20	400	20
20	45	5	25	−10	100	−50
17	70	2	4	15	225	30
16	80	1	1	25	625	25
19	65	4	16	10	100	40
18	70	3	9	15	225	45
$\sum x = 155$	$\sum y = 575$	$\sum dx = 7$	$\sum dx^2 = 85$	$\sum dy = 25$	$\sum dy^2 = 2725$	$\sum dxdy = 90$

$(\sum dx)^2 = 49$

$(\sum dy)^2 = 625$

$$\text{Correlation coefficient } r = \frac{\sum dxdy - \dfrac{(\sum dx)(\sum dy)}{10}}{\sqrt{\left(\sum dx^2 - \dfrac{(\sum dx)^2}{10}\right)\left(\sum dy^2 - \dfrac{(\sum dy)^2}{10}\right)}}$$

$$= \frac{90 - \dfrac{7 \times 25}{10}}{\sqrt{\left(85 - \dfrac{49}{10}\right)\left(2725 - \dfrac{625}{10}\right)}}$$

$$= \frac{90 - 17.5}{\sqrt{(85 - 4.9)(2725 - 62.5)}}$$

$$= \frac{90 - 17.5}{\sqrt{80.1 \times 2662.5}}$$

$$= \frac{72.5}{461.81} = 0.17$$

$$\text{Regression coefficient } b = \frac{\sum dxdy - \dfrac{\sum dx \sum dy}{10}}{\sum dx^2 - \dfrac{(\sum dx)^2}{10}} = \frac{72.5}{80.1} = 0.91$$

15. *The following data gives the number of flowers and the number of seeds per plant in one of the varieties of lentil. Calculate the regression coefficient and find its significance.*

Number of Flowers:	16	18	20	19	23	17	15	22
Number of Seeds:	35	39	42	48	52	41	60	51

x=No. of Flowers	y=No. of Seeds	x^2	y^2	xy
16	35	256	1225	560
18	39	324	1521	702
20	42	400	1764	840
19	48	361	2304	912
23	52	529	2704	1196
17	41	289	1681	697
15	60	225	3600	900
22	51	484	2601	1122
$\sum x = 150$	$\sum y = 368$	$\sum x^2 = 2868$	$\sum y^2 = 17400$	$\sum xy = 6929$

$(\sum x)^2 = 22500$ $\quad (\sum y)^2 = 135424$ $\quad \bar{x} = \frac{150}{8} = 18.75$ $\quad \bar{y} = \frac{368}{8} = 46$

$$\text{Regression Coefficient } b = \frac{\sum xy - \dfrac{\sum x \sum y}{8}}{\sum x^2 - \dfrac{(\sum x)^2}{8}} = \frac{6929 - 150 \times \dfrac{368}{8}}{2868 - 150 \times \dfrac{150}{8}}$$

$$= \frac{(6929 - 6900)}{(2868 - 2812.5)} = \frac{29}{55.5} = 0.52$$

$$\sum d_x^2 = \sum x^2 - \frac{(\sum x)^2}{n} = 2868 - \frac{150^2}{8} = 55.5$$

$$\sum d_y^2 = \sum y^2 - \frac{(\sum y)^2}{n} = 17400 - \frac{368^2}{8} = 472$$

$$\sum d_x d_y = \sum xy - \frac{(\sum x)(\sum y)}{n} = 6929 - 6900 = 29$$

Significance test

$$t = \frac{b}{SE \text{ of } b} \quad \text{SE of } b = \frac{\sqrt{\left(\frac{1}{n-2}\right)\left(\sum d_y^2 - \dfrac{(\sum d_x d_y)^2}{\sum d_x^2}\right)}}{\sqrt{\sum d_x^2}} = \frac{\sqrt{\dfrac{1}{6}\left(472 - \dfrac{29^2}{55.5}\right)}}{\sqrt{55.5}}$$

$$= \frac{\sqrt{\dfrac{1}{6}\left(472 - \dfrac{841}{55.5}\right)}}{\sqrt{55.5}} = \frac{\sqrt{\dfrac{1}{6}(472 - 15.15)}}{7.5}$$

$$= \frac{\sqrt{\dfrac{456.85}{6}}}{7.45} = \frac{\sqrt{76.14}}{7.45} = \frac{8.73}{7.45} = 1.17$$

$$t = \frac{b}{\text{SE of } b} = \frac{0.52}{1.17} = 0.44$$

t value from the table $= 2.45 (p = 0.05)$ for 6 d.f

$= 3.71 (p = 0.01)$ for 6 d.f

Thus the regression coefficient is not significant.

IX. Null Hypothesis

The statistical null hypothesis is "the hypothesis which is to be tested for possible rejection under the assumption it is true".

In a two group comparison it states that the population groups are the same.

The purpose of a hypothesis test is to allow generalization from a sample to a population from which it comes. In statistical hypothesis testing besides a null hypothesis, an alternate hypothesis is also stated. Then a statistical test is done to find which hypothesis should be accepted. Normally the aim is to disprove the null hypothesis and accept the alternative hypothesis.

p-Value

It is a measure of the probability of the observed result. The p-value is also described as the probability of the observed result being due to chance. It is to be noted that the p-value is calculated after performing the statistical test. It is a conditional probability as it is related to the null hypothesis. In other words if the p-value is small the null hypothesis is rejected and if it is large the null hypothesis is not rejected. The p-value always lies between 0 and 1.

Statistically significant result

A result is said to be statistically significant if the observed result is unlikely to be due to chance and that the observed result does not go with the null hypothesis. It also means that the sampling variation is not sufficient to explain the observed result and so the null hypothesis can be rejected and the alternate hypothesis can be accepted.

Significant level

It is the critical value for p to determine a statistically significant result. The most generally used values are $p < 0.05$ and $p < 0.01$. By critical value it is meant that the extreme (maximum or minimum) value that the test statistics must have in order to declare statistically significant result at a given level.

Confidence interval

Confidence interval is an interval estimate used to indicate the variability that the estimate would have in other samples. In other words it is a range of values about a sample estimate within which at a given confidence level, the true level of the corresponding population parameter can be found. Thus, confidence intervals can be established for any population parameter. The extreme end points of this confidence interval are known as confidence limits.

If a statistics is a measure of the degree of association between two variables, its confidence interval can be used to determine whether the observed association is due to chance.

If a confidence interval overlays the null value, it implies a nonsignificant result. On the other hand if a confidence interval excludes the null value it is equivalent to statistically significant result.

In general, significance test is performed in the following steps

- The null hypothesis H_0 is stated
- The significance level is selected
- The test statistic calculation is done
- The critical value of the distribution of the test statistic for the given (one tailed or two tailed) significant level is found.
- If the test statistic lies within the accepted interval, do not reject H_0.

On the other hand if it is in the critical region null hypothesis H_0 is rejected and a significant result is taken.

By convention, a significant level of 5% is generally assumed to be necessary for rejection of null hypothesis.

Type I and Type II errors

If the null hypothesis H_0 is true and a non-significant result is obtained, correct decision has been made. If the null hypothesis is true and a significant result is obtained, the decision to reject H_0 is incorrect and an error is made. This form of error is type I or α error. The probability of making a type I error α is defined as the probability of rejecting the null hypothesis when it is true.

A significant result is obtained if $p < 0.05$ and by definition p can be less than this value for 5% of the samples when the null hypothesis is true. Alternatively the p value can be taken as the chance of making type I error or α error.

Again if H_0 is false, a significant result implies a correct decision. But when H_0 is false and a non-significant result is obtained an error in decision is made and this error is called type II error or β error.

Hence for non-significant result it is necessary to calculate the probability of making this error.

Actual Situation

	H_0 is true	H_0 is false
Do not reject H_0	No error, correct decision	β or Type II error
Reject H_0	α or Type I error	No error correct decision

α-Probability of making type I error. Probability of rejecting H_0 when it is true

β- Probability of making type II error. Probability of not rejecting H_0 when it is false.

t-distribution

The student's t distribution or the t distribution is used in place of normal distribution when the population variance is unknown. The t test is the test statistic for testing one mean under this circumstance. It is the difference between sample mean and hypothesized mean divided by standard error.

i.e.

$$t = \frac{\bar{X} - \mu}{SE} = \frac{\bar{X} - \mu}{\frac{SD}{\sqrt{n}}}$$

where n is the sample size.

The t test is also the test of significance of difference between the two sample means (two sample t-test) when the population variances are equal and unknown. It is given by

$$t = \frac{|\bar{X}_1 - \bar{X}_2|}{\text{SE}}$$

$\bar{X}_1, \bar{X}_2$ being the means of the two samples.

t-Test for unpaired samples

This test is applied to unpaired data of independent observations of samples from two means. If n_1, n_2 are the sizes of two samples with means $\bar{X}_1$, $\bar{X}_2$, then the t value is given by

$$t = \frac{|\bar{X}_1 - \bar{X}_2|}{\text{SE}} = \frac{|\bar{X}_1 - \bar{X}_2|}{\text{S.D.}\sqrt{\frac{1}{n_1} + \frac{1}{n_2}}}$$

where, $$\text{SD} = \sqrt{\frac{\sum(X_1 - \bar{X}_1)^2 + \sum(X_2 - \bar{X}_2)^2}{n_1 + n_2 - 2}}$$

where X_1, X_2 are representative values from the two samples. The pooled degrees of freedom

$$= (n_1 - 1) + (n_2 - 1) = n_1 + n_2 - 2$$

t-Test for Paired Samples

It is applied to paired data of independent observations from a single sample only when each individual gives a pair of observations.

Here t is given by

$$t = \frac{\bar{D}}{SE} = \frac{\bar{D}}{\frac{SD}{\sqrt{n}}}$$

where $\bar{D}$ is the mean of the difference $X_1 - X_2$, X_1, X_2 being a set of paired observations For applying t test

- Random samples are drawn from normal population
- Sample size < 30
- The population variances are considered to be equal when testing the equality of 2 population means.

Like the normal distribution, the t distribution is also symmetrical, unimodal and bell shaped. It has greater spread than normal distribution. Its mean is zero. Due to the greater spread variability in the confidence interval also increases, when only the sample value of the standard deviation is known. The degrees of freedom increase with the sample size. The 't' distribution is similar to the normal distribution. As the degrees of freedom increases the 't' distribution gets very close to normal distribution.

Confidence interval

For a sample size of n, the t distribution has $n-1$ as the degrees of freedom as this is based on sample variance. The degree of freedom of the t distribution is related to the *df* for the sample estimate of the standard error. On the basis of this, the confidence interval of 95% for the mean can be calculated as

$$\bar{X} \pm t_C \frac{SD}{\sqrt{n}}$$

where t_C is the critical value corresponding to $n-1$ degrees of freedom. This is also called confidence coefficient. It is a number related to the level of confidence required. In fact for a very large sample size, the confidence interval is the same as the z-distribution.

As the t distribution is more spread out, the confidence interval is also larger, when compared to normal distribution. But with larger sample size ($n > 60$) the difference between t value and that for normal distribution is negligible.

The critical value t_C of t can be represented more precisely as $t_{\alpha(1),\nu}$ or $t_{\alpha(2),\nu}$ where $\alpha(1)$ and $\alpha(2)$ represent one tailed and two tailed values of probability and ν the number of degrees of freedom.

For a 2 tailed t test if $|t| \geq t_{\alpha(2),\nu}$ reject the null hypothesis.

X. Chi-square (χ^2) Test

χ^2-test is a statistical method suitable for the analysis of nominal data. It is a non-parametric test. χ^2-test is used to compare difference between two independent proportions. The test of association between two events in binomial and multinomial samples measures the probability of association between two discrete attributes.

The χ^2-test can be used to compare frequencies or proportions in two or more groups. Like the z-test, χ^2-test is also an approximate test. It is based on calculating expected values in the different categories of variable if the null hypothesis were true and comparing these values with the observed frequencies.

The data will be exhibited in a 2×2 table. The observed and expected frequencies in each of the four cells of the table will form the basis of this test statistic. It is important to note that the values in the cells are the actual observed values and not the percentages or proportions.

		Total	
	a	b	r_1 (row 1)
	c	d	r_2 (row 2)
Total	(Column 1)	(Column 2)	

The total number of observations in each row and the total number of observations in each column is taken to be predetermined or already known values. These values are also known as **Marginal frequencies.**

If the columns and rows values are assumed to be independent the number of observations expected to occur can be calculated. These values are called **Expected Frequencies.** These are obtained as the product of row total and column total divided by the overall sum.

The χ^2-Test is used to compare the observed values in each cell with the expected frequencies. After obtaining the expected numbers (E) in each category of the variable the χ^2 statistic is calculated as

$$\chi^2 = \frac{\sum(O - E)^2}{E}$$

where O represents the observed values in each category and the summation extends to all categories.

Note:

- Sum of $O - E$ values is always zero.
- χ^2-Test is used as a measure of how far the sample distribution is different or deviates from a theoretical distribution.
- It will be found that when the row and column variables are independent the difference between observed frequencies and expected frequencies will be very small and so also will the value of χ^2 statistic.
- If there is a relationship or the categories are dependent there will be suitable difference between observed frequencies and expected frequencies so that χ^2 value will also be large. For this reason, this type of calculation is termed as a measure of "Goodness of Fit"
- It may be noted that when the number of categories of data is more as the summation of all classes is involved in the calculation χ^2 value may be greater under this circumstance too. Hence while considering the significance of a calculated value of χ^2, the number of categories of data must also be taken in to account. For this the degrees of freedom df is considered.

Degree of freedom in χ^2-test

The degree of freedom depends upon the number of categories in the variable being examined, i.e. the number of rows and columns of data. It equals $(r-1)(c-1)$, where r represents the number of rows and c the number of columns in the **Contingency Table**, where the data are presented.

- The χ^2 distribution with one degree of freedom is the square of the standard normal distribution
- The mean of the χ^2 distribution is equal to the degrees of freedom. The more the number of degrees of freedom, the mean moves more towards the right. Also the standard deviation increases. When the number of degrees of freedom becomes very large χ^2 distribution becomes more like normal distribution.
- χ^2 distribution is always ≥ 0.
- The calculations involved in using χ^2 test are comparatively easy when applied to data presented in tables.

Short Cut χ^2 Formula for 2×2 Table

$$\chi^2 = \frac{n(ad-bc)^2}{(a+c)(b+d)(a+b)(c+d)}$$

where n represents the total of row and column values and a, b, c, d represent values in the 1st row and the 2nd row of the table (i.e. $n = a + b + c + d$).

Assumption

None of the expected numbers can be less than 5.

Null hypothesis

The two variables in the cells should be independent or not associated. For testing of hypothesis use critical value of χ^2, which is also different for different degrees of freedom.

Note down the critical value χ_C^2 of the Chi-Square Distribution, for their appropriate degree of freedom at the required two-sided significant level.

If $\chi^2 \geq \chi_C^2$ reject the null hypothesis

If $\chi^2 < \chi_C^2$ do not reject the null hypothesis

Continuity correction

For $df = 1$, if the expected value in any cell is < 5 using χ^2 test gives overestimate. In order to prevent this problem, "Yates correction" is applied. Expression for χ^2 with Yates correction is given by

$$\chi^2 = \frac{\sum(O - E - 0.5)^2}{E}$$

$$\chi^2 = \frac{[n(ad - bc) - \frac{n}{2}]^2}{(a + b)(c + d)(a + c)(b + d)}$$

This correction makes χ^2 value smaller and therefore the risk of type I error, that of rejecting null hypothesis is reduced.

χ^2 test for tables larger than 2×2

As in the case of table of 2×2, the expected numbers are calculated $\frac{(O-E)^2}{E}$ for each cell of the table is calculated and the Chi-square is given by

$$\chi_{df}^2 = \sum \frac{(O - E)^2}{E}$$

df being given by $(r - 1)(c - 1)$ and r and c being the number of rows and number of columns in the contingency table.

Assumption

Not more than 20% of the expected values in the cells should be < 5 and no cell should have an expected value < 1.

Null hypothesis

Two qualitative variables in a single variable sample are not associated.

Here too from the table of values look up the critical value of χ^2_C of the χ^2 distribution for the degrees of freedom $(r-1)(c-1)$ at the required two sided significance level.

If $\chi^2 \geq \chi^2_C$ reject the null hypothesis

If $\chi^2 < \chi^2_C$ do not reject the null Hypothesis

Worked out Examples

1. *In a random sample of 600 persons from Karnataka 350 are found to be vegetarians. In another sample of 800 persons from Tamil Nadu 500 are found to be vegetarians. Find whether there is any significant difference between Karnataka and Tamil Nadu as far as the proportion of vegetarians are concerned.*

Let us take the null hypothesis that no significant difference exists.

$$SE_{P_1-P_2} = \sqrt{pq\left(\frac{1}{n_1}+\frac{1}{n_2}\right)}$$

Here $n_1 = 600$, $n_2 = 800$

$$p = \frac{x_1+x_2}{n_1+n_2} = \frac{350+500}{600+800} = \frac{850}{1400} = 0.607 = 60.7\%$$

$$q = 1 - p = 0.393 = 39.3\%$$

$$\therefore \quad SE_{P_1-P_2} = \sqrt{60.7 \times 39.3\left(\frac{1}{600}+\frac{1}{800}\right)}$$

$$= \sqrt{60.7 \times 39.3 \times \frac{7}{2400}} = 2.64$$

Expected percentage of vegetarians in Karnataka $(p_1) = \dfrac{350}{600} \times 100 = 58.33$

Expected percentage of vegetarians in Tamil Nadu $(p_2) = \dfrac{500}{800} \times 100 = 62.5$

$$\frac{\text{Difference}}{\text{SE}} = \frac{62.5 - 58.33}{2.64} = \frac{4.17}{2.64} = 1.58$$

which is less than 1.96 at 5% significant level. Hence do not reject the null hypothesis.

2. *A sample of 90 students from a university has a mean weight of 65 kg and a standard deviation of 3 kg. Estimate the population mean with 95% confidence.*

$$n = 90;\ \bar{X} = 65,\ s = 3,\ SE_{\bar{X}} = \frac{s}{\sqrt{n}} = \frac{3}{\sqrt{90}} = \frac{1}{\sqrt{10}}$$

$$95\%\ \text{confidence limit} = \bar{X} \pm 1.96\ SE$$

$$= 75 \pm 1.96 \times \frac{1}{\sqrt{10}}$$

$$= 65 \pm 0.62$$

3. *An experiment on 2 sets of 80 and 120 animals was conducted for comparing 2 diets A and B. The average increase in weight due to diets A and B are respectively 12 and 8 kg. with SD of 2.00 kg and 2.5 kg. Find whether diet A is superior to diet B at 0.01 level of significance.*

H_0 : There is no significant difference between the diets.

$$SE_{\bar{X}_1-\bar{X}_2} = \sqrt{\frac{\sigma_1^2}{n_1} + \frac{\sigma_2^2}{n_2}} = \sqrt{\frac{2^2}{80} + \frac{2.5^2}{120}}$$

$$= \sqrt{\frac{4}{80} + \frac{6.25}{120}} = \sqrt{\frac{12.25}{120}}$$

$$= \sqrt{0.102} = 0.32$$

$$\frac{\text{difference}}{\text{SE}} = \frac{12-8}{0.32} = \frac{4}{0.32} = 12.5$$

The calculated value = 12.5

Tabulated value = 1.96 at 5% and = 2.576 at 1% level of significance.

As the calculated value is greater than the tabulated value even at 1% level of significance, reject the null hypothesis.

4. *Protein content in percentage for leaves of 10 potted plants was assessed before and after gamma irradiation and the following results are obtained.*

Plant	Before Exposure to gamma rays	After exposure to gamma rays
1	26	19
2	28	22
3	22	18
4	18	17

contd. from previous page

Plant	Before Exposure to gamma rays	After exposure to gamma rays
5	21	19
6	16	12
7	15	13
8	20	15
9	24	16
10	25	18

Find out whether gamma irradiation has any effect on leaf protein content of the plant.
H_0 : There is no effect of gamma irradiation on the protein content of the plant.

Plant	Before Exposure to gamma rays	After exposure to gamma rays	D	D_2
1	26	19	7	49
2	28	22	6	36
3	22	18	4	16
4	18	17	1	1
5	21	19	2	4
6	16	12	4	16
7	15	13	2	4
8	20	15	5	25
9	24	16	8	64
10	25	18	7	49
			$\sum D = 46$	$\sum D^2 = 264$

$n = 10$

$$\bar{D} = \frac{\sum D}{10} = \frac{46}{10} = 4.6$$

$$SE = \frac{\sqrt{\dfrac{\sum D^2 - \dfrac{(\sum D)^2}{n}}{n-1}}}{\sqrt{n}}$$

i.e. $SE = \dfrac{SD}{\sqrt{n}}$

$$SD = \sqrt{\frac{\sum D^2 - \frac{(\sum D)^2}{n}}{n-1}} = \sqrt{\frac{264 - 211.6}{9}}$$

$$= \sqrt{\frac{52.4}{9}} = \frac{7.239}{3} = 2.413$$

$$SE = \frac{2.413}{\sqrt{10}} = \frac{2.413}{3.162}$$

$$= 0.763 = 0.76$$

$$t = \frac{\bar{D}}{SE_D} = \frac{4.6}{0.76} = 6.05$$

$df = n - 1 = 9$

Tabulated t value (9 df) at 5% level of significance = 2.262

The computed value 6.05 is higher than this value. Hence the null hypothesis is rejected.

5. *The following data pertains to pod length (in cms) in 2 varieties of bengal gram. Find whether there is any significant difference in the pod length of the two varieties.*

Varieties	Pod Length (cm)									
A	6.2	6.7	4.8	5.9	4.9	5.4	5.0	7.1	6.9	6.0
B	4.9	5.8	6.1	6.3	6.9	6.5	5.6	5.3	7.0	6.8

H_0: There is no significant difference in pod length.

V_A		V_B	
x(cms)	x^2	y(cms)	y^2
6.2	38.44	4.9	24.01
6.7	44.89	5.8	33.64
4.8	23.04	6.1	37.21
5.9	34.81	6.3	39.69
4.9	24.01	6.9	47.61
5.4	29.16	6.5	42.25
5.0	25.00	5.6	31.36
7.1	50.41	5.3	28.09
6.9	47.61	7.0	49.00
6.0	36.00	6.8	46.24
$\sum x = 58.9$	$\sum x^2 = 353.37$	$\sum y = 61.2$	$\sum y^2 = 379.1$

$$(\sum x)^2 = 3469.21 \qquad (\sum y)^2 = 3745.44$$

$$\bar{x} = \frac{\sum x}{10} \qquad \bar{y} = \frac{\sum y}{10}$$

$$= 5.89 \qquad = 6.12$$

$$\sum x^2 - \frac{(\sum x)^2}{10} = 353.37 - 346.921 = 6.45$$

$$\sum y^2 - \frac{(\sum y)^2}{10} = 379.1 - 374.544 = 4.56$$

$$df = n_1 + n_2 - 2 = 10 + 10 - 2 = 18$$

$$\text{Pooled Variance SD}^2 = \frac{6.45 + 4.56}{18} = \frac{11.01}{18} = 0.61$$

$$\text{SD} = \sqrt{0.61} = 0.78$$

$$t = \frac{|\bar{x} - \bar{y}|}{\text{sd}\sqrt{\frac{1}{n_1} + \frac{1}{n_2}}} = \frac{|5.89 - 6.12|}{0.78\sqrt{\frac{1}{10} + \frac{1}{10}}} = \frac{0.23}{0.78 \times 0.45} = 0.66$$

Tabulated value at 0.05 level of significant (df 18) = 2.10
Computed value = 0.66 < 2.10
Hence do not reject the null hypothesis.

6. *The percentage of marks of 12 students from each of 2 schools in a region is given below.*

	Marks											
School I	89	75	56	67	58	45	90	77	62	60	48	85
School II	72	76	85	91	49	54	61	68	69	78	80	81

Find whether any significant difference exists in the students performance.

Sl.No.	Sch.I (x)	$d_1 = x - \bar{x}$	d_1^2	Sch.II (y)	$d_2 = y - \bar{y}$	d_2^2
1	89	21	441	72	0	0
2	75	−7	49	76	4	16
3	56	−12	144	85	13	169
4	67	0	0	91	19	361
5	58	−10	100	49	−23	529
6	45	−22	484	54	−18	324
7	90	22	484	61	−11	121

Sl.No.	Sch.I (x)	$d_1 = x - \bar{x}$	d_1^2	Sch.II (y)	$d_2 = y - \bar{y}$	d_2^2
8	77	9	81	68	−4	16
9	62	−4	16	69	−3	36
10	60	−8	64	78	6	36
11	48	−20	400	80	8	64
12	85	17	289	81	9	81
	$\sum x = 816$		$\sum d_1^2 = 2552$	$\sum y = 864$		$\sum d_2^2 = 1726$

$$\bar{x} = \frac{816}{12} = 68$$

$$\bar{y} = \frac{864}{12} = 72$$

$n_1 = 12;\ n_2 = 12;\ n_1 + n_2 - 2 = 22$

H_0: No significant difference in performance exists.

$$\text{SD}_{x,y} = \sqrt{\frac{\sum d_1^2 + \sum d_2^2}{n_1 + n_2 - 2}} = \sqrt{\frac{2552 + 1726}{22}}$$

$$= \sqrt{\frac{4278}{22}} = 13.94$$

$$\text{SE} = SD\sqrt{\frac{1}{n_1} + \frac{1}{n_2}} = 13.94\sqrt{\frac{1}{12} + \frac{1}{12}}$$

$$= 13.94\sqrt{0.167} = 13.94 \times 0.408$$

$$= 5.688$$

$$t = \frac{|\bar{x} - \bar{y}|}{SE} = \frac{|68 - 72|}{5.688} = \frac{4}{5.688} = 0.7$$

calculated t value at 5% level of significance $(df22) = 0.7$

tabulated t value $(t_{0.05,22}) = 2.07$

Calculated value is $<$ tabulated value.

∴ Null hypothesis is not rejected. There is no significant difference.

7. In a survey of 200 families with 5 children each in a village the following result is obtained.

Girls	0	1	2	3	4	5	Total
Boys	5	4	3	2	1	0	
No. of families	18	35	104	96	52	15	320

Test whether the result is consistent with the hypothesis that boy and girl child are equally probable.

H_0: The chances of female and male births are equal.

By the Binomial theorem the expected value of $0, 1, 2, 3, 4, 5$ female birth will be the terms of $N(p+q)^n$

$$\text{i.e.}\quad 320\left(\frac{1}{2}+\frac{1}{2}\right)^5 = 320\left\{\frac{1}{2^5}+5\frac{1}{2^4}\times\frac{1}{2}+{}^5C_2\left(\frac{1}{2}\right)^3\left(\frac{1}{2}\right)^2+{}^5C_3\left(\frac{1}{2}\right)^2\left(\frac{1}{2}\right)^3\right.$$

$$\left.+{}^5C_4\left(\frac{1}{2}\right)\left(\frac{1}{2}\right)^4+\left(\frac{1}{2}\right)^5\right\}$$

$$= 320\left(\frac{1}{2^5}+\frac{5}{2^5}+\frac{10}{2^5}+\frac{10}{2^5}+\frac{5}{2^5}+\frac{1}{2^5}\right)$$

$\therefore$ The expected values are

$$\frac{320}{32};\quad \frac{320\times 5}{32};\quad \frac{320\times 10}{32};\quad \frac{320\times 10}{32};\quad \frac{320\times 5}{32};\quad \frac{320}{32}$$

i.e. 10, 50, 100, 100, 50, 10.

No. of families (f_0)	No. of female children	f_e	$\lvert f_0 - f_e\rvert$	$\lvert f_0 - f_e\rvert^2$
18	0	10	8	64
35	1	50	15	225
104	2	100	4	16
96	3	100	4	16
52	4	50	2	4
15	5	10	5	25
				$df = 6-1 = 5$

$$\chi^2 = \sum \frac{(f_0 - f_e)^2}{f_e}$$

$$= \frac{64}{10} + \frac{225}{50} + \frac{16}{100} + \frac{16}{100} + \frac{4}{50} + \frac{25}{10}$$

$$= 6.4 + 4.5 + 0.16 + 0.16 + 0.08 + 2.5$$

$$= 13.8$$

Computed value of $\chi^2_{(0.05,5df)} = 13.8$

Tabulated value of $\chi^2_{(0.05,5df)} = 11.07$

As the computed value of χ^2 is greater than the tabulated value, binomial distribution with $p = q = \frac{1}{2}$ is not a good fit for the given data.

8. *Details of number of mistakes in a text book of statistics are recorded below. Fit a Poisson distribution and test the goodness of fit at 5% level of significance.*

No.of mistakes in a page	0	1	2	3	4	5	Total
No. of pages	150	30	15	1	3	1	200

The expected number of pages for a mistake is given by the terms of the Poisson series.

$$e^{-m},\ me^{-m},\ \frac{m^2e^{-m}}{2!},\ \frac{m^3e^{-3}}{3!};\ \frac{m^4e^{-m}}{4!},\ \frac{m^5e^{-m}}{5!}$$

m being the average number of pages for a mistake.

$$m = \frac{0 \times 150 + 1 \times 30 + 2 \times 15 + 3 \times 1 + 4 \times 3 + 5 \times 1}{200}$$

$$= \frac{30 + 30 + 3 + 12 + 5}{200}$$

$$= \frac{80}{200}$$

$$= 0.4$$

$$e^{-m} = e^{-0.4} = 0.6703$$

No. of mistakes	Poisson probability (p)		Expected frequency (200 × probability)
0	e^{-4}	$= 0.6703$	134.06
1	$0.4e^{-4}$	$= 0.2681$	53.62
2	$(0.4^2/2)e^{-0.4}$	$= 0.0536$	10.72
3	$(0.4^3/6)e^{-0.4}$	$= 0.0071$	1.42
4	$(0.4^4/24)e^{-0.4}$	$= 0.0007$	0.142
5	$(0.4^5/120)e^{-4}$	$= 0.00006$	0.012

No. of mistakes	Observed frequency(f_o)	Expected frequency(f_e)	$\lvert f_o - f_e\rvert$	$\lvert f_o - f_e\rvert^2$	$\frac{\lvert f_o - f_e\rvert^2}{f_e}$
0	150	134.06	15.94	254.08	1.89
1	30	53.62	23.62	557.90	10.40
2	15	10.72	4.28	18.32	1.71
3	1	1.42	0.42	0.18	0.127
4	3	0.142	2.858	8.19	57.68
5	1	0.012	0.988	0.98	81.67
					153.477

$$\chi^2 = \sum \frac{(f_O - f_e)^2}{f_e}$$
$$= 153.48$$

Tabulated value of χ^2 at df 5, 5% level of significance = 11.07 which is less than the calculated value of χ^2. The null hypothesis is therefore rejected. Poisson distribution does not give a goodness of fit to the given data.

9. *From the following table test whether the colour of the son's eye is associated with that of the father.*

Eye colour of father	Eye colour of son		
	Not Light	Light	Total
Not light	203	175	378
Light	178	444	622

H_o: Eye colours of father and son are independent. N = 1000

Expected frequency

(i) $= \dfrac{381 \times 378}{1000} = 144$ (ii) $= \dfrac{619 \times 378}{1000} = 234$

(iii) $= \dfrac{381 \times 622}{1000} = 237$ (iv) $\dfrac{619 \times 622}{1000} = 385$

f_O	f_e	$\lvert f_O - f_e \rvert$	$\lvert f_O - f_e \rvert$
203	144	59	3481
178	237	59	3481
175	234	59	3481
444	385	59	3481

$$\chi^2 = \sum \frac{(f_O - f_e)^2}{f_e}$$

$$= \frac{3481}{144} + \frac{3481}{237} + \frac{3481}{234} + \frac{3481}{385}$$

$$= 3481 \times (.00694 + 0.00422 + .00427 + .00259)$$

$$= 3481 \times 0.01802 = 62.73$$

No. of degrees of freedom $= (c-1)(r-1) = (2-1)(2-1) = 1$

The calculated χ^2 value $= 62.73$

and the tabulated value of χ^2 at 5% level of significance (1df) $= 3.84$

As the calculated value is higher than the tabulated value null hypothesis is rejected.

10. *In a survey of 250 girls of whom 90 were intelligent 50 had educated fathers while the fathers of 100 of the unintelligent girls were uneducated. Test whether educated fathers have intelligent children.*

Daughter	Father		
	Educated	Uneducated	Total
Intelligent	50 (a)	40 (b)	90
Unintelligent	60 (c)	100 (d)	160
Total	110	140	250

H_o: The two attributes are independent

$n = a + b + c + d = 250$

$$x^2 = \frac{n((ad - bc) - n/2)^2}{(a+b)(b+d)(c+d)(a+c)}$$

$$= \frac{250(5000 - 2400 - 125)^2}{90 \times 140 \times 160 \times 110}$$

$$= \frac{250 \times 2475^2}{90 \times 110 \times 140 \times 160} \qquad [df = (2-1)(2-1) = 1]$$

$$= 6.91$$

Tabulated value at 5% level of significance ($df = 1$) = 3.84

As the calculated value of $\chi^2 = 6.91$ is greater than the tabulated value, H_0 is rejected. The two attributes are not independent.

11. *A certain drug was administered to 600 persons out of 1000 persons in a locality to test its effectiveness against viral fever. The following table gives the results. Obtain the effectiveness of the drug against the disease.*

	Infection	No infection	Total
Drug	350	250	600
No drug	250	150	400
Total	600	400	1000

H_0: The drug is effective in controlling the fever.

Expected frequencies: $\dfrac{600 \times 600}{1000} = 360$

$$\frac{600 \times 400}{1000} = 240$$

$$\frac{400 \times 600}{1000} = 240$$

$$\frac{400 \times 400}{1000} = 160$$

		f_O	f_e	$\|f_O - f_e\|$	$\frac{\|f_O - f_e\|^2}{fe}$
	Infected	350	360	10	$\frac{100}{360} = 0.28$
Drug	Not infected	250	240	10	$\frac{100}{240} = 0.42$
	Infected	250	240	10	$\frac{100}{240} = 0.42$
No Drug	Not infected	150	160	10	$\frac{100}{160} = 0.63$

$\chi^2 = \sum \frac{(f_O - f_e)^2}{f_e} = 1.75$

$df = (2-1)(2-1) = 1$

At 5% level of significance ($df = 1$) tabulated value of $\chi^2 = 3.84$

Computed value of $\chi^2 = 1.75$

As the computed value is less than the tabulated value the null hypothesis need not be rejected.

XI. Analysis of Variance

Comparison of two groups can be done using hypothesis test and confidence intervals. The correct technique for comparing means in three or more groups is called Analysis of Variance or ANOVA.

It is one of the most powerful tools for the analysis of experimental data, particularly laboratory experiments. It is a parametric test like t-test.

In ANOVA, comparison of three means is considered.

One way ANOVA is like an extension of a t-test. In a t-test the null hypothesis testing is to check whether two populations have the same mean. In one way ANOVA the null hypothesis test is to see whether all the populations have the same mean. One way ANOVA can be used to compare just two populations and the results would be consistent with that of a t-test.

One way ANOVA

When there is only one qualitative variable one-way ANOVA is performed. It is a technique for comparing independent means in more than 2 groups.

Assumptions in ANOVA

1. The population variance is the same in each group
2. Within each group, the values of the samples/variables are normally distributed.
3. The observations are independent random samples.

The null hypothesis H_0 and the alternative hypothessis H_1 for one way ANOVA are:

H_0: all the samples come from populations with the same mean

H_1: all the samples do not come from population within the same mean

So a significant result does not mean that all of the mean differ from one another, just they are all not all the same.

Note:

1. Deviations from the above assumptions especially that of normality will not affect the F test particularly when the number of observations in each group is very large.
2. Differences in variance will affect F test. Homogeneity is almost a necessity but when this assumption is not satisfied this can be over looked if the sample sizes are equal or almost the same in each of group. When this is not possible in certain experiments, equal variance is obtained by making necessary transformation of data within each group. Otherwise samples of equal size can be selected at random from each group.
3. It is of utmost importance to have independent observations.

Because of the randomness the one-way ANOVA is said to represent a completely randomized experimental design.

Before describing the procedure an explanation of the notations used will be in order.

K = Number of experimental groups (Independent Random samples)

$\mu_1, \mu_2, \ldots, \mu_k$ = Corresponding population means

N $= \sum_{j=1}^{k} nj$ total number of observations

X_{ij} - the ith observation in the jth group

$\bar{X}_j$ - Mean of group j

$\bar{X}$ - Grand mean of all observations

Null hypothesis H_0: The k population means μ_i's are equal.

Between group and within group variations

To determine whether or not group means are significantly different, comparison of between groups and within group variability can be done. With the assumption of equal sample sizes, the best estimate of the common population variance from within the sample values will be the mean of all the groups. This is called the Within Group Variance . It is also called the error of the mean square and denoted by

$$S_W^2 = \frac{\sum\sum(X_{ij} - \bar{X}_j)^2}{N - k}$$

Thus error sum of squares $SS_W = \sum_{j=i}^{k}\sum_{i=1}^{nj}(X_{ij} - \bar{X}_i)^2$

Within Group $DF = \sum_{j=i}^{k}(n_j - 1) = N - k$ (SS- Squared Sum)

Between group mean square

The variation of the sample means is obtained and the summation is taken over k groups and we get $\dfrac{\sum(\bar{X}_j - \bar{X})^2}{k-1}$

The best estimate of the population variance is given by $S_B^2 = \dfrac{\sum(X_j - \bar{X})^2}{k-1}$ (Sample sizes being equal).

This is called Between Group Mean Square

Between group $DF = k - 1$

The total sum of squares $SS_T = \sum(Xij - \bar{X})^2$

$$= \sum X_{ij}^2 - \frac{(\sum X_{ij})^2}{N}$$

$$SS_B = \sum(\bar{X}_j - \bar{X})^2$$

$$= \sum n_j\bar{X}_j^2 - \left[\frac{(\sum X_{ij})^2}{N}\right]$$

Also SS_W is found simply from the relation

$$SS_W = SS_T - SS_B$$

Which results from the relation that

$$\sum(X_{ij} - \bar{X})^2 = \sum(X_{ij} - \bar{X}_j)^2 + \sum(\bar{X}_j - \bar{X})^2$$

$$\text{i.e.} \quad SS_T = SS_W + SS_B$$

Now the mean squares are given by

$$MS_W = \frac{SS_W}{N-k}$$

$$\text{and,} \qquad MS_B = \frac{SS_B}{k-1}$$

It is to be noted here that the measure of variability consists of 2 parts: a sum of squares and degree of freedom associated with this sum.

F test

For the ANOVA test the F Test, which is the variance ratio test is used. F is the most appropriate hypothesis test for comparing more than two means and it is based on between group sum of squares and within group sum of squares.

The F test is very sensitive to data that are not normal. To use the F test the larger variance must be in the numerator and the smaller in the denominator of the F ratio so that F thus calculated may be ≥ 1. When the ratio is significantly greater than 1, it will be concluded that the variances are unequal. No interpretation will be made to the F-ratio if it is less than 1. Such a situation can be only a chance occurrence. An important property of F distribution is that unlike t and χ^2 distribution, it has two degrees of freedom. These correspond to the degrees of freedom of the two variances in the ratio.

Steps for calculations in ANOVA

- Square each observation, add to obtain the sum $\sum\limits_i\sum\limits_j X_{ij}^2$
- Find the square of the sum of the observations and divide this sum by N to get $\frac{(\sum\sum X_{ij})^2}{N}$
- Obtain $\sum_{j=1}^{k} nj\bar{X}_j^2$ by squaring each mean $\bar{X}_j$ multiplying these by the corresponding number of observations n_j in the j-th group and then summing all these values.
- Form the following sums of squares.

$$SS_T = \sum\sum X_{ij}^2 - \frac{(\sum\sum X_{ij})^2}{N}$$

Between Group Squares $SS_B = \sum nj\bar{X}_j^2 - \frac{(\sum\sum X_{ij})^2}{N}$

Within Group Squares $SS_W = SS_T - SS_B$

- Calculate the mean squares
$$MS_W = \frac{SS_W}{N-k}, \quad MS_B = \frac{SS_B}{k-1}$$
- Finally the F ratio is calculated as $F = \dfrac{MS_B}{MS_W}$
- This value is compared with the critical value Fc from the table corresponding to $k-1, N-k$ degrees of freedom at the required level of significance.
- If $F \geq Fc$ reject the Null hypothesis If $F < Fc$ do not reject the Null hypothesis

The ANOVA table is prepared as follows

Source of Variation	DF	Sum of squares (SS)	Mean sum of squares (MSS)	F ratio
Between Groups	$k-1$	$SS_B = \sum\sum njX_j^2 - \frac{(\sum\sum X_{ij})^2}{N}$	$MS_B = \frac{SS_B}{k-1}$	$F = \frac{MS_B}{MS_W}$
Within groups (Error)	$N-k$	$SS_W = SS_T - SS_B$	$MS_W = \frac{SS_W}{N-k}$	
Total	$N-1$	$SS_T = \sum\sum X_{ij}^2 - \frac{(\sum\sum X_{ij})^2}{N}$		

Interpreting the results

(2) In practice most people carry out the calculations on a computer but it is improtant to know how to interpet the output. The results of ANOVA are usually presented in ANOVA table as follows:

Sl. No.	Source of Variation	df	SS	MS	F	P-value
1.	Between groups	5	0.2131	0.0607	4.089	0.071
2.	Within groups	10	0.1214	0.0112		
3.	Total	15	0.3345			

df = degrees of freedom. SS: Sum of squares

df between groups = number of groups -1

df for total = total number of measurements -1

df within groups = df for total − df between groups

MS = Sum of squares/degrees of freedom = Mean square

$F = MS$ for between groups/MS for within groups

MS are the estimates of variance and what is compared in the test

$P = 0.0171$ is the P- value for $F = 4.08$ with 5 to 10 degrees of freedom.

F values have two associated degrees of freedom df for between groups and df for within groups.

If the result is P value > 0.05 there is no significant differences between any of the populations means. If the P value is ≤ 0.05 then at least two of the populations have significantly different means.

Multiway ANOVA

Multiway ANOVA is an extensive subject. It is theoreticaly possible to use multiway ANOVA to study the effect of a large number of factors at once. Eg. pH, Salinity, temperature, BOD, humidity, time of the year. But it is very difficult to interpret the results of multifactor studies, if we want to know not only which factors have an effect but also what their effect is. In such cases *one will* usually get clear results by limiting the study to no more than three factors. The most common use of multiway ANOVA in biological or environmental sciences is to determine whether two or three factors interact together to produce a greater or lesser effect than are expected from their effects individually.

Null hypothesis

For multiway ANOVA there are several null hypotheses (H_0) and alternative hypotheses (H_1), all of which will be tested together.

For two way ANOVA three null hypotheses are tested:

H_0 Factor 1 has **no** effect on the population mean
H_1 Factor 1 does have an effect on the population mean
H_0 Factor 2 has no effect on the population mean
H_1 Factor 2 does have an effect on the population mean
H_0 There is no interaction between factor 1 and factor 2
H_1 There is an interaction between factor 1 and factor 2

In similar way three way ANOVA can have upto seven null hypotheses. The complexity increases by including extra factors. A Significant result for any of these hypotheses usually means that one has to carry out further comparisons to decide how the factors interact. With four factors, a full interpretation of the result becomes even more complex and this test would include upto 15 null hypotheses.

In short for one way ANOVA, the test makes an assessment of how likely it is that samples come from populations with the same mean by considering the variation between sample means relative to the variation within the sample. For two way ANOVA there will be two factors and each factor may have more than two levels. The two way ANOVA table is prepared in the same way as one way ANOVA, only the source of variation will have more entities. An example to interpret a multiway ANOVA table.

1. If the result is a P-value > 0.05 for any of the main effects the conclusion is: The factor had no significant effect.
2. If the result is P-value ≤ 0.05 for any of the main effects, the conclusion is the factor had a significant effect.
3. If the result is a P value > 0.05 for an interactions one could conclude. there was no significant interactions between those factors.
4. If the result is a P value ≤ 0.05 for an interaction one could conclude; there was a significant interaction between these factors.

In summary the technique can be extended to analyse studies with a variety of different blockery designs and to compare groups of treatment with other groups of treatments in the design rather than just to compare individual means.

Multivariate ANOVA or MANOVA

An extension of the ideas behind t-test and ANOVA is to test whether populations differ with several variables at the same time. The method used to test this is called multivariate ANOVA or MANOVA. MANOVA is one of the class of methods that consider together several types of measurements made on each individual. These are called multivariate methods. MANOVA is less commonly used in biological sciences than t-test or ANOVA because the investigations are usually aimed at identifying which aspects of populations are affected by their environments. MANOVA often gives too general an answer. Simply establishing that populations differ is often not very helpful; one also wants to know how they differ.

In summary MANOVA compares two groups in terms of several variables.

Worked out Example

1. *In an experiment different concentrations of plant cytokine were applied to the leaves and the emergence of roots was tested. The following results were obtained.*

Replicates	No. of emergent roots under different concentration of cytokine (ppm)			
	0	5	10	15
1	1	2	3	5
2	2	3	3	4
3	1	3	4	5
4	0	4	3	4
5	1	3	4	4

Apply ANOVA to test whether different concentrations have any effect on root emergence $(Gn.F_{0.01}(3,16) = 5.3)$

H_0: There is no significant difference under different concentrations

Repli-cates	(x) No. of emergent roots under Different concentration							
	0	5	10	15	X^2_{c1}	X^2_{c2}	X^2_{c3}	X^2_{c4}
	x_{c_1}	x_{c_2}	x_{c_3}	x_{c_4}				
1	1	2	3	5	1	4	9	25
2	2	3	3	4	4	9	9	16
3	1	3	4	5	1	9	16	25
4	0	4	3	4	0	16	9	16
5	1	3	4	4	1	9	16	16
	$\sum x_{c_1} = 5$	$\sum x_{c_2} = 15$	$\sum x_{c_3} = 17$	$\sum x_{c_4} = 22$	$\sum x^2_{c_1} = 7$	$\sum x^2_{c_2} = 47$	$\sum x^2_{c_3} = 59$	$\sum x^2_{c_4} = 98$

$n = 5 \times 4 = 20$

n_c = no. of elements in each column

$$\sum x = 5 + 15 + 17 + 22 = 59$$

$$\sum x^2 = 7 + 47 + 59 + 98 = 211 \qquad (1)$$

$$\frac{(\sum x_c)^2}{n_c} = \frac{5^2}{5} + \frac{15^2}{5} + \frac{17^2}{5} + \frac{22^2}{5}$$

$$= \frac{1023}{5} = 204.6 \qquad (2)$$

$$= \frac{(\sum x)^2}{n} = \frac{59^2}{20} = 174.5 \qquad (3)$$

Variations

Between concentration $= 204.6 - 174.05 = 30.55$

Within concentrations $= 211 - 204.6 \quad = 6.4$

Total $= 211 - 174.05 \quad = 36.95$

Anova Table

Sources of variation \ *df*	*DF*	*SS*	*MS*	F
Between concentrations	$(c-1) = 4-1 = 3$	30.55	$\frac{30.55}{3} = 10.18$	$\frac{10.18}{0.4} = 25.45$
Within concentration	$c(r-1) = 4(5-1) = 16$	6.40	$\frac{6.4}{16} = 0.4$	
Total	$cr - 1 = 20 - 1 = 19$	36.95		

Calculated value of $F = 25.45$ at *df* (3,16) 0.01 level of significance

tabulated value of $F = 5.3$

As the calculated value of F is greater than the tabulated value, the null hypothesis is rejected.

2. *Data recorded on seedling height (cm) of a pulse crop in gamma irradiation material are given below. Calculate the F value and conclude your result.*

Replicates	Seedling ht (cm)		
	10 KR	20 KR	30 KR
1	12	15	13
2	18	13	11
3	23	18	19
4	14	12	10
5	20	11	16

H_0 : The effect of the three levels of gamma rays on seed germination is the same.

Replicates	Seeding's ht (cm)			
	10 KR xc_1	20 KR xc_2	30 KR xc_3	Total
1	12	15	13	40
2	18	13	11	42
3	23	18	19	60

contd. from previous page

Replicates	Seeding's ht (cm)			
	10 KR xc_1	20 KR xc_2	30 KR xc_3	Total
4	14	12	10	36
5	20	11	16	47
Total	87	69	69	225

$$\sum x = \sum_c x = 87 + 69 + 69 = 225 \qquad n = 5 \times 3 = 15$$

	$x_{c_1}^2$	$x_{c_2}^2$	$x_{c_3}^2$	
	144	225	169	
	324	169	121	
	529	324	361	
	196	144	100	
	400	121	256	
$\sum x^2 =$	1593	983	1007	$= 3591$

$$\frac{(\sum x)^2}{n_c} = \frac{87^2 + 69^2 + 69^2}{5}$$

$$= \frac{7659 + 4761 + 4761}{5}$$

$$= \frac{17181}{5} = 3436.2$$

$$= \frac{(\sum x)^2}{n} = \frac{225^2}{15} = 3375$$

Variation

Between varieties = $3436.2 - 3375 = 60.8$

Within varieties = $3591 - 3436.2 = 154.8$

Total 215.6

Anova Table

Sources of variation	*df*	*S.S*	*M.S*	*F*
Between varieties	$c - 1 = 3 - 1 = 2$	60.8	$\frac{60.8}{2} = 30.4$	$\frac{30.4}{12.9} = 2.36$
Within varieties	$c(r - 1) = 3(5 - 1) = 12$	154.8	$\frac{154.8}{12} = 12.9$	
Total	14	215.6		

Calculated F value = 2.36

Tabulated vlaue of F at 1% level of significance with df (2,12) = 6.9

As the calculated F value is less than the tabulated value do not reject the null hypothesis.

3. *The following data give the production (in kg) of 3 varieties, A,B,C of wheat sown in 12 plots.*

A	20	18	25	21	
B	20	28	23		
C	17	22	16	15	32

Find if there is any significant difference in the production of these three varieties.

H_0 : There is no significant difference in the production of these varieties.

V_A (x_1)	V_B (x_2)	V_C (x_3)
20	20	17
18	28	22
25	23	16
21		15
		32

$\sum x_1 = 84$, $\sum x_2 = 71$; $\sum x_3 = 102$

$$\bar{X}_1 = \frac{\sum x_1}{4} = \frac{84}{4} = 21$$

$$\bar{X}_2 = \frac{71}{3} = 23.67$$

$$\bar{X}_3 = \frac{102}{5} = 20.4$$

$$\text{Grand mean } \bar{X} = \frac{\bar{X}_1 + \bar{X}_2 + \bar{X}_3}{3}$$

$$= \frac{21 + 23.67 + 20.4}{3}$$

$$= 21.69$$

$(\bar{X}_1-\bar{X})^2$	$(\bar{X}_2-\bar{X})^2$	$(\bar{X}_3-\bar{X})^2$
$.69^2 = .4761$	$2.02^2 = 4.0804$	$1.29^2 = 1.6641$
.4761	4.0804	1.6641
.4761	4.0804	1.6641
.4761		1.6641
$a=\sum(\bar{X}_1-\bar{X}^2)=$ $4\times 0.4761=1.9044$	$b=\sum(\bar{X}_2-\bar{X}^2)=$ $3\times 4.0804=12.2412$	$c=\sum(\bar{X}_3-\bar{X})^2=$ $5\times 1.6641=8.3205$

$$SS \text{ between varieties} = a+b+c = 1.9044+12.2412+8.3205$$

$$= 22.4661 = 22.47$$

$$\text{Mean square between varieties} = \frac{a+b+c}{(3-1)} = \frac{22.47}{2} = 11.24$$

S.S. within samples

	X_1	$(X_1-\bar{X}_1)^2$	X_2	$(X_2-\bar{X}_2)^2$	X_3	$(X_3-\bar{X}_3)^2$
	20	$1^2=1$	20	$3.67^2=13.4689$	17	$3.4^2=11.56$
	18	$2^2=4$	28	$4.33^2=18.7489$	22	$1.6^2=2.56$
	25	$4^2=16$	23	$0.67^2=0.4489$	16	$4.4^2=19.36$
	21	0			15	$5.4^2=29.16$
					32	$11.6^2=134.56$
Total		21		$32.6667=32.67$		197.20

$$SS \text{ within varieties} = 21+32.67+197.2 = 250.87$$

$$MS \text{ within varieties} = \frac{250.87}{12-3} = 27.87$$

Anova table

Sources of variation	Df	$S.S$	$M.S$	F
Between samples	2	22.47	11.24	$\frac{11.24}{27.7}=0.4$
Within samples	9	250.87	27.87	
Total	11	273.34		

Tabulated value of $F_{0.05}$ $(df\ 2,9) = 4.26$

As the calculated value is less than the tabulated value do not reject the null hypothesis.

4. *From the following data determine if there is any significant difference in the means of the 4 samples*

Sample I	Sample II	Sample III	Sample IV
57	52	58	42
49	46	67	53
54	55	68	65

H_0: There is no significant difference in the mean.

Sample I (x_1)	Sample II (x_2)	Sample III (x_3)	Sample IV(x_4)	Total
57	52	58	42	209
49	46	67	53	215
54	55	68	65	242
Total : 160	153	193	160	666

(x_1^2)	(x_2^2)	(x_3^2)	(x_4^2)
3249	2704	3364	1764
2401	2116	4489	2809
2916	3025	4624	4225
8566	7845	12477	8798

$$\sum x^2 = \sum x_1^2 + \sum x_2^2 + \sum x_3^2 + \sum x_4^2 = 37686 \qquad (1)$$

$$\begin{aligned}\frac{(\sum x)^2}{n_c} &= \frac{160^2 + 153^2 + 193^2 + 160^2}{3} \\ &= \frac{25600 + 23409 + 37249 + 25600}{3} \\ &= \frac{111858}{3} \\ &= 37286 \qquad (2)\end{aligned}$$

$$\begin{aligned}\frac{(\sum x)^2}{n} &= \frac{(\sum x_1 + \sum x_2 + \sum x_3 + \sum x_4)^2}{n} \\ &= \frac{(160 + 153 + 193 + 160)^2}{12} \\ &= \frac{666^2}{12} = 36963 \qquad (3)\end{aligned}$$

S. S. between samples = 37286 – 36963 = 323

S. S. within samples = 37686 – 37286 = 400

Total 723

Anova Table

Sources of variation	Df	SS	MS	F
Between samples	$(4-1)=3$	323	107.67	$\frac{107.67}{50} = 2.1534$
Within samples	$(12-4)=8$	400	50	
Total	11	723		i.e. 2.15

Calculated value of $F = 2.15$

Tabulated value 5% level of significance df $(3, 8) = 4.07$

As the calculated value is less than the tabulated value do not reject the null hypothesis.

5. *The rating of 5 candidates by 4 experts is listed below. Examine whether there is any significant difference between the candidates and between the experts.*

Candidates	**Experts**			
	A	B	C	D
1	9	8	7	7
2	6	9	8	7
3	5	4	3	4
4	7	6	5	8
5	3	4	4	3

H_0 : There is no significant difference

Candidates	**Experts**				
	A	B	C	D	Total
1	9	8	7	7	31
2	6	9	8	7	30
3	5	4	3	4	16
4	7	6	5	8	26
5	3	4	4	3	14
$\sum x$	30	31	27	29	117

	x_1^2	x_2^2	x_3^2	x_4^2
	81	64	49	49
	36	81	64	49
	25	16	9	16
	49	36	25	64
	9	16	16	9
Total	200	213	163	187

$$\sum x^2 = 200 + 213 + 163 + 187 = 763 \quad (1)$$

$$\sum \frac{(\sum x)^2}{n_c} = \frac{30^2}{5} + \frac{31^2}{5} + \frac{27^2}{5} + \frac{29^2}{5} = \frac{3431}{5} = 686.2 \quad (2)$$

$$\sum \frac{(\sum x)^2}{n_r} = \frac{31^2}{4} + \frac{30^2}{4} + \frac{16^2}{4} + \frac{26^2}{4} + \frac{14^2}{4} = \frac{2989}{4} = 747.25 \quad (3)$$

$$\frac{(\sum \sum x)^2}{n} = \frac{117^2}{20} = 684.45 \quad (4)$$

Variation

Between columns = 686.20 – 684.45 = 1.75 (2–4)

Between rows = 747.25 – 684.45 = 62.80 (3–4)

Residual = (763 – 684.45) – 64.55

{(1-4) - (2-4 + 3-4)} = 78.55 – 64.55 = 14

Total (1 – 4) = 78.55

Two way ANOVA Table

Sources of Variation	*df*	*SS*	*MS*	*F*
Between (Rows)	$r-1=5-1=4$	62.80	15.7	27.07
Between Experts (Column)	$c-1=4-1=3$	1.75	0.58	0.49
Residual	$(c-1)(r-1)=12$	14	1.17	
Total	$cr-1=19$			

Conclusion:

	Computed Value	Tabulated Value	*Df*	Level of Significance
F for candidates	27.07	3.26	4, 12	
F for experts	0.49	3.49	3, 12	

Candidates: F (Calculated) > F critical
Null hypothesis is rejected
There is significant variation between candidates.

Experts: F (Calculated) < F (Tabulated)
Do not reject the null hypothesis
There is no significant variation in the ratings by experts.

6. *Three colleges which have started 4 new courses get the percentage of passes for the first batch of students as given in the following table. Apply two way classification of analysis of variance and interpret your result.*

College	Courses			
	B1	B2	B3	B4
A1	55	65	60	62
A2	70	66	65	68
A3	62	52	49	48

Colleges	Courses				
	B1	B2	B3	B4	Total
A1	55	65	60	62	242
A2	70	66	65	68	269
A3	62	52	49	48	211
Total	187	183	174	178	722

H_{01}: There is no significant difference between the college.
H_{02}: There is no significant difference between performances in the subjects.

$$\sum x = 187 + 183 + 174 + 178 = 722$$

$$\frac{(\sum x)^2}{12} = \frac{(722)^2}{12} = 43440.33$$

$$\begin{aligned}\sum x^2 &= \text{Sum of the squares of all data}\\ &= (55^2 + 70^2 + 62^2) + (65^2 + 66^2 + 52^2) +\\ &\quad (60^2 + 65^2 + 49^2) + (62^2 + 68^2 + 48^2)\\ &= 11769 + 9285 + 10226 + 10772 = 44052.00 \qquad (1)\end{aligned}$$

$$\frac{(\sum x_c)^2}{n_c} = \frac{187^2}{3} + \frac{183^2}{3} + \frac{174^2}{3} + \frac{178^2}{3} = 43472.67 \qquad (2)$$

$$\frac{(\sum x_r)^2}{n_r} = \frac{242^2}{4} + \frac{269^2}{4} + \frac{211^2}{4} + = 43861.5 \quad (3)$$

$$\frac{(\sum x)^2}{12} = 43440.33 \quad (4)$$

$$(1) - (4) = 611.67 \quad (5)$$

Sources of variations	Df	SS	MS	F	Tabulated $F_{0.05}$
Between Colleges (Rows)	$3 - 1 = 2$	(3) – (4) 421.17	210.59	7.99	5.14 at (2, 6) df
Between Subjects (Columns)	$4 - 1 = 3$	(2) – (4) 32.34	10.78	0.41	4.76 at (3,6) df
Residual	$(3 - 1)(4 - 1)$ =6	(5) – {(2) – (4) + (3 – 4)} = 158.16	26.36		

1. **Colleges:** As the calculated F Value is greater than the tabulated value reject the null hypothesis
2. **Students:** Since the computed value of F is less than the tabulated value, do not reject the null hypothesis.

XII. Computer Packages

Though all the statistical problems are carried out on a computer it is important be able to interpret the output. For this reason the basic knowledge in the subject is essential which has been extensively dealt with in this book.

This appendix contains points about how to carry out most of the analysis described in this book using Excel. Excel is common programme available in all MS packages. The students need not spend to have special packages for doing basic statistical calculation in computer. This notes relate specifically to Microsoft Excel 2000.

Excel includes some usefule statistical functions which can be entered into a cell as part of a formula.

Click fx on the standard toolbar for a list of functions. For example the list is as follows:

AVERAGE — calculates the mean

CHIDIST — Gives P-values for given values of $\chi^2 df$.

CHITEST: Carries out χ^2 test for associates when both the observed and expected values are given. The function gives a P-value directly.

FDIST: This gives the P-value for given values of F and df. It can be used to see the result of ANOVA or an F-test. For the F-test multiply the P-value by 2.

STDEV: Calculates the standard deviation.

TDIST: Gives P-values for given values of t and df. It gives two tailed P-value

TINV: Gives t-values for a given probability and df. It can be used to obtain the t value required to calculate a least significant difference or a confidence interval.

VAR: Calculates the variance. Most of the statistical tests are accessed by opening the tools menu and selecting DATA analysis. Most of the tests in Excel require the data for each sample to be in a separate column or row. Non parametric tests and multivariate methods are not available as functions or commands in Excel.

Appendix

Statistical Calculations at a Glance

s_1	Standard deviation of Sample 1
s_1^2	Variance of sample 1
n_1	Sample size for sample 1
$\bar{X}_1$	Mean of sample 1
$\|x\|$	absolute values of x
df	degrees of freedom
$\sum x_1$	all the values in sample 1 are added together
$\sum x_1^2$	each value in sample 1 is squared then these squares are added together
$(\sum x_1)^2$	all the value in sample 1 are added together and then the total of these is squared
$(\sum xy)$	Each pair of X and Y values are multiplied together then these values are added together
$(\sum xy)^2$	Each pair of X and Y values are multiplied togther then these values are added to together and then the total of these is squared.

Mean, variance, SD and coefficient of variation.

It is often useful to be able to characterize a population in terms of a few well-chosen statistics. These allow us to summarize a large number of measurements in order to present results and also to compare the measuerements with one another.

Example:

Mean: $\bar{x} = \dfrac{\sum x}{n}$

Variance: $s^2 = \dfrac{\sum x^2 - \frac{(\sum x)^2}{n}}{n-1}$

Standard deviation $= SD = \sqrt{s^2}$

Standard error $= SE = \dfrac{S}{\sqrt{n}}$

95% confidence interval: $\bar{x} \pm (t_{n-1,\ 0.05}) \times \dfrac{s}{\sqrt{n}}$;

Here $t_{n-1,\ 0.05}$ is value obtained from t-table with $(n-1)$ degrees of freedom and $P = 0.05$.

95% confidence interval is used to quote a margin of error and to be more confident to show where the population mean really lies. The 95% CI gives approximately twice as wide a range of values as the standard error.

Unpaired t-test

Unpaired t-test is comparing means of two populations, Sample 1 and Sample 2. Calculate $\bar{x}_1$, $\bar{x}_2$, s_1^2 and s_2^2.

Calculate a pooled estimate of variance $S_p^2 = \dfrac{(n_1-1)s_1^2 + (n_2-1)s_2^2}{n_1+n_2-2}$

Calculate the t-value with df,

$$df = (n_1 + n_2 - 2)$$

$$t_{n_1+n_2-2} = \frac{(\bar{x}_1 - \bar{x}_2)}{\sqrt{S_p^2(\frac{1}{n_1} + \frac{1}{n_2})}}$$

from t-table find out p- value for significant results.

Paired t-test

Paired t-test is comparing means of two populations using paired samples, Sample 1 and Sample 2.

1. Calculate pairwise differences X_d
2. Calculate mean and variance of pairwise differences like the same way to calculate the mean and variance of a sample. $\bar{x}_d$ and S_d^2.

Calculate the t-value with df $n_d - 1$ where n_d is the number of pairwise difference.

$$t_{n_d-1} = \frac{(\bar{x}_d)}{\sqrt{\frac{s_d^2}{n_d}}}$$

Calculate the p value from t-table.

t-test:

It is comparing the mean of a population with a fixed value. Calculate, $\bar{x}$, s^2, μ_0 (hypothesized mean) for a given sample.

Calculate the t-value with $df = (n-1)$

$$t_{n-1} = \frac{(\bar{x}\mu_0)}{\sqrt{\frac{s^2}{n}}}$$

From t-tables find the p-value with t_{n-1}

F-test

It is comparing variances of two populations.

Calculate $\bar{x}_1$, $\bar{x}_2$, s_1^2 and s_2^2.

Calculate F with $df = (n_1 - 1)$ and $df = (n_2 - 1)$

$F_{n_1-1,n_2-1} = \frac{S_1^2}{S_2^2}$, Always divide the larger variance by the smaller.

If s_2^2 is larger the formula is $F_{n_2-1,\ n_1-1} = \dfrac{s_2^2}{s_1^2}$. Calculate P-values from F-table.

One way ANOVA

Comparing means of several populations. Consider five different samples (several population).

Calculate

$(\sum x_1)^2 + (\sum x_2)^2 + (\sum x_3)^2 + (\sum x_4)^2 + (\sum x_5)^2 = (\sum x_T)^2$

Arrive at $(\sum x_T)^2$, n_T and $\sum x_T^2$

$\sum x_T$ is calculated by adding all sample values together and then squaring the total.

$\sum x_T^2$ is calculated by squaring each of the values and then adding those squares together.

Calculate sums of squares

Total sum of squares $SS_T = \sum x_T^2 - \frac{(\sum X_T)^2}{n_T}$

Between groups sum of squares

$$SS_B = \frac{(\sum x_1)^2}{n_1} + \frac{(\sum x_2)^2}{n_2} + \frac{(\sum x_3)^2}{n_3} + \frac{(\sum x_4)^2}{n_4} + \frac{(\sum x_5)^2}{n_5} - \frac{(\sum x_T)^2}{n_T}$$

within the groups sum of squares

$SS_W = SS_T - SS_G$

From ANOVA table calculations of degrees of freedom mean squares and the F-value are taken. the F-value takes its degrees of freedom from the df for between groups and the df for 'within groups' in that order. From F-tables look for p-values.

Two way ANOVA

Testing for the effect of two different factors and whether they interact. The samples involve two factors.

Factor A and Factor B

Calculate sums of squares, total sum of squares $SS_T = \sum x_T^2 - \frac{(\sum x_T)^2}{n_T}$

Factor A sum squares

$SS_A = \frac{(\sum x_a)^2}{n_a} + \frac{(\sum x_b)^2}{n_b} + \ldots + \frac{(\sum x_d)^2}{n_d} - \frac{(\sum x_T)^2}{n_T}$

Factor B sum of squares

$SS_B = \frac{(\sum x_1)^2}{n_1} + \frac{(\sum x_2)^2}{n_2} + \ldots - \frac{(\sum x_T)^2}{n_T}$

Calculate SS_G as in one way ANOVA.

Instead of n_1, n_2, use n_{a_1}, n_{a_2}, n_{b_1}, n_{b_2} etc.

Interaction sum of squares.

$SS_1 = SS_G - SS_A - SS_B$

within groups sum of squares.

$SS_R = SS_T - SS_A - SS_B - SS_T$

From F table calculate P-values.

Correlation: Testing whether there is a straight line relationship between two variables.

Sample is with two variables X and Y. Calculate XY for samples.

Then calculate $\sum x$, $\sum y$, $\sum xy$, $(\sum x)^2$, $(\sum y)^2$, $\sum x^2$ and $\sum y^2$ know the n value which is number of pairs of measurements.

Calculate the correlations co-efficient r with $df = n - 2$

$$r_{n-2} = \frac{\sum xy - \frac{\sum x \sum y}{n}}{\sqrt{\left(\sum x^2 - \frac{(\sum x)^2}{n}\right)\left(\sum y^2 - \frac{(\sum y)^2}{n}\right)}}$$

From r table get P values.

Linear Regression: It is determining an equation to describe the relationship between two variables. A multiple regression is determining an equation to describe how values of one variable can be estimated from two or more other variables. In simple linear regressions the variable 'y' should be the one that is responding to changes in x.

Calculate:

$\sum x$, $(\sum x)^2$, $\sum x^2$, $\bar{x}$
$\sum y$, $(\sum y)^2$, $\sum y^2$, $\bar{y}$

and note n the number of pairs of measurements.

1. Calculate the equation of regression line

   $\beta = \frac{n\sum xy - (\sum x)(\sum y)}{n\sum x^2 - (\sum x)^2}$ $\alpha = \bar{y} - \beta\bar{x}$

   The equation of the line of best fit is

   $y = \alpha + \beta x$

2. Calculate the sum of squares

   Total sum of squares $SS_T = \sum y^2 - \frac{(\sum y)^2}{n}$

   Regression sum of squares

   $SS_{reg} = \frac{\sum xy - \frac{(\sum x)(\sum y)}{n}}{\sum x^2 - \frac{(\sum x)^2}{n}}$

Residual sum of squares

$$SS_{res} = SS_T - SS_{reg}$$

The calculation of mean squares and the F-value is the same as one way ANOVA. From F tables calculate P.

Calculate the coefficient of determination, r^2

$$r^2 = \frac{SS_{reg}}{SS_T}.$$

Statistical Table

Logarithms

	0	1	2	3	4	5	6	7	8	9	Mean Differences 1	2	3	4	5	6	7	8	9
10	0000	0043	0086	0128	0170	0212	0253	0294	0334	0374	4	8	12	17	21	25	29	33	37
11	0414	0453	0492	0531	0569	0607	0645	0682	0719	0755	4	8	11	15	19	23	26	30	34
12	0792	0828	0864	0899	0934	0969	1004	1038	1072	1106	3	7	10	14	17	21	24	28	31
13	1139	1173	1206	1239	1271	1303	1335	1367	1399	1430	3	6	10	13	16	19	23	26	29
14	1461	1492	1523	1553	1584	1614	1644	1673	1703	1732	3	6	9	12	15	18	21	24	27
15	1761	1790	1818	1847	1875	1903	1931	1959	1987	2014	3	6	8	11	14	17	20	22	25
16	2041	2068	2095	2122	2148	2175	2201	2227	2253	2279	3	5	8	11	13	16	18	21	24
17	2304	2330	2355	2380	2405	2430	2455	2480	2504	2529	2	5	7	10	12	15	17	20	22
18	2553	2577	2601	2625	2648	2672	2695	2718	2742	2765	2	5	7	9	12	14	16	19	21
19	2788	2810	2833	2856	2878	2900	2923	2945	2967	2989	2	4	7	9	11	13	16	18	20
20	3010	3032	3054	3075	3096	3118	3139	3160	3181	3201	2	4	6	8	11	13	15	17	19
21	3222	3243	3263	3284	3304	3324	3345	3365	3385	3404	2	4	6	8	10	12	14	16	18
22	3424	3444	3464	3483	3502	3522	3541	3560	3579	3598	2	4	6	8	10	12	14	15	17
23	3617	3636	3655	3674	3692	3711	3729	3747	3766	3784	2	4	6	7	9	11	13	15	17
24	3802	3820	3838	3856	3874	3892	3909	3927	3945	3962	2	4	5	7	9	11	12	14	16
25	3979	3997	4014	4031	4048	4065	4082	4099	4116	4133	2	3	5	7	9	10	12	14	15
26	4150	4166	4183	4200	4216	4232	4249	4265	4281	4298	2	3	5	7	8	10	11	13	15
27	4314	4330	4346	4362	4378	4393	4409	4425	4440	4456	2	3	5	6	8	9	11	13	14
28	4472	4487	4502	4518	4533	4548	4564	4579	4594	4609	2	3	5	6	8	9	11	12	14
29	4624	4639	4654	4669	4683	4698	4713	4728	4742	4757	1	3	4	6	7	9	10	12	13
30	4771	4786	4800	4814	4829	4843	4857	4871	4886	4900	1	3	4	6	7	9	10	11	13
31	4914	4948	4942	4955	4969	4983	4997	5011	5024	5038	1	3	4	6	7	8	10	11	12
32	5051	5065	5079	5092	5105	5119	5132	5145	5159	5172	1	3	4	5	7	8	9	11	12
33	5185	5198	5211	5224	5237	5250	5263	5276	5289	5302	1	3	4	5	6	8	9	10	12
34	5315	5328	5340	5353	5366	5378	5391	5403	5416	5428	1	3	4	5	6	8	9	10	11
35	5441	5453	5465	5478	5490	5502	5514	5527	5539	5551	1	2	4	5	6	7	9	10	11
36	5563	5575	5587	5599	5611	5623	5635	5647	5658	5670	1	2	4	5	6	7	8	10	11
37	5682	5694	5705	5717	5729	5740	5752	5763	5775	5786	1	2	3	5	6	7	8	9	10
38	5798	5809	5821	5832	5843	5855	5866	5877	5888	5899	1	2	3	5	6	7	8	9	10
39	5911	5922	5933	5944	5955	5966	5977	5988	5999	6010	1	2	3	4	5	7	8	9	10
40	6021	6031	6042	6053	6064	6075	6085	6096	6107	6117	1	2	3	4	5	6	8	9	10
41	6128	6138	6149	6160	6170	6180	6191	6201	6212	6222	1	2	3	4	5	6	7	8	9
42	6232	6243	6253	6263	6274	6284	6294	6304	6314	6325	1	2	3	4	5	6	7	8	9
43	6335	6345	6355	6365	6375	6385	6395	6405	6415	6425	1	2	3	4	5	6	7	8	9
44	6435	6444	6454	6464	6474	6484	6493	6503	6513	6522	1	2	3	4	5	6	7	8	9
45	6535	6542	6551	6561	6571	6580	6590	6599	6609	6618	1	2	3	4	5	6	7	8	9
46	6628	6637	6646	6656	6665	6675	6684	6693	6702	6712	1	2	3	4	5	6	7	7	8
47	6721	6730	6739	6749	6758	6767	6776	6785	6794	6803	1	2	3	4	5	5	6	7	8
48	6812	6821	6830	6839	6848	6857	6866	6875	6884	6893	1	2	3	4	4	5	6	7	8
49	6902	6911	6920	6928	6937	6946	6955	6964	6972	6981	1	2	3	4	4	5	6	7	8
50	6990	6998	7007	7016	7024	7033	7042	7050	7059	7067	1	2	3	3	4	5	6	7	8
51	7076	7084	7093	7101	7110	7118	7126	7135	7143	7152	1	2	3	3	4	5	6	7	8
52	7160	7168	7177	7185	7193	7202	7210	7218	7226	7235	1	2	2	3	4	5	6	7	7
53	7243	7251	7259	7267	7275	7284	7292	7300	7306	7316	1	2	2	3	4	5	6	6	7
54	7324	7332	7340	7348	7356	7364	7372	7380	7388	7396	1	2	2	3	4	5	6	6	7
55	7404	7412	7419	7427	7435	7443	7451	7459	7466	7474	1	2	2	3	4	5	5	6	7
56	7482	7490	7497	7505	7513	7520	7528	7536	7543	7551	1	2	2	3	4	5	5	6	7
57	7559	7566	7574	7582	7589	7597	7604	7612	7619	7627	1	2	2	3	4	5	5	6	7
58	7634	7642	7649	7657	7664	7672	7679	7686	7694	7701	1	1	2	3	4	4	5	6	7
59	7709	7716	7723	7731	7738	7745	7752	7760	7767	7774	1	1	2	3	4	4	5	6	7
60	7782	7789	7796	7803	7810	7818	7825	7832	7839	7846	1	1	2	3	4	4	5	6	6

Logarithms

											Mean Differences								
	0	1	2	3	4	5	6	7	8	9	1	2	3	4	5	6	7	8	9
61	7853	7860	7868	7875	7882	7889	7896	7903	7910	7917	1	1	2	3	4	4	5	6	6
62	7924	7931	7938	7945	7952	7959	7966	7973	7980	7987	1	1	2	3	3	4	5	6	6
63	7993	8000	8007	8014	8021	8028	8035	8041	8048	8055	1	1	2	3	3	4	5	5	6
64	8062	8069	8075	8082	8089	8096	8102	8109	8116	8122	1	1	2	3	3	4	5	5	6
65	8129	8136	8142	8149	8156	8162	8169	8176	8182	8189	1	1	2	3	3	4	5	6	6
66	8195	8202	8209	8215	8222	8228	8235	8241	8248	8254	1	1	2	3	3	4	5	5	6
67	8261	8267	8274	8280	8287	8293	8299	8306	8312	8319	1	1	2	3	3	4	5	5	6
68	8325	8331	8338	8344	8351	8357	8363	8370	8376	8382	1	1	2	3	3	4	4	5	6
69	8388	8395	8401	8407	8414	8420	8426	8432	8439	8445	1	1	2	2	3	4	4	5	6
70	8451	8457	8463	8470	8476	8482	8488	8494	8500	8506	1	1	2	2	3	4	4	5	6
71	8513	8519	8525	8531	8537	8543	8549	8555	8561	8567	1	1	2	2	3	4	4	5	5
72	8573	8579	8585	8591	8597	8603	8609	8615	8621	8627	1	1	2	2	3	4	4	5	5
73	8633	8639	8645	8651	8657	8663	8669	8675	8681	8686	1	1	2	2	3	4	4	5	5
74	8692	8698	8704	8710	8716	8722	8727	8733	8739	8745	1	1	2	2	3	4	4	5	5
75	8751	8756	8762	8768	8774	8779	8785	8791	8797	8802	1	1	2	2	3	3	4	5	5
76	8808	8814	8820	8825	8831	8837	8842	8848	8854	8859	1	1	2	2	3	3	4	5	5
77	8865	8871	8876	8882	8887	8893	8899	8904	8910	8915	1	1	2	2	3	3	4	4	5
78	8921	8927	8932	8938	8943	8949	8954	8960	8965	8971	1	1	2	2	3	3	4	4	5
79	8976	8982	8987	8993	8998	9004	9009	9015	9020	9025	1	1	2	2	3	3	4	4	5
80	9031	9036	9042	9047	9053	9058	9063	9069	9074	9079	1	1	2	2	3	3	4	4	5
81	9085	9090	9096	9101	9106	9112	9117	9122	9128	9133	1	1	2	2	3	3	4	4	5
82	9138	9143	9149	9154	9159	9165	9170	9175	9180	9186	1	1	2	2	3	3	4	4	5
83	9191	9196	9201	9206	9212	9217	9222	9227	9232	9238	1	1	2	2	3	3	4	4	5
84	9243	9248	9253	9258	9263	9269	9274	9279	9284	9289	1	1	2	2	3	3	4	4	5
85	9294	9299	9304	9309	9315	9320	9325	9330	9335	9340	1	1	2	2	3	3	4	4	5
86	9345	9350	9355	9360	9365	9370	9375	9380	9385	9390	1	1	2	2	3	3	4	4	5
87	9395	9400	9405	9410	9415	9420	9425	9430	9435	9440	0	1	1	2	2	3	3	4	4
88	9445	9450	9455	9460	9465	9469	9474	9479	9484	9489	0	1	1	2	2	3	3	4	4
89	9494	9499	9504	9509	9513	9518	9523	9528	9533	9538	0	1	1	2	2	3	3	4	4
90	9542	9547	9552	9557	9562	9566	9571	9576	9581	9586	0	1	1	2	2	3	3	4	4
91	9590	9595	9600	9605	9609	9614	9619	9624	9628	9633	0	1	1	2	2	3	3	4	4
92	9638	9643	9647	9653	9657	9661	9666	9671	9675	9680	0	1	1	2	2	3	3	4	4
93	9685	9689	9694	9699	9703	9708	9713	9717	9722	9727	0	1	1	2	2	3	3	4	4
94	9731	9736	9741	9745	9750	9754	9759	9763	9768	9773	0	1	1	2	2	3	3	4	4
95	9777	9782	9786	9791	9795	9800	9805	9809	9814	9818	0	1	1	2	2	3	3	4	4
96	9823	9827	9832	9836	9841	9845	9850	9854	9859	9863	0	1	1	2	2	3	3	4	4
97	9868	9872	9877	9881	9886	9890	9894	9899	9903	9908	0	1	1	2	2	3	3	4	4
98	9912	9917	9921	9926	9930	9934	9939	9943	9948	9952	0	1	1	2	2	3	3	4	4
99	9956	9961	9965	9969	9974	9978	9983	9987	9991	9996	0	1	1	2	2	3	3	3	4

Antilogarithms

	0	1	2	3	4	5	6	7	8	9	Mean Differences 1	2	3	4	5	6	7	8	9
.00	1000	1002	1006	1007	1009	1012	1014	1016	1019	1021	0	0	1	1	1	1	2	2	2
.01	1023	1026	1028	1030	1033	1035	1038	1040	1042	1045	0	0	1	1	1	1	2	2	2
.02	1047	1050	1052	1054	1057	1059	1062	1064	1067	1069	0	0	1	1	1	1	2	2	2
.03	1072	1074	1076	1079	1081	1084	1086	1089	1091	1094	0	0	1	1	1	1	2	2	2
.04	1096	1099	1102	1104	1107	1109	1112	1114	1117	1119	0	1	1	1	1	2	2	2	2
.05	1122	1125	1127	1130	1132	1135	1138	1140	1143	1146	0	1	1	1	1	2	2	2	2
.06	1148	1151	1153	1156	1159	1161	1164	1167	1169	1172	0	1	1	1	1	2	2	2	2
.07	1175	1178	1180	1183	1186	1189	1191	1194	1197	1199	0	1	1	1	1	2	2	2	2
.08	1202	1205	1208	1211	1213	1216	1219	1222	1225	1227	0	1	1	1	2	2	2	2	3
.09	1230	1233	1236	1239	1242	1245	1247	1250	1253	1256	0	1	1	1	2	2	2	2	3
.10	1259	1262	1265	1268	1271	1274	1276	1279	1282	1285	0	1	1	1	2	2	2	2	3
.11	1288	1291	1294	1297	1300	1303	1306	1309	1312	1315	0	1	1	1	2	2	2	2	3
.12	1318	1321	1324	1327	1330	1334	1337	1340	1343	1346	0	1	1	1	2	2	2	2	3
.13	1349	1352	1355	1358	1361	1365	1368	1371	1374	1377	0	1	1	1	2	2	2	3	3
.14	1380	1384	1387	1390	1393	1396	1400	1403	1406	1409	0	1	1	1	2	2	2	3	3
.15	1413	1416	1419	1422	1426	1429	1432	1435	1439	1442	0	1	1	1	2	2	2	3	3
.16	1445	1449	1452	1455	1459	1462	1466	1469	1472	1476	0	1	1	1	2	2	2	3	3
.17	1479	1483	1486	1489	1493	1496	1500	1503	1507	1510	0	1	1	1	2	2	2	3	3
.18	1514	1517	1521	1524	1528	1531	1535	1538	1542	1545	0	1	1	1	2	2	2	3	3
.19	1549	1552	1556	1560	1563	1567	1570	1574	1578	1581	0	1	1	1	2	2	3	3	3
.20	1585	1589	1592	1596	1600	1603	1607	1611	1614	1618	0	1	1	1	2	2	3	3	3
.21	1622	1626	1629	1633	1637	1641	1644	1648	1652	1656	0	1	1	1	2	2	3	3	3
.22	1660	1663	1667	1671	1675	1679	1683	1687	1690	1694	0	1	1	1	2	2	3	3	3
.23	1698	1702	1706	1710	1714	1718	1722	1726	1730	1734	0	1	1	1	2	2	3	3	4
.24	1738	1742	1746	1750	1754	1758	1762	1766	1770	1774	0	1	1	1	2	2	3	3	4
.25	1778	1782	1786	1791	1795	1799	1803	1807	1811	1816	0	1	1	1	2	2	3	3	4
.26	1820	1824	1828	1832	1837	1841	1845	1849	1854	1858	0	1	1	1	2	2	3	3	4
.27	1862	1866	1871	1875	1879	1884	1888	1892	1897	1901	0	1	1	1	2	2	3	3	4
.28	1905	1910	1914	1919	1923	1928	1932	1936	1941	1945	0	1	1	2	2	3	3	4	4
.29	1950	1954	1959	1963	1968	1972	1977	1982	1986	1991	0	1	1	2	2	3	3	4	4
.30	1995	2000	2004	2009	2014	2018	2023	2028	2032	2037	0	1	1	2	2	3	3	4	4
.31	2042	2046	2051	2056	2061	2065	2070	2075	2080	2084	0	1	1	2	2	3	3	4	4
.32	2089	2094	2099	2104	2109	2113	2118	2123	2128	2133	0	1	1	2	2	3	3	4	4
.33	2138	2143	2148	2153	2158	2163	2168	2173	2178	2183	0	1	1	2	2	3	3	4	4
.34	2188	2193	2198	2203	2208	2213	2218	2223	2228	2234	1	1	2	2	3	3	4	4	5
.35	2239	2244	2249	2254	2259	2265	2270	2275	2280	2286	1	1	2	2	3	3	4	4	5
.36	2291	2296	2301	2307	2312	2317	2323	2328	2333	2339	1	1	2	2	3	3	4	4	5
.37	2344	2350	2355	2360	2366	2371	2377	2382	2388	2393	1	1	2	2	3	3	4	4	5
.38	2399	2404	2410	2415	2421	2427	2432	2438	2443	2449	1	1	2	2	3	3	4	4	5
.39	2455	2460	2466	2472	2477	2483	2489	2495	2500	2506	1	1	2	2	3	3	4	5	5
.40	2512	2518	2523	2529	2535	2541	2547	2553	2559	2564	1	1	2	2	3	4	4	5	5
.41	2570	2576	2582	2588	2594	2600	2606	2612	2618	2624	1	1	2	2	3	4	4	5	5
.42	2630	2636	2642	2649	2655	2661	2667	2673	2679	2685	1	1	2	2	3	4	4	5	6
.43	2692	2696	2704	2710	2716	2723	2729	2735	2742	2748	1	1	2	3	3	4	4	5	6
.44	2754	2761	2767	2773	2780	2786	2793	2799	2805	2812	1	1	2	3	3	4	4	5	6
.45	2818	2825	2831	2838	2844	2851	2858	2864	2871	2877	1	1	2	3	3	4	5	5	6
.46	2884	2891	2897	2904	2911	2917	2924	2931	2938	2944	1	1	2	3	3	4	5	5	6
.47	2951	2958	2965	2972	2979	2985	2992	2999	3006	3013	1	1	2	3	3	4	5	5	6
.48	3020	3027	3034	3041	3048	3055	3062	3069	3076	3083	1	1	2	3	4	4	5	6	6
.49	3090	3097	3105	3112	3119	3126	3133	3141	3148	3155	1	1	2	3	4	4	5	6	6
.50	3162	3170	3177	3184	3192	3199	3206	3214	3221	3228	1	1	2	3	4	4	5	6	7

Antilogarithms

	0	1	2	3	4	5	6	7	8	9	Mean Differences								
											1	2	3	4	5	6	7	8	9
.51	3236	3243	3251	3258	3266	3273	3281	3289	3296	3304	1	2	2	3	4	5	5	6	7
.52	3311	3319	3327	3334	3342	3350	3357	3365	3373	3381	1	2	2	3	4	5	5	6	7
.53	3388	3396	3404	3412	3420	3428	3436	3443	3451	3459	1	2	2	3	4	5	5	6	7
.54	3467	3475	3783	3491	3499	3508	3516	3524	3532	3540	1	2	2	3	4	5	6	6	7
.55	3548	3556	3565	3573	3581	3589	3597	3606	3614	3622	1	2	2	3	4	5	6	7	7
.56	3631	3639	3648	3656	3664	6673	3681	3690	3698	3707	1	2	3	3	4	5	6	7	8
.57	3715	3724	3733	3741	3750	3758	3767	3776	3784	3793	1	2	3	3	4	5	6	7	8
.58	3802	3811	3819	3828	3837	3846	3855	3864	3873	3882	1	2	3	4	4	5	6	7	8
.59	3890	3899	3908	3917	3926	3936	3945	3954	3963	3972	1	2	3	4	5	5	6	7	8
.60	3981	3990	3999	4009	4018	4027	4036	4046	4055	4064	1	2	3	4	5	6	6	7	8
.61	4074	4083	4093	4102	4111	4121	4130	4140	4150	4159	1	2	3	4	5	6	7	8	9
.62	4169	4178	4188	4198	4207	4217	4227	4236	4246	4256	1	2	3	4	5	6	7	8	9
.63	4266	4276	4285	4295	4305	4315	4325	4335	4345	4355	1	2	3	4	5	6	7	8	9
.64	4365	4375	4385	4395	4406	4416	4426	4436	4446	4457	1	2	3	4	5	6	7	8	9
.65	4467	4477	4487	4498	4508	4519	4529	4539	4550	4560	1	2	3	4	5	6	7	8	9
.66	4571	4581	4592	4603	4613	4624	4634	4645	4656	4667	1	2	3	4	5	6	7	9	10
.67	4677	4688	4699	4710	4721	4732	4742	4753	4764	4775	1	2	3	4	5	7	8	9	10
.68	4786	4797	4808	4819	4831	4842	4853	4864	4875	4887	1	2	3	4	6	7	8	9	10
.69	4898	4909	4920	4932	4943	4955	4966	4977	4989	5000	1	2	3	5	6	7	8	9	10
.70	5012	5023	5035	5047	5058	5070	5082	5093	5105	5117	1	2	4	5	6	7	8	9	11
.71	5129	5140	5152	5164	5176	5188	5200	5212	5224	5236	1	2	4	5	6	7	8	10	11
.72	5248	5260	5272	5284	5297	5309	5321	5333	5346	5358	1	2	4	5	6	7	9	10	11
.73	5370	5383	5395	5408	5420	5433	5445	5458	5470	5883	1	3	4	5	6	8	9	10	11
.74	5495	5508	5521	5534	5546	5559	5572	5585	5598	5610	1	3	4	5	6	8	9	10	11
.75	5623	5636	5649	5662	5675	5689	5702	5715	5728	5741	1	3	4	5	7	8	9	10	11
.76	5754	5768	5781	5794	5808	5821	5834	5848	5861	5875	1	3	4	5	7	8	9	11	11
.77	5888	5902	5916	5929	5943	5957	5970	5984	5998	6012	1	3	4	5	7	8	10	11	12
.78	6026	6039	6053	6067	6081	6095	6109	6124	6138	6152	1	3	4	6	7	8	10	11	13
.79	6166	6180	6194	6209	6223	6237	6252	6266	6281	6295	1	3	4	6	7	9	10	11	13
.80	6310	6424	6339	6353	6368	6383	6397	6412	6427	6442	1	3	4	6	7	9	10	12	13
.81	6457	6471	6486	6501	6516	6531	6546	6561	6577	6592	2	3	5	6	8	9	11	12	14
.82	6607	6622	6637	6653	6668	6683	6699	6714	6730	6745	2	3	5	6	8	9	11	12	14
.83	6761	6776	6792	6808	6823	6839	6855	6871	6887	6902	2	3	5	6	8	9	11	13	14
.84	6918	6934	6950	6966	6982	6998	7015	7031	7047	7063	2	3	5	6	8	10	11	13	15
.85	7079	7096	7112	7129	7145	7161	7178	7194	7211	7228	2	3	5	7	8	10	12	13	15
.86	7244	7261	7278	7295	7311	7328	7345	7362	7379	7396	2	3	5	7	8	10	12	13	15
.87	7413	7430	7447	7464	7482	7499	7516	7534	7551	7568	2	3	5	7	9	10	12	14	16
.88	7586	7603	7621	7638	7656	7674	7691	7709	7727	7745	2	4	5	7	9	11	12	14	16
.89	7762	7780	7798	7816	7834	7852	7870	7889	7907	7925	2	4	5	7	9	11	13	14	16
.90	7943	7962	7980	7998	8017	8035	8054	8072	8091	8110	2	4	6	7	9	11	13	15	17
.91	8128	8147	8166	8185	8204	8222	8241	8260	8279	8299	2	4	6	8	9	11	13	15	17
.92	8318	8337	8356	8375	8395	8414	8433	8453	8472	8492	2	4	6	8	10	12	14	15	17
.93	8511	8531	8551	8570	8590	8610	8630	8650	8670	8690	2	4	6	8	10	12	14	16	18
.94	8710	8730	8750	8770	8790	8810	8831	8851	8872	8892	2	4	6	8	10	12	14	16	18
.95	8913	8933	8954	8974	8995	9016	9036	9057	9078	9099	2	4	6	8	10	12	15	17	19
.96	9120	9141	9162	9183	9204	9226	9247	9268	9290	9311	2	4	6	8	11	13	15	17	19
.97	9333	9354	9376	9397	9419	9441	9462	9484	9506	9528	2	4	7	9	11	13	15	17	20
.98	9550	9572	9594	9616	9638	9661	9683	9705	9727	9750	2	4	7	9	11	13	16	18	20
.99	9772	9795	9817	9840	9863	9886	9908	9931	9954	9977	2	5	7	9	11	14	16	18	20

Example: 1

Find logarithm for 543.7

1. It is a three digit integer, hence the characteristic of log 543.7 is 2
2. From table refer to row headed 54 under column headed 3 = 7348
3. The mean difference for 7 in the same is 6(0,0006)
4. 7348 + 0.0006 = 77354
5. log for 543.7 = 2.7354

Example: 2

Find log for 0.007654

1. The number has 2 zeros between decimal point and the first significant figure. So the characteristic of log 0.007654 is $\bar{3}$, i.e -3

Follow the rest of the steps as above

log for 0.007654 = $\bar{3}$.8839

Antilogarithm table is used to obtain the number from logarithm of number. Find antilogarithm of 3.4567. The mantissa of logarithm is 0.4567. Refer to the row headed 0.4567. The mantissa of logarithm is 0.4567. Refer to the row headed 0.45 under the column headed 6. The value is 2863 by adding the mean difference for 7 in the same row. Since the characteristic is 3, the number of digits in the integral part is 4. Thus the number.

Table 1: Critical Values of the Chi-Square Distribution

ν	α : 0.999	0.995	0.99	0.975	0.95	0.90	0.75	0.50	0.25	0.10	0.05	0.025	0.01	0.005	0.001
1	0.000	0.000	0.000	0.001	0.004	0.016	0.102	0.455	1.323	2.706	3.841	5.024	6.635	7.879	10.828
2	0.002	0.010	0.020	0.051	0.103	0.211	0.575	1.386	2.773	4.605	5.991	7.378	9.210	10.597	13.816
3	0.024	0.072	0.115	0.216	0.352	0.584	1.213	2.366	4.108	6.251	7.815	9.348	11.345	12.838	16.266
4	0.091	0.207	0.297	0.484	0.711	1.064	1.923	3.357	5.385	7.779	9.488	11.143	13.277	14.860	18.467
5	0.210	0.412	0.554	0.831	1.145	1.610	2.675	4.351	6.626	9.236	11.070	12.833	15.086	16.750	20.515
6	0.381	0.676	0.872	1.237	1.635	2.204	3.455	5.348	7.841	10.645	12.592	14.449	16.812	18.548	22.458
7	0.599	0.989	1.239	1.690	2.167	2.833	4.255	6.346	9.037	12.017	14.067	16.013	18.475	20.278	24.322
8	0.857	1.344	1.646	2.180	2.733	3.490	5.071	7.344	10.219	13.362	15.507	17.535	20.090	21.955	26.124
9	1.152	1.735	2.088	2.700	3.325	4.168	5.899	8.343	11.389	14.684	16.919	19.023	21.666	23.589	27.877
10	1.479	2.156	2.558	3.247	3.940	4.865	6.737	9.342	12.549	15.987	18.307	20.483	23.209	25.188	29.588
11	1.834	2.603	3.053	3.816	4.575	5.578	7.584	10.341	13.701	17.275	19.675	21.920	24.725	26.757	31.264
12	2.214	3.074	3.571	4.404	5.226	6.304	8.438	11.340	14.845	18.549	21.026	23.337	26.217	28.300	32.909
13	2.617	3.565	4.107	5.009	5.892	7.042	9.299	12.340	15.984	19.812	22.362	24.736	27.688	29.819	34.528
14	3.041	4.075	4.660	5.629	6.571	7.790	10.165	13.339	17.117	21.064	23.685	26.119	29.141	31.319	36.123
15	3.483	4.601	5.229	6.262	7.261	8.547	11.037	14.339	18.245	22.307	24.996	27.488	30.578	32.801	37.697
16	3.942	5.142	5.812	6.908	7.962	9.312	11.912	15.338	19.369	23.542	26.296	28.845	32.000	34.267	39.252
17	4.416	5.697	6.408	7.564	8.672	10.085	12.792	16.338	20.489	24.769	27.587	30.191	33.409	35.718	40.790
18	4.905	6.265	7.015	8.231	9.390	10.865	13.675	17.338	21.605	25.989	28.869	31.526	34.805	37.156	42.312
19	5.407	6.844	7.633	8.907	10.117	11.651	14.562	18.338	22.718	27.204	30.144	32.852	36.191	38.582	43.820
20	5.921	7.434	8.260	9.591	10.851	12.443	15.452	19.337	23.828	28.412	31.410	34.170	37.566	39.997	45.315
21	6.447	8.034	8.897	10.283	11.591	13.240	16.344	20.337	24.935	29.615	32.671	35.479	38.932	41.401	46.797
22	6.983	8.643	9.542	10.982	12.338	14.041	17.240	21.337	26.039	30.813	33.924	36.781	40.289	42.796	48.268
23	7.529	9.260	10.196	11.689	13.091	14.848	18.137	22.337	27.141	32.007	35.172	38.076	41.638	44.181	49.728
24	8.085	9.886	10.856	12.401	13.848	15.659	19.037	23.337	28.241	33.196	36.415	39.364	42.980	45.559	51.179
25	8.649	10.520	11.524	13.120	14.611	16.473	19.939	24.337	29.339	34.382	37.652	40.646	44.314	46.928	52.620
26	9.222	11.160	12.198	13.844	15.379	17.292	20.843	25.336	30.435	35.563	38.885	41.923	45.642	48.290	54.052
27	9.803	11.808	12.879	14.573	16.151	18.114	21.749	26.336	31.528	36.741	40.113	43.195	46.963	49.645	55.476
28	10.391	12.461	13.565	15.308	16.928	18.939	22.657	27.336	32.620	37.916	41.337	44.461	48.278	50.993	56.892
29	10.986	13.121	14.256	16.047	17.708	19.768	23.567	28.336	33.711	39.087	42.557	45.722	49.588	52.336	58.301
30	11.588	13.787	14.953	16.791	18.493	20.599	24.478	29.336	34.800	40.256	43.773	46.979	50.892	53.672	59.703
31	12.196	14.458	15.655	17.539	19.281	21.434	25.390	30.336	35.887	41.422	44.985	48.232	52.191	55.003	61.098
32	12.811	15.134	16.362	18.291	20.072	22.271	26.304	31.336	36.973	42.585	46.194	49.480	53.486	56.328	62.487
33	13.431	15.815	17.074	19.047	20.867	23.110	27.219	32.336	38.058	43.745	47.400	50.725	54.776	57.648	63.870
34	14.057	16.501	17.789	19.806	21.664	23.952	28.136	33.336	39.141	44.903	48.602	51.966	56.061	58.964	65.247
35	14.688	17.192	18.509	20.569	22.465	24.797	29.054	34.336	40.223	46.059	49.802	53.203	57.342	60.275	66.619

Table 2: Critical Values of the Chi-Square Distribution

ν	α : 0.999	0.995	0.99	0.975	0.95	0.90	0.75	0.50	0.25	0.10	0.05	0.025	0.01	0.005	0.001
36	15.324	17.887	19.233	21.336	23.269	25.643	29.973	35.336	41.304	47.212	50.998	54.437	58.619	61.581	67.985
37	15.965	18.586	19.960	22.106	24.075	26.492	30.893	36.336	42.383	48.363	52.192	55.668	59.893	62.883	69.346
38	16.611	19.289	20.691	22.878	24.884	27.343	31.815	37.335	43.462	49.513	53.384	56.896	61.162	64.181	70.703
39	17.262	19.996	21.426	23.654	25.695	28.196	32.737	38.335	44.539	50.660	54.572	58.120	62.428	65.476	72.055
40	17.916	20.707	22.164	24.433	26.509	29.051	33.660	39.335	45.616	51.805	55.758	59.342	63.691	66.766	73.402
41	18.576	21.421	22.906	25.215	27.326	29.907	34.585	40.335	46.692	52.949	56.942	60.561	64.950	68.053	74.745
42	19.239	22.138	23.650	25.999	28.144	30.765	35.510	41.335	47.766	54.090	58.124	61.777	66.206	69.336	76.084
43	19.906	22.859	24.398	26.785	28.965	31.625	36.436	42.335	48.840	55.230	59.304	62.990	67.459	70.616	77.419
44	20.576	23.584	25.148	27.575	29.787	32.487	37.363	43.335	49.913	56.369	60.481	64.201	68.710	71.893	78.750
45	21.251	24.311	25.901	28.366	30.612	33.350	38.291	44.335	50.985	57.505	61.656	65.410	69.957	73.166	80.077
46	21.929	25.041	26.657	29.160	31.439	34.215	39.220	45.335	52.056	58.641	62.830	66.617	71.201	74.437	81.400
47	22.610	25.775	27.416	29.956	32.268	35.081	40.149	46.335	53.127	59.774	64.001	67.821	72.443	75.704	82.720
48	23.295	26.511	28.177	30.755	33.098	35.949	41.079	47.335	54.196	60.907	65.171	69.023	73.683	76.969	84.037
49	23.983	27.249	28.941	31.555	33.930	36.818	42.010	48.335	55.265	62.038	66.339	70.222	74.919	78.231	85.351
50	24.674	27.991	29.707	32.357	34.764	37.689	42.942	49.335	56.334	63.167	67.505	71.420	76.154	79.490	86.661
51	25.368	28.735	30.475	33.162	35.600	38.560	43.874	50.335	57.401	64.295	68.669	72.616	77.386	80.747	87.968
52	26.065	29.481	31.246	33.968	36.437	39.433	44.808	51.335	58.468	65.422	69.832	73.810	78.616	82.001	89.272
53	26.765	30.230	32.019	34.776	37.276	40.308	45.741	52.335	59.534	66.548	70.993	75.002	79.843	83.253	90.753
54	27.468	30.981	32.793	35.586	38.116	41.183	46.676	53.335	60.600	67.673	72.153	76.192	81.069	84.502	91.872
55	28.173	31.735	33.570	36.398	38.958	42.060	47.610	54.335	61.665	68.796	73.311	77.380	82.292	85.749	93.168
56	28.881	32.491	34.350	37.212	39.801	42.937	48.546	55.335	62.729	69.919	74.468	78.567	83.513	86.994	94.461
57	29.592	33.248	35.131	38.027	40.646	43.816	49.482	56.335	63.793	71.040	75.624	79.752	84.733	88.236	95.751
58	30.305	34.008	35.913	38.844	41.492	44.696	50.419	57.335	64.857	72.160	76.778	80.936	85.950	89.477	97.039
59	31.021	34.770	36.698	39.662	42.339	45.577	51.356	58.335	65.919	73.279	77.931	82.117	87.166	90.715	98.324
60	31.738	35.535	37.485	40.482	43.188	46.459	52.294	59.335	66.981	74.397	79.082	83.298	88.379	91.952	99.607
61	32.459	36.301	38.273	41.303	44.038	47.342	53.232	60.335	68.043	75.514	80.232	84.476	89.591	93.186	100.888
62	33.181	37.068	39.063	42.126	44.889	48.226	54.171	61.335	69.104	76.630	81.381	85.654	90.802	94.419	102.166
63	33.906	37.838	39.855	42.950	45.741	49.111	55.110	62.335	70.165	77.745	82.529	86.830	92.010	95.649	103.442
64	34.633	38.610	40.649	43.776	46.595	49.996	56.050	63.335	71.225	78.860	83.675	88.004	93.217	96.878	104.716
65	35.362	39.383	41.444	44.603	47.450	50.883	56.990	64.335	72.285	79.973	84.821	89.177	94.422	98.105	105.988
66	36.093	40.158	42.240	45.431	48.305	51.770	57.931	65.335	73.344	81.085	85.965	90.349	95.626	99.330	107.258
67	36.826	40.935	43.038	46.261	49.162	52.659	58.872	66.335	74.403	82.197	87.108	91.519	96.828	100.554	108.526
68	37.561	41.713	43.838	47.092	50.020	53.548	59.814	67.335	75.461	83.308	88.250	92.689	98.028	101.776	109.791
69	38.298	42.494	44.639	47.924	50.879	54.438	60.756	68.334	76.519	84.418	89.391	93.856	99.228	102.996	111.055
70	39.036	43.275	45.442	48.758	51.739	55.329	61.698	69.334	77.577	87.527	90.531	95.023	100.425	104.215	112.317

Table 3: Critical Values of the Chi-Square Distribution

ν	α : 0.999	0.995	0.99	0.975	0.95	0.90	0.75	0.50	0.25	0.10	0.05	0.025	0.01	0.005	0.001
71	39.777	44.058	46.246	49.592	52.600	56.221	62.641	70.334	78.634	86.365	91.670	96.189	101.621	105.432	113.577
72	40.520	44.843	47.051	50.428	53.462	57.113	63.585	71.334	79.690	87.743	92.808	97.353	102.816	106.648	114.835
73	41.264	45.629	47.858	51.265	54.325	58.006	64.528	72.334	80.747	88.850	93.945	98.516	104.010	107.862	116.092
74	42.010	46.417	48.666	52.103	55.189	58.900	65.472	73.334	81.803	89.956	95.081	99.678	105.202	109.074	117.346
75	42.757	47.206	49.475	52.942	56.054	59.795	66.417	74.334	82.858	91.061	96.217	100.839	106.393	110.286	118.599
76	43.507	47.997	50.286	53.782	56.920	60.690	67.362	75.334	83.913	92.166	97.351	101.999	107.583	111.495	119.850
77	44.258	48.788	51.097	54.623	57.786	61.586	68.307	76.334	84.968	93.270	98.484	103.158	108.771	112.704	121.100
78	45.010	49.582	51.910	55.466	58.654	62.483	69.252	77.334	86.022	94.374	99.617	104.316	109.958	113.911	122.348
79	45.764	50.376	52.725	56.309	59.522	63.380	70.198	78.334	87.077	95.476	100.749	105.473	111.144	115.117	123.594
80	46.520	51.172	53.540	57.153	60.391	64.278	71.145	79.334	88.130	96.578	101.879	106.629	112.329	116.321	124.839
81	47.277	51.969	54.357	57.998	61.261	65.176	72.091	80.334	89.184	97.680	103.010	107.783	113.512	117.524	126.083
82	48.036	52.767	55.174	58.845	62.132	66.076	73.038	81.334	90.237	98.780	104.139	108.937	114.695	118.726	127.324
83	48.796	53.567	55.993	59.692	63.004	66.976	73.985	82.334	91.289	99.880	105.267	110.090	115.876	119.927	128.565
84	49.557	54.368	56.813	60.540	63.876	67.876	74.933	83.334	92.342	100.980	106.395	111.242	117.057	121.126	129.804
85	50.320	55.170	57.634	61.389	64.749	68.777	75.881	84.334	93.394	102.079	107.522	112.393	118.236	122.325	131.041
86	51.085	55.973	58.456	62.239	65.623	69.679	76.829	85.334	94.446	103.177	108.648	113.544	119.414	123.522	132.277
87	51.850	56.777	59.279	63.089	66.498	70.581	77.777	86.334	95.497	104.275	109.773	114.693	120.591	124.718	133.512
88	52.617	57.582	60.103	63.941	67.373	71.484	78.726	87.334	96.548	105.372	110.898	115.841	121.767	125.913	134.745
89	53.386	58.389	60.928	64.793	68.249	72.387	79.675	88.334	97.599	106.469	112.022	116.989	122.942	127.106	135.978
90	54.155	59.196	61.754	65.647	69.126	73.291	80.625	89.334	98.650	107.565	113.145	118.136	124.116	128.299	137.208
91	54.926	60.005	62.581	66.501	70.003	74.196	81.574	90.334	99.700	108.661	114.268	119.282	125.289	129.491	138.438
92	55.698	60.815	63.409	67.356	70.882	75.100	82.524	91.334	100.750	109.756	115.390	120.427	126.462	130.681	139.666
93	56.472	61.625	64.238	68.211	71.760	76.006	83.474	92.334	101.800	110.850	116.511	121.571	127.633	131.871	140.893
94	57.246	62.437	65.068	69.068	72.640	76.912	84.425	93.334	102.850	111.944	117.632	122.715	128.803	133.059	142.119
95	58.022	63.250	65.898	69.925	73.520	77.818	85.376	94.344	103.899	113.038	118.752	123.858	129.973	134.247	143.344
96	58.799	64.063	66.730	70.783	74.401	78.725	86.327	95.334	104.948	114.131	119.871	125.000	131.141	135.433	144.567
97	59.577	64.878	67.562	71.642	75.282	79.633	87.278	96.334	105.997	115.223	120.990	126.141	132.309	136.619	145.789
98	60.356	65.694	68.396	72.501	76.164	80.541	88.229	97.334	107.045	116.315	122.108	127.282	133.476	137.803	147.010
99	61.137	66.510	69.230	73.361	77.046	81.449	89.181	98.334	108.093	117.407	123.225	128.422	134.642	138.987	148.230
100	61.918	67.328	70.065	74.222	77.929	82.358	90.133	99.334	109.141	118.498	124.342	129.561	135.807	140.169	149.449
101	62.701	68.146	70.901	75.083	78.813	83.267	91.085	100.334	110.189	119.589	125.458	130.700	136.971	141.351	150.667
102	63.484	68.965	71.737	75.946	79.697	84.177	92.038	101.334	111.236	120.679	126.574	131.838	138.134	142.532	151.884
103	64.269	69.785	72.575	76.809	80.582	85.088	92.991	102.334	112.284	121.769	127.689	132.975	139.297	143.712	153.099
104	65.055	70.606	73.413	77.672	81.468	85.998	93.944	103.334	113.331	122.858	128.804	134.111	140.459	144.891	154.314
105	65.841	71.428	74.252	78.536	82.354	86.909	94.897	104.334	114.378	123.947	129.918	135.247	141.620	146.070	155.528

Table 4: Critical Values of the Chi-Square Distribution

ν	α : 0.999	0.995	0.99	0.975	0.95	0.90	0.75	0.50	0.25	0.10	0.05	0.025	0.01	0.005	0.001
106	66.629	72.251	75.092	79.401	83.240	87.821	95.850	105.334	115.424	125.035	131.031	136.382	142.780	147.247	156.740
107	67.418	73.075	75.932	80.267	84.127	88.733	96.804	106.334	116.741	126.123	132.144	137.517	143.940	148.424	157.952
108	68.207	73.899	76.774	81.133	85.015	89.645	97.758	107.334	117.517	127.211	133.257	138.651	145.099	149.599	159.162
109	68.998	74.724	77.616	82.000	85.903	90.558	98.712	108.334	118.563	128.298	134.369	139.784	146.257	150.774	160.372
110	69.790	75.550	78.458	82.867	86.792	91.471	99.666	109.334	119.608	129.385	135.480	140.917	147.414	151.948	161.581
111	70.582	76.377	79.302	83.735	87.681	92.385	100.620	110.334	120.654	130.472	136.591	142.049	148.571	153.122	162.788
112	71.376	77.204	80.146	84.604	88.570	93.299	101.575	111.334	121.699	131.558	137.701	143.180	149.727	154.294	163.995
113	72.170	78.033	80.991	85.473	89.461	94.213	102.530	112.334	122.744	132.643	138.811	144.311	150.882	155.466	165.201
114	72.965	78.862	81.836	86.342	90.351	95.128	103.485	113.343	123.789	133.729	139.921	145.441	152.037	156.637	166.406
115	73.761	79.692	82.682	87.213	91.242	96.043	104.440	114.334	124.834	134.813	141.030	146.571	153.191	157.808	167.610
116	74.558	80.522	83.529	88.084	92.134	96.958	105.396	115.334	125.878	135.898	142.138	147.700	154.344	158.977	168.813
117	75.356	81.353	84.377	88.955	93.026	97.874	106.352	116.334	126.923	136.982	143.246	148.829	155.496	160.146	170.016
118	76.155	82.185	85.225	89.827	93.918	98.790	107.307	117.334	127.967	138.066	144.354	149.957	156.648	161.314	171.217
119	76.955	83.018	86.074	90.700	94.811	99.707	108.263	118.334	129.011	139.149	145.461	151.084	157.800	162.481	172.418
120	77.755	83.852	86.923	91.573	95.705	100.624	109.220	119.334	130.055	140.233	146.567	152.211	158.950	163.648	173.617
121	78.557	84.686	87.773	92.446	96.598	101.541	110.176	120.334	131.098	141.315	147.674	153.338	160.100	164.814	174.816
122	79.359	85.521	88.624	93.320	97.493	102.458	111.133	121.334	132.142	142.398	148.779	154.464	161.250	165.980	176.014
123	80.162	86.356	89.475	94.195	98.387	103.376	112.089	122.334	133.185	143.480	149.885	155.589	162.398	167.144	177.212
124	80.965	87.192	90.327	95.070	99.283	104.295	113.046	123.334	134.228	144.562	150.989	156.714	163.546	168.308	178.408
125	81.770	88.029	91.180	95.946	100.178	105.213	114.004	124.334	135.271	145.643	152.094	157.839	164.694	169.471	179.604
126	82.575	88.866	92.033	96.822	101.074	106.132	114.961	125.334	136.313	146.724	153.198	158.962	165.841	170.634	180.799
127	83.381	89.704	92.887	97.698	101.971	107.051	115.918	126.334	137.356	147.805	154.302	160.086	166.987	171.796	181.993
128	84.188	90.543	93.741	98.576	102.867	107.971	116.876	127.334	138.398	148.885	155.405	161.209	168.133	172.957	183.186
129	84.996	91.383	94.596	99.453	103.765	108.891	117.834	128.334	139.440	149.965	156.508	162.331	169.278	174.118	184.379
130	85.804	92.223	95.451	100.331	104.662	109.811	118.792	129.334	140.482	151.045	157.610	163.453	170.423	175.278	185.571
131	86.613	93.063	96.307	101.210	105.560	110.732	119.750	130.334	141.524	152.125	158.712	164.575	171.567	176.438	186.762
132	87.423	93.904	97.163	102.089	106.459	111.652	120.708	131.334	142.566	153.204	159.814	165.696	172.711	177.597	187.953
133	88.233	94.746	98.021	102.968	107.357	112.573	121.667	132.334	143.608	154.283	160.915	166.816	173.854	178.755	189.142
134	89.044	95.588	98.878	103.848	108.257	113.495	122.625	133.334	144.649	155.361	162.016	167.936	174.996	179.913	190.331
135	89.856	96.431	99.736	104.729	109.156	114.417	123.584	134.334	145.690	156.440	163.116	169.056	176.138	181.070	191.520
136	90.669	97.275	100.595	105.609	110.056	115.338	124.543	135.334	146.731	157.518	164.216	170.175	177.280	182.226	192.707
137	91.482	98.119	101.454	106.491	110.956	116.261	125.502	136.334	147.772	158.595	165.316	171.294	178.421	183.382	193.894
138	92.296	98.964	102.314	107.372	111.857	117.183	126.461	137.334	148.813	159.673	166.415	172.412	179.561	184.538	195.080
139	93.111	99.809	103.174	108.254	112.758	118.106	127.421	138.334	149.854	160.750	167.514	173.530	180.701	185.693	196.266
140	93.926	100.655	104.034	109.137	113.659	119.029	128.380	139.334	150.894	161.827	168.613	174.648	181.840	186.847	197.451

Table 5: Critical Values of the t distribution

ν	$\alpha(2)$: 0.50 $\alpha(1)$: 0.25	0.20 0.10	0.10 0.05	0.05 0.025	0.02 0.01	0.01 0.005	0.005 0.0025	0.002 0.001	0.001 0.0005
1	1.000	3.078	6.314	12.706	31.821	63.657	127.321	318.309	636.619
2	0.816	1.886	2.920	4.303	6.965	9.925	14.089	22.327	31.599
3	0.765	1.638	2.353	3.182	4.541	5.841	7.453	10.215	12.924
4	0.741	1.533	2.132	2.776	3.747	4.604	5.598	7.173	8.610
5	0.727	1.476	2.015	2.571	3.365	4.032	4.773	5.893	6.869
6	0.718	1.440	1.943	2.447	3.143	3.707	4.317	5.208	5.959
7	0.711	1.415	1.895	2.365	2.998	3.499	4.029	4.785	5.408
8	0.706	1.397	1.860	2.306	2.896	3.355	3.833	4.501	5.041
9	0.703	1.383	1.833	2.262	2.821	3.250	3.690	4.297	4.781
10	0.700	1.372	1.812	2.228	2.764	3.169	3.581	4.144	4.587
11	0.697	1.363	1.796	2.201	2.718	3.106	3.497	4.025	4.437
12	0.695	1.356	1.782	2.179	2.681	3.055	3.428	3.930	4.318
13	0.694	1.350	1.771	2.160	2.650	3.012	3.372	3.852	4.221
14	0.692	1.345	1.761	2.145	2.624	2.977	3.326	3.787	4.140
15	0.691	1.341	1.753	2.131	2.602	2.947	3.286	3.733	4.073
16	0.690	1.337	1.746	2.120	2.583	2.921	3.252	3.686	4.015
17	0.689	1.333	1.740	2.110	2.567	2.898	3.222	3.646	3.965
18	0.688	1.330	1.734	2.101	2.552	2.878	3.197	3.610	3.922
19	0.688	1.328	1.729	2.093	2.539	2.861	3.174	3.579	3.883
20	0.687	1.325	1.725	2.086	2.528	2.845	3.153	3.552	3.850
21	0.686	1.323	1.721	2.080	2.518	2.831	3.135	3.527	3.819
22	0.686	1.321	1.717	2.074	2.508	2.819	3.119	3.505	3.792
23	0.685	1.319	1.714	2.069	2.500	2.807	3.104	3.485	3.768
24	0.685	1.318	1.711	2.064	2.492	2.797	3.091	3.467	3.745
25	0.684	1.316	1.708	2.060	2.485	2.787	3.078	3.450	3.725
26	0.684	1.315	1.706	2.056	2.479	2.779	3.067	3.435	3.707
27	0.684	1.314	1.703	2.052	2.473	2.771	3.057	3.421	3.690
28	0.683	1.313	1.701	2.048	2.467	2.763	3.047	3.408	3.674
29	0.683	1.311	1.699	2.045	2.462	2.756	3.038	3.396	3.659
30	0.683	1.310	1.697	2.042	2.457	2.750	3.030	3.385	3.646
31	0.682	1.309	1.696	2.040	2.453	2.744	3.022	3.375	3.633
32	0.682	1.309	1.694	2.037	2.449	2.738	3.015	3.365	3.622
33	0.682	1.308	1.692	2.035	2.445	2.733	3.008	3.356	3.611
34	0.682	1.307	1.691	2.032	2.441	2.728	3.002	3.348	3.601
35	0.682	1.306	1.690	2.030	2.438	2.724	2.996	3.340	3.591
36	0.681	1.306	1.688	2.028	2.434	2.719	2.990	3.333	3.582
37	0.681	1.305	1.687	2.026	2.431	2.715	2.985	3.326	3.574
38	0.681	1.304	1.686	2.024	2.429	2.712	2.980	3.319	3.566
39	6.681	1.304	1.685	2.023	2.426	2.708	2.976	3.313	3.558
40	0.681	1.303	1.684	2.021	2.423	2.704	2.971	3.307	3.551
41	0.681	1.303	1.683	2.020	2.421	2.701	2.967	3.301	3.544
42	0.680	1.302	1.682	2.018	2.418	2.698	2.963	3.296	3.538
43	0.680	1.302	1.681	2.017	2.416	2.695	2.959	3.291	3.532
44	0.680	1.301	1.680	2.015	2.414	2.692	2.956	3.286	3.526
45	0.680	1.301	1.679	2.014	2.412	2.690	2.952	3.281	3.520
46	0.680	1.300	1.679	2.013	2.410	2.687	2.949	3.277	3.515
47	0.680	1.300	1.678	2.012	2.408	2.685	2.946	3.273	3.510
48	0.680	1.299	1.677	2.011	2.407	2.682	2.943	3.269	3.505
49	0.680	1.299	1.677	2.010	2.405	2.680	2.940	3.265	3.500
50	0.679	1.299	1.676	2.009	2.403	2.678	2.937	3.261	3.496
52	0.679	1.298	1.675	2.007	2.400	2.674	2.932	3.255	3.488
54	0.679	1.297	1.674	2.005	2.397	2.670	2.927	3.248	3.480
56	0.679	1.297	1.673	2.003	2.395	2.667	2.923	3.242	3.473
58	0.679	1.296	1.672	2.002	2.392	2.663	2.918	3.237	3.466
60	0.679	1.296	1.671	2.000	2.390	2.660	2.915	3.232	3.460

Table 6: Critical Values of the t distribution

ν	$\alpha(2)$: 0.50	0.20	0.10	0.05	0.02	0.01	0.005	0.002	0.001
	$\alpha(1)$: 0.25	0.10	0.05	0.025	0.01	0.005	0.0025	0.001	0.0005
62	0.678	1.295	1.670	1.999	2.388	2.657	2.911	3.227	3.454
64	0.678	1.295	1.669	1.998	2.386	2.655	2.908	3.223	3.449
66	0.678	1.295	1.668	1.997	2.384	2.652	2.904	3.218	3.444
68	0.678	1.294	1.668	1.995	2.382	2.650	2.902	3.214	3.439
70	0.678	1.294	1.667	1.994	2.381	2.648	2.899	3.211	3.435
72	0.678	1.293	1.666	1.993	2.379	2.646	2.896	3.207	3.431
74	0.678	1.293	1.666	1.993	2.378	2.644	2.894	3.204	3.427
76	0.678	1.293	1.665	1.992	2.376	2.642	2.891	3.201	3.423
78	0.678	1.292	1.665	1.991	2.375	2.640	2.889	3.198	3.420
80	0.678	1.292	1.664	1.990	2.374	2.639	2.887	3.195	3.416
82	0.677	1.292	1.664	1.989	2.373	2.637	2.885	3.193	3.413
84	0.677	1.292	1.663	1.989	2.372	2.636	2.883	3.190	3.410
86	0.677	1.291	1.663	1.988	2.370	2.634	2.881	3.188	3.407
88	0.677	1.291	1.662	1.987	2.369	2.633	2.880	3.185	3.405
90	0.677	1.291	1.662	1.987	2.368	2.632	2.878	3.183	3.402
92	0.677	1.291	1.662	1.986	2.368	2.630	2.876	3.181	3.399
94	0.677	1.291	1.661	1.986	2.367	2.629	2.875	3.179	3.397
96	0.677	1.290	1.661	1.985	2.366	2.628	2.873	3.177	3.395
98	0.677	1.290	1.661	1.984	2.365	2.627	2.872	3.175	3.393
100	0.677	1.290	1.660	1.984	2.364	2.626	2.871	3.174	3.390
105	0.677	1290	1.659	1.983	2.362	2.623	2.868	3.170	3.386
110	0.677	1.289	1.659	1.982	2.361	2.621	2.865	3.166	3.381
115	0.677	1.289	1.658	1.981	2.359	2.619	2.862	3.163	3.377
120	0.677	1.289	1.658	1.980	2.358	2.617	2.860	3.160	3.373
125	0.676	1.288	1.657	1.979	2.357	2.616	2.858	3.157	3.370
130	0.676	1.288	1.657	1.978	2.355	2.614	2.856	3.154	3.367
135	0.676	1.288	1.656	1.978	2.354	2.613	2.854	3.152	3.364
140	0.676	1.288	1.656	1.977	2.353	2.611	2.852	3.149	3.361
145	0.676	1.287	1.655	1.976	2.352	2.610	2.851	3.147	3.359
150	0.676	1.287	1.655	1.976	2.351	2.609	2.849	3.145	3.357
160	0.676	1.287	1.654	1.975	2.350	2.607	2.846	3.142	3.352
170	0.676	1.287	1.654	1.974	2.348	2.605	2.844	3.139	3.349
180	0.676	1.286	1.653	1.973	2.347	2.603	2.842	3.136	3.345
190	0.676	1.286	1.653	1.973	2.346	2.602	2.840	3.134	3.342
200	0.676	1.286	1.653	1.972	2.345	2.601	2.839	3.131	3.340
250	0.675	1.285	1.651	1.969	2.341	2.596	2.832	3.123	3.330
300	0.675	1.284	1.650	1.968	2.339	2.592	2.828	3.118	3.323
350	0.675	1.284	1.649	1.967	2.337	2.590	2.825	3.114	3.319
400	0.675	1.284	1.649	1.966	2.336	2.588	2.823	3.111	3.315
450	0.675	1.283	1.648	1.965	2.335	2.587	2.821	3.108	3.312
500	0.675	1.283	1.648	1.965	2.334	2.586	2.820	3.107	3.310
600	0.675	1.283	1.647	1.964	2.333	2.584	2.817	3.104	3.307
700	0.675	1.283	1.647	1.963	2.332	2.583	2.816	3.102	3.304
800	0.675	1.283	1.647	1.963	2.331	2.582	2.815	3.100	3.303
900	0.675	1.282	1.647	1.963	2.330	2.581	2.814	3.099	3.301
1000	0.675	1.282	1.646	1.962	2.330	2.581	2.813	3.098	3.300
∞	0.6745	1.2816	1.6449	1.9600	2.3263	2.5758	2.8070	3.0902	3.2905

Table 7: Critical Values of the F Distribution

Numerator DF = 1

ν	$\alpha(2)$: 0.50 $\alpha(1)$: 0.25	0.20 0.10	0.10 0.05	0.05 0.025	0.02 0.01	0.01 0.005	0.005 0.0025	0.002 0.001	0.001 0.0005
1.	5.83	39.9	161.0	648.0	4050.0	16200.0	64800.0	405000.0	1620000.0
2	2.57	8.53	18.5	38.5	98.5	199.0	399.0	999.0	2000.0
3	2.02	5.54	10.1	17.4	34.1	55.6	89.6	167.0	267.00
4	1.81	4.54	7.71	12.2	21.2	31.3	45.7	74.1	106.0
5	1.69	4.06	6.61	10.0	16.3	22.8	31.4	47.2	63.6
6	1.62	3.78	5.99	8.81	13.7	18.6	24.8	35.5	46.1
7	1.57	3.59	5.59	8.07	12.2	16.2	21.1	29.2	37.0
8	1.54	3.46	5.32	7.57	11.3	14.7	18.8	25.4	31.6
9	1.51	3.36	5.12	7.21	10.6	13.6	17.2	22.9	28.0
10	1.49	3.29	4.96	6.94	10.0	12.8	16.0	21.0	25.5
11	1.47	3.23	4.84	6.72	9.65	12.2	15.2	19.7	23.7
12	1.46	3.18	4.75	6.55	9.33	11.8	14.5	18.6	22.2
13	1.45	3.14	4.67	6.41	9.07	11.4	13.9	17.8	21.1
14	1.44	3.10	4.60	6.30	8.86	11.1	13.5	17.1	20.2
15	1.43	3.07	4.54	6.20	8.68	10.8	13.1	16.6	19.5
16	1.42	3.05	4.49	6.12	8.53	10.6	12.8	16.1	18.9
17	1.42	3.03	4.45	6.04	8.40	10.4	12.6	15.7	18.4
18	1.41	3.01	4.41	5.98	8.29	10.2	12.3	15.4	17.9
19	1.41	2.99	4.38	5.92	8.18	10.1	12.1	15.1	17.5
20	1.40	2.97	4.35	5.87	8.10	9.94	11.9	14.8	17.2
21	1.40	2.96	4.32	5.83	8.02	9.83	11.8	14.6	16.9
22	1.40	2.95	4.30	5.79	7.95	9.73	11.6	14.4	16.6
23	1.39	2.94	4.28	5.75	7.88	9.63	11.5	14.2	16.4
24	1.39	2.93	4.26	5.72	7.82	9.55	11.4	14.0	16.2
25	1.39	2.92	4.24	5.69	7.77	9.48	11.3	13.9	16.0
26	1.38	2.91	4.23	5.66	7.72	9.41	11.2	13.7	15.8
27	1.38	2.90	4.21	5.63	7.68	9.34	11.1	13.6	15.6
28	1.38	2.89	4.20	5.61	7.64	9.28	11.0	13.5	15.5
29	1.38	2.89	4.18	5.59	7.60	9.23	11.0	13.4	15.3
30	1.38	2.88	4.17	5.57	7.56	9.18	10.9	13.3	15.2
35	1.37	2.85	4.12	5.48	7.42	8.98	10.6	12.9	14.7
40	1.36	2.84	4.08	5.42	7.31	8.83	10.4	12.6	14.4
45	1.36	2.82	4.06	5.38	7.23	8.71	10.3	12.4	14.1
50	1.35	2.81	4.03	5.34	7.17	8.63	10.1	12.2	13.9
60	1.35	2.79	4.00	5.29	7.08	8.49	9.96	12.0	13.5
70	1.35	2.78	3.98	5.25	7.01	8.40	9.84	11.8	13.3
80	1.34	2.77	3.96	5.22	6.96	8.33	9.75	11.7	13.2
90	1.34	2.76	3.95	5.20	6.93	8.28	9.68	11.6	13.0
100	1.34	2.76	3.94	5.18	6.90	8.24	9.62	11.5	12.9
120	1.34	2.75	3.92	5.15	6.85	8.18	9.54	11.4	12.8
140	1.33	2.74	3.91	5.13	6.82	8.14	9.48	11.3	12.7
160	1.33	2.74	3.90	5.12	6.80	8.10	9.44	11.2	12.6
180	1.33	2.73	3.89	5.11	6.78	8.08	9.40	11.2	12.6
200	1.33	2.73	3.89	5.10	6.76	8.06	9.38	11.2	12.5
300	1.33	2.72	3.87	5.07	6.72	8.00	9.30	11.0	12.4
500	1.33	2.72	3.86	5.05	6.69	7.95	9.23	11.0	12.3
∞	1.32	2.71	3.84	5.02	6.64	7.88	9.14	10.8	12.1

Table 8: Critical Values of the F Distribution

Numerator DF = 2

ν	$\alpha(2)$: 0.50 $\alpha(1)$: 0.25	0.20 0.10	0.10 0.05	0.05 0.025	0.02 0.01	0.01 0.005	0.005 0.0025	0.002 0.001	0.001 0.0005
1	7.50	49.5	200.0	800.0	5000.0	20000.0	80000.0	500000.0	2000000.0
2	3.00	9.00	19.0	39.0	99.0	199.0	399.0	999.0	2000.0
3	2.28	5.46	9.55	16.0	30.8	49.8	79.9	149.0	237.0
4	2.00	4.32	6.94	10.6	18.0	26.3	38.0	61.2	87.4
5	1.85	3.78	5.79	8.43	13.3	18.3	25.0	37.1	49.8
6	1.76	3.46	5.14	7.26	10.9	14.5	19.1	27.0	34.8
7	1.70	3.26	4.74	6.54	9.55	12.4	15.9	21.7	27.2
8	1.66	3.11	4.46	6.06	8.65	11.0	13.9	18.5	22.7
9	1.62	3.01	4.26	5.71	8.02	10.1	12.5	16.4	19.9
10	1.60	2.92	4.10	5.46	7.56	9.43	11.6	14.9	17.9
11	1.58	2.86	3.98	5.26	7.21	8.91	10.8	13.8	16.4
12	1.56	2.81	2.89	5.10	6.93	8.51	10.3	13.0	15.3
13	1.55	2.76	3.81	4.97	6.70	8.19	9.84	12.3	14.4
14	1.53	2.73	3.74	4.86	6.51	7.92	9.47	11.8	13.7
15	1.52	2.70	3.68	4.77	6.36	7.70	9.17	11.3	13.2
16	1.51	2.67	3.63	4.69	6.23	7.51	8.92	11.0	12.7
17	1.51	2.64	3.59	4.62	6.11	7.35	8.70	10.7	12.3
18	1.50	2.62	3.55	4.56	6.01	7.21	8.51	10.4	11.9
19	1.49	2.61	3.52	4.51	5.93	7.09	8.35	10.2	11.6
20	1.49	2.59	3.49	4.46	5.85	6.99	8.21	9.95	11.4
21	1.48	2.57	3.47	4.42	5.78	6.89	8.08	9.77	11.2
22	1.48	2.56	3.44	4.38	5.72	6.81	7.96	9.61	11.0
23	1.47	2.55	3.42	4.35	5.66	6.73	7.86	9.47	10.8
24	1.47	2.54	3.40	4.32	5.61	6.66	7.77	9.34	10.6
25	1.47	2.53	3.39	4.29	5.57	6.60	7.69	9.22	10.5
26	1.46	2.52	3.37	4.27	5.53	6.54	7.61	9.12	10.3
27	1.46	2.51	3.35	4.24	5.49	6.49	7.54	9.02	10.2
28	1.46	2.50	3.34	4.22	5.45	6.44	7.48	8.93	10.1
29	1.45	2.50	3.33	4.20	5.42	6.40	7.42	8.85	9.99
30	1.45	2.49	3.32	4.18	5.39	6.35	7.36	8.77	9.90
35	1.44	2.46	3.27	4.11	5.27	6.19	7.14	8.47	9.52
40	1.44	2.44	3.23	4.05	5.18	6.07	6.99	8.28	9.25
45	1.43	2.42	2.20	4.01	5.11	5.97	6.86	8.09	9.04
50	1.43	2.41	3.18	3.97	5.06	5.90	6.77	7.96	8.88
60	1.42	2.39	3.15	3.93	4.98	5.79	6.63	7.77	8.65
70	1.41	2.38	3.13	3.89	4.92	5.72	6.53	7.64	8.49
80	1.41	2.37	3.11	3.86	4.88	5.67	6.46	7.54	8.37
90	1.41	2.36	3.10	3.84	4.85	5.62	6.41	7.47	8.28
100	1.41	2.36	3.09	3.83	4.82	5.59	6.37	7.41	8.21
120	1.40	2.35	3.07	3.80	4.79	5.54	6.30	7.32	8.10
140	1.40	2.34	3.06	3.79	4.76	5.50	6.26	7.26	8.03
160	1.40	2.34	3.05	3.78	4.74	5.48	6.22	7.21	7.97
180	1.40	2.33	3.05	3.77	4.73	5.46	6.20	7.18	7.93
200	1.40	2.33	3.04	3.76	4.71	5.44	6.17	7.15	7.90
300	1.39	2.32	3.03	3.73	4.68	5.39	6.11	7.07	7.80
500	1.39	2.31	3.01	3.72	4.65	5.35	6.06	7.00	7.72
∞	1.39	2.30	3.00	3.69	4.61	5.30	5.99	6.91	7.60

Table 9: Critical Values of the F Distribution

Numerator DF = 3

ν	$\alpha(2)$: 0.50	0.20	0.10	0.05	0.02	0.01	0.005	0.002	0.001
	$\alpha(1)$: 0.25	0.10	0.05	0.025	0.01	0.005	0.0025	0.001	0.0005
1	8.20	53.6	216.0	864.0	5400.0	21600.0	86500.0	540000.0	2160000.0
2	3.15	9.16	19.2	39.2	99.2	199.0	399.0	999.0	2000.0
3	2.36	5.39	9.28	15.4	29.5	47.5	76.1	141.0	225.0
4	2.05	4.19	6.59	9.98	16.7	24.3	35.0	56.2	80.1
5	1.88	3.62	5.41	7.76	12.1	16.5	22.4	33.2	44.4
6	1.78	3.29	4.76	6.60	9.78	12.9	16.9	23.7	30.5
7	1.72	3.07	4.35	5.89	8.45	10.9	13.8	18.8	23.5
8	1.67	2.92	4.07	5.42	7.59	9.60	12.0	15.8	19.4
9	1.63	2.81	3.86	5.08	6.99	8.72	10.7	13.9	16.8
10	1.60	2.73	3.71	4.83	6.55	8.08	9.83	12.6	15.0
11	1.58	2.66	3.59	4.63	6.22	7.60	9.17	11.6	13.7
12	1.56	2.61	3.49	4.47	5.95	7.23	8.65	10.8	12.7
13	1.55	2.56	3.41	4.35	5.74	6.93	8.28	10.2	11.9
14	1.53	2.52	3.34	4.24	5.56	6.68	7.91	9.73	11.3
15	1.52	2.49	3.29	4.15	5.42	6.48	7.63	9.34	10.8
16	1.51	2.46	2.24	4.08	5.29	6.30	7.40	9.01	10.3
17	1.50	2.44	3.20	4.01	5.19	6.16	7.21	8.73	9.93
18	1.49	2.42	3.16	3.95	5.09	6.03	7.04	8.49	9.69
19	1.49	2.40	3.13	3.90	5.01	5.92	6.89	8.28	9.42
20	1.48	2.38	3.10	3.86	4.94	5.82	6.76	8.10	9.20
21	1.48	2.36	3.07	3.82	4.87	5.73	6.64	7.94	8.99
22	1.47	2.35	3.05	3.78	4.82	5.65	6.54	7.80	8.82
23	1.47	2.34	3.03	3.75	4.76	5.58	6.45	7.67	8.66
24	1.46	2.33	3.01	3.72	4.72	5.52	6.36	7.55	8.51
25	1.46	2.32	2.99	3.69	4.68	5.46	6.29	7.45	8.39
26	1.45	2.31	2.98	3.67	4.64	5.41	6.22	7.36	8.27
27	1.45	2.30	2.96	3.65	4.60	5.36	6.16	7.27	8.16
28	1.45	2.29	2.95	3.63	4.57	5.32	6.10	7.19	8.07
29	1.45	2.28	2.93	3.61	4.54	5.28	6.05	7.12	7.98
30	1.44	2.28	2.92	3.59	4.51	5.24	6.00	7.05	7.89
35	1.43	2.25	2.87	3.52	4.40	5.09	5.80	6.79	7.56
40	1.42	2.23	2.84	3.46	4.31	4.98	5.66	6.59	7.33
45	1.42	2.21	2.81	3.42	4.25	4.89	5.59	6.45	7.15
50	1.41	2.20	2.79	3.39	4.20	4.83	5.47	6.34	7.01
60	1.41	2.18	2.76	3.34	4.13	4.73	5.34	6.17	6.81
70	1.40	2.16	2.74	3.31	4.07	4.66	5.26	6.06	6.67
80	1.40	2.15	2.72	3.28	4.04	4.61	5.19	5.97	6.57
90	1.39	2.15	2.71	3.26	4.01	4.57	5.14	5.91	6.49
100	1.39	2.14	2.70	3.25	3.98	4.54	5.11	5.86	6.43
120	1.39	2.13	2.68	3.23	3.95	4.50	5.05	5.78	6.34
140	1.38	2.12	2.67	3.21	3.92	4.47	5.01	5.73	6.28
160	1.38	2.12	2.66	3.20	3.91	4.44	4.98	5.69	6.23
180	1.38	2.11	2.65	3.19	3.89	4.42	4.95	5.66	6.19
200	1.38	2.11	2.65	3.18	3.88	4.41	4.94	5.63	6.16
300	1.38	2.10	2.63	3.16	3.85	4.36	4.88	5.56	6.08
500	1.37	2.09	2.62	3.14	3.82	4.33	4.84	5.51	6.01
∞	1.37	2.08	2.61	3.12	3.78	4.28	4.77	5.42	5.91

Table 10: Critical Values of the F Distribution

Numerator DF = 4

ν	$\alpha(2)$: 0.50 $\alpha(1)$: 0.25	0.20 0.10	0.10 0.05	0.05 0.025	0.02 0.01	0.01 0.005	0.005 0.0025	0.002 0.001	0.001 0.0005
1	8.58	55.5	225.0	900.0	5620.0	22500.0	90000.0	562000.0	2250000.0
2	3.23	9.24	19.2	39.2	99.2	199.0	399.0	999.0	2000.0
3	2.39	5.34	9.12	15.1	28.7	46.2	73.9	137.0	218.0
4	2.06	4.11	6.39	9.60	16.0	23.2	33.3	53.4	76.1
5	1.89	3.52	5.19	7.39	11.4	15.6	21.0	31.1	41.5
6	1.79	3.18	4.53	6.23	9.15	12.0	15.7	21.9	28.1
7	1.72	2.96	4.12	5.52	7.85	10.1	12.7	17.2	21.4
8	1.66	2.81	3.84	5.05	7.01	8.81	10.9	14.4	17.6
9	1.63	2.69	3.63	4.72	6.42	7.96	9.74	12.6	15.1
10	1.59	2.61	3.48	4.47	5.99	7.34	8.89	11.3	13.4
11	1.57	2.54	3.36	4.28	5.67	6.88	8.25	10.3	12.2
12	1.55	2.48	3.26	4.12	5.41	6.52	7.76	9.63	11.2
13	1.53	2.43	3.18	4.00	5.21	6.23	7.37	9.07	10.5
14	1.52	2.39	3.11	3.89	5.04	6.00	7.06	8.62	9.95
15	1.51	2.36	3.06	3.80	4.89	5.80	6.80	8.25	9.48
16	1.50	2.33	3.01	3.73	4.77	5.64	6.58	7.94	9.08
17	1.49	2.31	2.96	3.66	4.67	5.50	6.39	7.68	8.75
18	1.48	2.29	2.93	3.61	4.58	5.37	6.23	7.46	8.47
19	1.47	2.27	2.90	3.56	4.50	5.27	6.09	7.27	8.23
20	1.47	2.25	2.87	3.51	4.43	5.17	5.97	7.10	8.02
21	1.46	2.23	2.84	3.48	4.37	5.09	5.86	6.95	7.83
22	1.45	2.22	2.82	3.44	4.31	5.02	5.76	6.81	7.67
23	1.45	2.21	2.80	3.41	4.26	4.95	5.67	6.70	7.52
24	1.44	2.19	2.78	3.38	4.22	4.89	5.60	6.59	7.39
25	1.44	2.18	2.76	3.35	4.18	4.84	5.53	6.49	7.27
26	1.44	2.17	2.74	3.33	4.14	4.79	5.46	6.41	7.16
27	1.43	2.17	2.73	3.31	4.11	4.74	5.40	6.33	7.06
28	1.43	2.16	2.71	3.29	4.07	4.70	5.35	6.25	6.97
29	1.43	2.15	2.70	3.27	4.04	4.66	5.30	6.19	6.89
30	1.42	2.14	2.69	3.25	4.02	4.62	5.25	6.12	6.82
35	1.41	2.11	2.64	3.18	3.91	4.48	5.07	5.88	6.51
40	1.40	2.09	2.61	3.13	3.83	4.37	4.93	5.70	6.30
45	1.40	2.07	2.58	3.09	3.77	4.29	4.83	5.56	6.13
50	1.39	2.06	2.56	3.05	3.72	4.23	4.75	5.46	6.01
60	1.38	2.04	2.53	3.01	3.65	4.14	4.64	5.31	5.82
70	1.38	2.03	2.50	2.97	3.60	4.08	4.56	5.20	5.70
80	1.38	2.02	2.49	2.95	3.56	4.03	4.50	5.12	5.60
90	1.37	2.01	2.47	2.93	3.53	3.99	4.45	5.06	5.53
100	1.37	2.00	2.46	2.92	3.51	3.96	4.42	5.02	5.48
120	1.37	1.99	2.45	2.89	3.48	3.92	4.36	4.95	5.39
140	1.36	1.99	2.44	2.88	3.46	3.89	4.32	4.90	5.33
160	1.36	1.98	2.43	2.87	3.44	3.87	4.30	4.86	5.29
180	1.36	1.98	2.42	2.86	3.43	3.85	4.27	4.83	5.26
200	1.36	1.97	2.42	2.85	3.41	3.84	4.26	4.81	5.23
300	1.35	1.96	2.40	2.83	3.38	3.80	4.21	4.75	5.15
500	1.35	1.96	2.39	2.81	3.36	3.76	4.17	4.69	5.09
∞	1.35	1.94	2.37	2.79	3.32	3.72	4.11	4.62	5.00

Table 11: Critical Values of the F Distribution

Numerator DF = 5

ν	$\alpha(2)$: 0.50 $\alpha(1)$: 0.25	0.20 0.10	0.10 0.05	0.05 0.025	0.02 0.01	0.01 0.005	0.005 0.0025	0.002 0.001	0.001 0.0005
1	8.82	57.2	230.0	922.0	5760.0	23100.0	92200.0	576000.0	2310000.0
2	3.28	9.29	19.3	39.3	99.3	199.0	399.0	999.0	2000.0
3	2.41	5.31	9.01	14.9	28.2	45.4	72.6	135.0	214.0
4	2.07	4.05	6.26	9.36	15.5	22.5	32.3	51.7	73.6
5	1.89	3.45	5.05	7.15	11.0	14.9	20.2	29.8	39.7
6	1.79	3.11	4.39	5.99	8.75	11.5	14.9	20.8	26.6
7	1.71	2.88	3.97	5.29	7.46	9.52	12.0	16.2	20.2
8	1.66	2.73	3.69	4.82	6.63	8.30	10.3	13.5	16.4
9	1.62	2.61	3.48	4.48	6.06	7.47	9.12	11.7	14.1
10	1.59	2.52	3.33	4.24	5.64	6.87	8.29	10.5	12.4
11	1.56	2.45	3.20	4.04	5.32	6.42	7.67	9.58	11.2
12	1.54	2.39	3.11	3.89	5.06	6.07	7.20	8.89	10.4
13	1.52	2.35	3.03	3.77	4.86	5.79	6.82	8.35	9.66
14	1.51	2.31	2.96	3.66	4.69	5.56	6.51	7.92	9.11
15	1.49	2.27	2.90	3.58	4.56	5.37	6.26	7.57	8.66
16	1.48	2.24	2.85	3.50	4.44	5.21	6.05	7.27	8.29
17	1.47	2.22	2.81	3.44	4.34	5.07	5.87	7.02	7.98
18	1.46	2.20	2.77	3.38	4.25	4.96	5.72	6.81	7.71
19	1.46	2.18	2.74	3.33	4.17	4.85	5.58	6.62	7.48
20	1.45	2.16	2.71	3.29	4.10	4.76	5.46	6.46	7.27
21	1.44	2.14	2.68	3.25	4.04	4.68	5.36	6.32	7.10
22	1.44	2.13	2.66	3.22	3.99	4.61	5.26	6.19	6.94
23	1.43	2.11	2.64	3.18	3.94	4.54	5.18	6.08	6.80
24	1.43	2.10	2.62	3.15	3.90	4.49	5.11	5.98	6.68
25	1.42	2.09	2.60	3.13	3.85	4.43	5.04	5.89	6.56
26	1.42	2.08	2.59	3.10	3.82	4.38	4.98	5.80	6.46
27	1.42	2.07	2.57	3.08	3.78	4.34	4.92	5.73	6.37
28	1.41	2.06	2.56	3.06	3.75	4.30	4.87	5.66	6.28
29	1.41	2.06	2.55	3.04	3.73	4.26	4.82	5.59	6.21
30	1.41	2.05	2.53	3.03	3.70	4.23	4.78	5.53	6.13
35	1.40	2.02	2.49	2.96	3.59	4.09	4.60	5.30	5.85
40	1.39	2.00	2.45	2.90	3.51	3.99	4.47	5.13	5.64
45	1.38	1.98	2.42	2.86	3.45	3.91	4.37	5.00	5.49
50	1.37	1.97	2.40	2.83	3.41	3.85	4.30	4.90	5.37
60	1.37	1.95	2.37	2.79	3.34	3.76	4.19	4.76	5.20
70	1.36	1.93	2.35	2.75	3.29	3.70	4.11	4.66	5.08
80	1.36	1.92	2.33	2.73	3.26	3.65	4.05	4.58	4.99
90	1.35	1.91	2.32	2.71	3.23	3.62	4.01	4.53	4.92
100	1.35	1.91	2.31	2.70	3.21	3.59	3.97	4.48	4.87
120	1.35	1.90	2.29	2.67	3.17	3.55	3.92	4.42	4.79
140	1.34	1.89	2.28	2.66	3.15	3.52	3.89	4.37	4.74
160	1.34	1.88	2.27	2.65	3.13	3.50	3.86	4.33	4.69
180	1.34	1.88	2.26	2.64	3.12	3.48	3.84	4.31	4.66
200	1.34	1.88	2.26	2.63	3.11	3.47	3.82	4.29	4.64
300	1.33	1.87	2.24	2.61	3.08	3.43	3.77	4.22	4.56
500	1.33	1.86	2.23	2.59	3.05	3.40	3.73	4.18	4.51
∞	1.33	1.85	2.21	2.57	3.02	3.35	3.68	4.10	4.42

Table 12: Critical Values of the F Distribution

Numerator DF = 6

ν	$\alpha(2)$: 0.50 $\alpha(1)$: 0.25	0.20 0.10	0.10 0.05	0.05 0.025	0.02 0.01	0.01 0.005	0.005 0.0025	0.002 0.001	0.001 0.0005
1	8.98	58.2	234.0	937.0	5860.0	23400.0	93700.0	586000.0	2340000.0
2	3.31	9.33	19.3	39.3	99.3	199.0	399.0	999.0	2000.0
3	2.42	5.28	8.94	14.7	27.9	44.8	71.7	133.0	211.0
4	2.08	4.01	6.16	9.20	15.2	22.0	31.5	50.5	71.9
5	1.89	3.40	4.95	6.98	10.7	14.5	19.6	28.8	38.5
6	1.78	3.05	4.28	5.82	8.47	11.1	14.4	20.0	25.6
7	1.71	2.83	3.87	5.12	7.19	9.16	11.5	15.5	19.3
8	1.65	2.67	3.58	4.65	6.37	7.95	9.83	12.9	15.7
9	1.61	2.55	3.37	4.32	5.80	7.13	8.68	11.1	13.3
10	1.58	2.46	3.22	4.07	5.39	6.54	7.87	9.93	11.7
11	1.55	2.39	3.09	3.88	5.07	6.10	7.27	9.05	10.6
12	1.53	2.33	3.00	3.73	4.82	5.76	6.80	8.38	9.74
13	1.51	2.28	2.92	3.60	4.62	5.48	6.44	7.86	9.07
14	1.50	2.24	2.85	3.50	4.46	5.26	6.14	7.44	8.53
15	1.48	2.21	2.79	3.41	4.32	5.07	5.89	7.09	8.10
16	1.47	2.18	2.74	3.34	4.20	4.91	5.68	6.80	7.74
17	1.46	2.15	2.70	3.28	4.10	4.78	5.51	6.56	7.43
18	1.45	2.13	2.66	3.22	4.01	4.66	5.36	6.35	7.18
19	1.44	2.11	2.63	3.17	3.94	4.56	5.21	6.18	6.95
20	1.44	2.09	2.60	3.13	3.87	4.47	5.11	6.02	6.76
21	1.43	2.08	2.57	3.09	3.81	4.39	5.01	5.88	6.59
22	1.42	2.06	2.55	3.05	3.76	4.32	4.92	5.76	6.44
23	1.42	2.05	2.53	3.02	3.71	4.26	4.84	5.65	6.30
24	1.41	2.04	2.51	2.99	3.67	4.20	4.76	5.55	6.18
25	1.41	2.02	2.49	2.97	3.63	4.15	4.70	5.46	6.07
26	1.41	2.01	2.47	2.94	3.59	4.10	4.64	5.38	5.98
27	1.40	2.00	2.46	2.92	3.56	4.06	4.58	5.31	5.89
28	1.40	2.00	2.45	2.90	3.53	4.02	4.53	5.24	5.80
29	1.40	1.99	2.43	2.88	3.50	3.98	4.48	5.18	5.73
30	1.39	1.98	2.42	2.87	3.47	3.95	4.44	5.12	5.66
35	1.38	1.95	2.37	2.80	3.37	3.81	4.27	4.89	5.39
40	1.37	1.93	2.34	2.74	3.29	3.71	4.14	4.73	5.19
45	1.36	1.91	2.31	2.70	3.23	3.64	4.05	4.61	5.04
50	1.36	1.90	2.29	2.67	3.19	3.58	3.98	4.51	4.93
60	1.35	1.87	2.25	2.63	3.12	3.49	3.87	4.37	4.76
70	1.34	1.86	2.23	2.59	3.07	3.43	3.79	4.28	4.64
80	1.34	1.85	2.21	2.57	3.04	3.39	3.74	4.20	4.56
90	1.33	1.84	2.20	2.55	3.01	3.35	3.70	4.15	4.50
100	1.33	1.83	2.19	2.54	2.99	3.33	3.66	4.11	4.45
120	1.33	1.82	2.18	2.52	2.96	3.28	3.61	4.04	4.37
140	1.32	1.82	2.16	2.50	2.93	3.26	3.58	4.00	4.32
160	1.32	1.81	2.16	2.49	2.92	3.24	3.55	3.97	4.28
180	1.32	1.81	2.15	2.48	2.90	3.22	3.53	3.94	4.25
200	1.32	1.80	2.14	2.47	2.89	3.21	3.52	3.92	4.22
300	1.32	1.79	2.13	2.45	2.86	3.17	3.47	3.86	4.15
500	1.31	1.79	2.12	2.43	2.84	3.14	3.43	3.81	4.10
∞	1.31	1.77	2.10	2.41	2.80	3.09	3.37	3.74	4.02

Table 13: Critical Values of the F Distribution

Numerator DF = 7

ν	$\alpha(2)$: 0.50	0.20	0.10	0.05	0.02	0.01	0.005	0.002	0.001
	$\alpha(1)$: 0.25	0.10	0.05	0.025	0.01	0.005	0.0025	0.001	0.0005
1	9.10	58.9	237.0	948.0	5930.0	23700.0	94900.0	593000.0	2370000.0
2	3.34	9.35	19.4	39.4	99.4	199.0	399.0	999.0	2000.0
3	2.43	5.27	8.89	14.6	27.7	44.4	71.0	132.0	209.0
4	2.08	3.98	6.09	9.07	15.0	21.6	31.0	49.7	70.7
5	1.89	3.37	4.88	6.85	10.5	14.2	19.1	28.2	37.6
6	1.78	3.01	4.21	5.70	8.26	10.8	14.0	19.5	24.9
7	1.70	2.78	3.79	4.99	6.99	8.89	11.2	15.0	18.7
8	1.64	2.62	3.50	4.53	6.18	7.69	9.49	12.4	15.1
9	1.60	2.51	3.29	4.20	5.61	6.88	8.36	10.7	12.8
10	1.57	2.41	3.14	3.95	5.20	6.30	7.56	9.52	11.2
11	1.54	2.34	3.01	3.76	4.89	5.86	6.97	8.66	10.1
12	1.52	2.28	2.91	3.61	4.64	5.52	6.51	8.00	9.28
13	1.50	2.23	2.83	3.48	4.44	5.25	6.15	7.49	8.63
14	1.49	2.19	2.76	3.38	4.28	5.03	5.86	7.08	8.11
15	1.47	2.16	2.71	3.29	4.14	4.85	5.62	6.74	7.68
16	1.46	2.13	2.66	3.22	4.03	4.69	5.41	6.46	7.33
17	1.45	2.10	2.61	3.16	3.93	4.56	5.24	6.22	7.04
18	1.44	2.08	2.58	3.10	3.84	4.44	5.09	6.02	6.78
19	1.43	2.06	2.54	3.05	3.77	4.34	4.96	5.85	6.57
20	1.43	2.04	2.51	3.01	3.70	4.26	4.85	5.69	6.38
21	1.42	2.02	2.49	2.97	3.64	4.18	4.75	5.56	6.21
22	1.41	2.01	2.46	2.93	3.59	4.11	4.66	5.44	6.07
23	1.41	1.99	2.44	2.90	3.54	4.05	4.58	5.33	5.94
24	1.40	1.98	2.42	2.87	3.50	3.99	4.51	5.23	5.82
25	1.40	1.97	2.40	2.85	3.46	3.94	4.44	5.15	5.71
26	1.39	1.96	2.39	2.82	3.42	3.89	4.38	5.07	5.62
27	1.39	1.95	2.37	2.80	3.39	3.85	4.33	5.00	5.53
28	1.39	1.94	2.36	2.78	3.36	3.81	4.28	4.93	5.45
29	1.38	1.93	2.35	2.76	3.33	3.77	4.24	4.87	5.38
30	1.38	1.93	2.33	2.75	3.30	3.74	4.19	4.82	5.31
35	1.37	1.90	2.29	2.68	3.20	3.61	4.02	4.59	5.04
40	1.36	1.87	2.25	2.62	3.12	3.51	3.90	4.44	4.85
45	1.35	1.85	2.22	2.58	3.07	3.43	3.81	4.32	4.71
50	1.34	1.84	2.20	2.55	3.02	3.38	3.74	4.22	4.60
60	1.33	1.82	2.17	2.51	2.95	3.29	3.63	4.09	4.44
70	1.33	1.80	2.14	2.47	2.91	3.23	3.56	3.99	4.32
80	1.32	1.79	2.13	2.45	2.87	3.19	3.50	3.92	4.24
90	1.32	1.78	2.11	2.43	2.84	3.15	3.46	3.87	4.18
100	1.32	1.78	2.10	2.42	2.82	3.13	3.43	3.83	4.13
120	1.31	1.77	2.09	2.39	2.79	3.09	3.38	3.77	4.06
140	1.31	1.76	2.08	2.38	2.77	3.06	3.35	3.72	4.01
160	1.31	1.75	2.07	2.37	2.75	3.04	3.32	3.69	3.97
180	1.31	1.75	2.06	2.36	2.74	3.02	3.30	3.67	3.94
200	1.30	1.75	2.06	2.35	2.73	3.01	3.29	3.65	3.92
300	1.30	1.74	2.04	2.33	2.70	2.97	3.24	3.59	3.85
500	1.30	1.73	2.03	2.31	2.68	2.94	3.20	3.54	3.80
∞	1.29	1.72	2.01	2.29	2.64	2.90	3.15	3.47	3.72

Table 14: Critical Values of the F Distribution

Numerator DF = 8

ν	$\alpha(2)$: 0.50 $\alpha(1)$: 0.25	0.20 0.10	0.10 0.05	0.05 0.025	0.02 0.01	0.01 0.005	0.005 0.0025	0.002 0.001	0.001 0.0005
1	9.19	59.4	239.0	957.0	5980.0	23900.0	95700.0	598000.0	2390000.0
2	3.35	9.37	19.4	39.4	99.4	199.0	399.0	999.0	2000.0
3	2.44	5.25	8.85	14.5	27.5	44.1	70.5	131.0	208.0
4	2.08	3.95	6.04	8.98	14.8	21.4	30.6	49.0	69.7
5	1.89	3.34	4.82	6.76	10.3	14.0	18.8	27.6	36.9
6	1.78	2.98	4.15	5.60	8.10	10.6	13.7	19.0	24.3
7	1.70	2.75	3.73	4.90	6.84	8.68	10.9	14.6	18.2
8	1.64	2.59	3.44	4.43	6.03	7.50	9.24	12.0	14.6
9	1.60	2.47	3.23	4.10	5.47	6.69	8.12	10.4	12.4
10	1.56	2.38	3.07	3.85	5.06	6.12	7.33	9.20	10.9
11	1.53	2.30	2.95	3.66	4.74	5.68	6.74	8.35	9.76
12	1.51	2.24	2.85	3.51	4.50	5.35	6.29	7.71	8.94
13	1.49	2.20	2.77	3.39	4.30	5.08	5.93	7.21	8.29
14	1.48	2.15	2.70	3.29	4.14	4.86	5.64	6.80	7.78
15	1.46	2.12	2.64	3.20	4.00	4.67	5.40	6.47	7.37
16	1.45	2.09	2.59	3.12	3.89	4.52	5.20	6.19	7.02
17	1.44	2.06	2.55	3.06	3.79	4.39	5.03	5.96	6.73
18	1.43	2.04	2.51	3.01	3.71	4.28	4.89	5.76	6.48
19	1.42	2.02	2.48	2.96	3.63	4.18	4.76	5.59	6.27
20	1.42	2.00	2.45	2.91	3.56	4.09	4.65	5.44	6.09
21	1.41	1.98	2.42	2.87	3.51	4.01	4.55	5.31	5.92
22	1.40	1.97	2.40	2.84	3.45	3.94	4.46	5.19	5.78
23	1.40	1.95	2.37	2.81	3.41	3.88	4.38	5.09	5.65
24	1.39	1.94	2.36	2.78	3.36	3.83	4.31	4.99	5.54
25	1.39	1.93	2.34	2.75	3.32	3.78	4.25	4.91	5.43
26	1.38	1.92	2.32	2.73	3.29	3.73	4.19	4.83	5.34
27	1.38	1.91	2.31	2.71	3.26	3.69	4.14	4.76	5.25
28	1.38	1.90	2.29	2.69	3.23	3.65	4.09	4.69	5.18
29	1.37	1.89	2.28	2.67	3.20	3.61	4.04	4.64	5.11
30	1.37	1.88	2.27	2.65	3.17	3.58	4.00	4.58	5.04
35	1.36	1.85	2.22	2.58	3.07	3.45	3.83	4.36	4.78
40	1.35	1.83	2.18	2.53	2.99	3.35	3.71	4.21	4.59
45	1.34	1.81	2.15	2.49	2.94	3.28	3.62	4.09	4.45
50	1.33	1.80	2.13	2.46	2.89	3.22	3.55	4.00	4.34
60	1.32	1.77	2.10	2.41	2.82	3.13	3.45	3.86	4.19
70	1.32	1.76	2.07	2.38	2.78	3.08	3.37	3.77	4.08
80	1.31	1.75	2.06	2.35	2.74	3.03	3.32	3.70	4.00
90	1.31	1.74	2.04	2.34	2.72	3.00	3.28	3.65	3.94
100	1.30	1.73	2.03	2.32	2.69	2.97	3.25	3.61	3.89
120	1.30	1.72	2.02	2.30	2.66	2.93	3.20	3.55	3.82
140	1.30	1.71	2.01	2.28	2.64	2.91	3.17	3.51	3.77
160	1.29	1.71	2.00	2.27	2.62	2.88	3.14	3.48	3.73
180	1.29	1.70	1.99	2.26	2.61	2.87	3.12	3.45	3.70
200	1.29	1.70	1.98	2.26	2.60	2.86	3.11	3.43	3.68
300	1.29	1.69	1.97	2.23	2.57	2.82	3.06	3.38	3.61
500	1.28	1.68	1.96	2.22	2.55	2.79	3.03	3.33	3.56
∞	1.28	1.67	1.94	2.19	2.51	2.74	2.97	3.27	3.48

Table 15: Critical Values of the F Distribution

Numerator DF = 9

ν	$\alpha(2)$: 0.50 $\alpha(1)$: 0.25	0.20 0.10	0.10 0.05	0.05 0.025	0.02 0.01	0.01 0.005	0.005 0.0025	0.002 0.001	0.001 0.0005
1	9.26	59.9	241.0	963.0	6020.0	24100.0	96400.0	602000.0	2410000.0
2	3.37	9.38	19.4	39.4	99.4	199.0	399.0	999.0	2000.0
3	2.44	5.24	8.81	14.5	27.3	43.9	70.1	130.0	207.0
4	2.08	3.94	6.00	8.90	14.7	21.1	30.3	48.5	69.0
5	1.89	3.32	4.77	6.68	10.2	13.8	18.5	27.2	36.3
6	1.77	2.96	4.10	5.52	7.98	10.4	13.4	18.7	23.9
7	1.69	2.72	3.68	4.82	6.72	8.51	10.7	14.3	17.8
8	1.63	2.56	3.39	4.36	5.91	7.34	9.03	11.8	14.3
9	1.59	2.44	3.18	4.03	5.35	6.54	7.92	10.1	12.1
10	1.56	2.35	3.02	3.78	4.94	5.97	7.14	8.96	10.6
11	1.53	2.27	2.90	3.59	4.63	5.54	6.56	8.12	9.48
12	1.51	2.21	2.80	3.44	4.39	5.20	6.11	7.48	8.66
13	1.49	2.16	2.71	3.31	4.19	4.94	5.76	6.98	8.03
14	1.47	2.12	2.65	3.21	4.03	4.72	5.47	6.58	7.52
15	1.46	2.09	2.59	3.12	3.89	4.54	5.23	6.26	7.11
16	1.44	2.06	2.54	3.05	3.78	4.38	5.04	5.98	6.77
17	1.43	2.03	2.49	2.98	3.68	4.25	4.87	5.75	6.49
18	1.42	2.00	2.46	2.93	3.60	4.14	4.72	5.56	6.24
19	1.41	1.98	2.42	2.88	3.52	4.04	4.60	5.39	6.03
20	1.41	1.96	2.39	2.84	3.46	3.96	4.49	5.24	5.85
21	1.40	1.95	2.37	2.80	3.40	3.88	4.39	5.11	5.69
22	1.39	1.93	2.34	2.76	3.35	3.81	4.30	4.99	5.55
23	1.39	1.92	2.32	2.73	3.30	3.75	4.22	4.89	5.43
24	1.38	1.91	2.30	2.70	3.26	3.69	4.15	4.80	5.31
25	1.38	1.89	2.28	2.68	3.22	3.64	4.09	4.71	5.21
26	1.37	1.88	2.27	2.65	3.18	3.60	4.03	4.64	5.12
27	1.37	1.87	2.25	2.63	3.15	3.56	3.98	4.57	5.04
28	1.37	1.87	2.24	2.61	3.12	2.52	3.93	4.50	4.96
29	1.36	1.86	2.22	2.59	3.09	3.48	3.89	4.45	4.89
30	1.36	1.85	2.21	2.57	3.07	3.45	3.85	4.39	4.82
35	1.35	1.82	2.16	2.50	2.96	3.32	3.68	4.18	4.57
40	1.34	1.79	2.12	2.45	2.89	3.22	3.56	4.02	4.38
45	1.33	1.77	2.10	2.41	2.83	3.15	3.47	3.91	4.25
50	1.32	1.76	2.07	2.38	2.78	3.09	3.40	3.82	4.14
60	1.31	1.74	2.04	2.33	2.72	3.01	3.30	3.69	3.98
70	1.31	1.72	2.02	2.30	2.67	2.95	3.23	3.60	3.88
80	1.30	1.71	2.00	2.28	2.64	2.91	3.17	3.53	3.80
90	1.30	1.70	1.99	2.26	2.61	2.87	3.13	3.48	3.74
100	1.29	1.69	1.97	2.24	2.59	2.85	3.10	3.44	3.69
120	1.29	1.68	1.96	2.22	2.56	2.81	3.06	3.38	3.62
140	1.29	1.68	1.95	2.21	2.54	2.78	3.02	3.34	3.57
160	1.28	1.67	1.94	2.19	2.52	2.76	3.00	3.31	3.54
180	1.28	1.67	1.93	2.19	2.51	2.74	2.98	3.28	3.51
200	1.28	1.66	1.93	2.18	2.50	2.73	2.96	3.26	3.49
300	1.27	1.65	1.91	2.16	2.47	2.69	2.92	3.21	3.42
500	1.27	1.64	1.90	2.14	2.44	2.66	2.88	3.16	3.37
∞	1.27	1.63	1.88	2.11	2.41	2.62	2.83	3.10	3.30

Table 16: Critical Values of the F Distribution

Numerator DF = 10

ν	$\alpha(2)$: 0.50 $\alpha(1)$: 0.25	0.20 0.10	0.10 0.05	0.05 0.025	0.02 0.01	0.01 0.005	0.005 0.0025	0.002 0.001	0.001 0.0005
1	9.32	60.2	242.0	969.0	6060.0	24200.0	96900.0	606000.0	2420000.0
2	3.38	9.39	19.4	39.4	99.4	199.0	399.0	999.0	2000.0
3	2.44	5.23	8.79	14.4	27.2	43.7	69.8	129.0	206.0
4	2.08	3.92	5.96	8.84	14.5	21.0	30.0	48.1	68.3
5	1.89	3.30	4.74	6.62	10.1	13.6	18.3	26.9	35.9
6	1.77	2.94	4.00	5.46	7.87	10.3	13.2	18.4	23.5
7	1.69	2.70	3.64	4.76	6.62	8.38	10.5	14.1	17.5
8	1.63	2.54	3.35	4.30	5.81	7.21	8.87	11.5	14.0
9	1.59	2.42	3.14	3.96	5.26	6.42	7.77	9.89	11.8
10	1.55	2.32	2.98	3.72	4.85	5.85	6.99	8.75	10.3
11	1.52	2.25	2.85	3.53	4.54	5.42	6.41	7.92	9.24
12	1.50	2.19	2.75	3.37	4.30	5.09	5.97	7.29	8.43
13	1.48	2.14	2.67	3.25	4.10	4.82	5.62	6.80	7.81
14	1.46	2.10	2.60	3.15	3.94	4.60	5.33	6.40	7.31
15	1.45	2.06	2.54	3.06	3.80	4.42	5.10	6.08	6.91
16	1.44	2.03	2.49	2.99	3.69	4.27	4.90	5.81	6.57
17	1.43	2.00	2.45	2.92	3.59	4.14	4.73	5.58	6.29
18	1.42	1.98	2.41	2.87	3.51	4.03	4.59	5.39	6.05
19	1.41	1.96	2.38	2.82	3.43	3.93	4.46	5.22	5.84
20	1.40	1.94	2.35	2.77	3.37	3.85	4.35	5.08	5.66
21	1.39	1.92	2.32	2.73	3.31	3.77	4.26	4.95	5.50
22	1.39	1.90	2.30	2.70	3.26	3.70	4.17	4.83	5.36
23	1.38	1.89	2,27	2.67	3.21	3.64	4.09	4.73	5.24
24	1.38	1.88	2.25	2.64	3.17	3.59	4.03	4.64	5.13
25	1.37	1.87	2.24	2.61	3.13	3.54	3.96	4.56	5.03
26	1.37	1.86	2.22	2.59	3.09	3.49	3.91	4.48	4.94
27	1.36	1.85	2.20	2.57	3.06	3.45	3.85	4.41	4.86
28	1.36	1.84	2.19	2.55	3.03	3.41	3.81	4.35	4.78
29	1.35	1.83	2.18	2.53	3.00	3.38	3.76	4.29	4.71
30	1.35	1.82	2.16	2.51	2.98	3.34	3.72	4.24	4.65
35	1.34	1.79	2.11	2.44	2.88	3.21	3.56	4.03	4.39
40	1.33	1.76	2.08	2.39	2.80	3.12	3.44	3.87	4.21
45	1.32	1.74	2.05	2.35	2.74	3.04	3.35	3.76	4.08
50	1.31	1.73	2.03	2.32	2.70	2.99	3.28	3.67	3.97
60	1.30	1.71	1.99	2.27	2.63	2.90	3.18	3.54	3.82
70	1.30	1.69	1.97	2.24	2.59	2.85	3.11	3.45	3.71
80	1.29	1.68	1.95	2.21	2.55	2.80	3.05	3.39	3.64
90	1.29	1.67	1.94	2.19	2.52	2.77	3.01	3.34	3.58
100	1.28	1.66	1.93	2.18	2.50	2.74	2.98	3.30	3.53
120	1.28	1.65	1.91	2.16	2.47	2.71	2.94	3.24	3.46
140	1.28	1.64	1.90	2.14	2.45	2.68	2.90	3.20	3.42
160	1.27	1.64	1.89	2.13	2.43	2.66	2.88	3.17	3.38
180	1.27	1.63	1.88	2.12	2.42	2.64	2.86	3.14	3.35
200	1.27	1.63	1.88	2.11	2.41	2.63	2.84	3.12	3.33
300	1.26	1.62	1.86	2.09	2.38	2.59	2.80	3.07	3.27
500	1.26	1.61	1.85	2.07	2.36	2.56	2.76	3.02	3.22
∞	1.25	1.60	1.83	2.05	2.32	2.52	2.71	2.96	3.14

Table 17: Critical Values of the F Distribution

Numerator DF = 11

ν	$\alpha(2)$: 0.50	0.20	0.10	0.05	0.02	0.01	0.005	0.002	0.001
	$\alpha(1)$: 0.25	0.10	0.05	0.025	0.01	0.005	0.0025	0.001	0.0005
1	9.37	60.5	243.0	973.0	6080.0	24300.0	97300.0	608000.0	2430000.0
2	3.39	9.40	19.4	39.4	99.4	199.0	399.0	999.0	2000.0
3	2.45	5.22	8.76	14.4	27.1	43.5	69.5	129.0	205.0
4	2.08	3.91	5.94	8.79	14.5	20.8	29.8	47.7	67.8
5	1.89	3.28	4.70	6.57	9.96	13.5	18.1	26.6	35.5
6	1.77	2.92	4.03	5.41	7.79	10.1	13.1	18.2	23.2
7	1.69	2.68	3.60	4.71	6.54	8.27	10.4	13.9	17.2
8	1.63	2.52	3.31	4.24	5.73	7.10	8.73	11.4	13.8
9	1.58	2.40	3.10	3.91	5.18	6.31	7.63	9.72	11.6
10	1.55	2.30	2.94	3.66	4.77	5.75	6.86	8.59	10.1
11	1.52	2.23	2.82	3.47	4.46	5.32	6.29	7.76	9.05
12	1.49	2.17	2.72	3.32	4.22	4.99	5.85	7.14	8.25
13	1.47	2.12	2.63	3.20	4.02	4.72	5.50	6.65	7.63
14	1.46	2.07	2.57	3.09	3.86	4.51	5.21	6.26	7.13
15	1.44	2.04	2.51	3.01	3.73	4.33	4.98	5.94	6.73
16	1.43	2.01	2.46	2.93	3.62	4.18	4.79	5.67	6.40
17	1.42	1.98	2.41	2.87	3.52	4.05	4.62	5.44	6.12
18	1.41	1.95	2.37	2.81	3.43	3.94	4.48	5.25	5.89
19	1.40	1.93	2.34	2.76	3.36	3.84	4.35	5.08	5.68
20	1.39	1.91	2.31	2.72	3.29	3.76	4.24	4.94	5.50
21	1.39	1.90	2.28	2.68	3.24	3.68	4.15	4.81	5.35
22	1.38	1.88	2.26	2.65	3.18	3.61	4.06	4.70	5.21
23	1.37	1.87	2.24	2.62	3.14	3.55	3.99	4.60	5.09
24	1.37	1.85	2.22	2.59	3.09	3.50	3.92	4.51	4.98
25	1.36	1.84	2.20	2.56	3.06	3.45	3.85	4.42	4.88
26	1.36	1.83	2.18	2.54	3.02	3.40	3.80	4.35	4.79
27	1.35	1.82	2.17	2.51	2.99	3.36	3.75	4.28	4.71
28	1.35	1.81	2.15	2.49	2.96	3.32	3.70	4.22	4.63
29	1.35	1.80	2.14	2.48	2.93	3.29	3.66	4.16	4.56
30	1.34	1.79	2.13	2.46	2.91	3.25	3.61	4.11	4.50
35	1.33	1.76	2.07	2.39	2.80	3.12	3.45	3.90	4.25
40	1.32	1.74	2.04	2.33	2.73	3.03	3.33	3.75	4.07
45	1.31	1.72	2.01	2.29	2.67	2.96	3.25	3.64	3.94
50	1.30	1.70	1.99	2.26	2.63	2.90	3.18	3.55	3.83
60	1.29	1.68	1.95	2.22	2.56	2.82	3.08	3.42	3.68
70	1.29	1.66	1.93	2.18	2.51	2.76	3.00	3.33	3.58
80	1.28	1.65	1.91	2.16	2.48	2.72	2.95	3.27	3.50
90	1.28	1.64	1.90	2.14	2.45	2.68	2.91	3.22	3.44
100	1.27	1.64	1.89	2.12	2.43	2.66	2.88	3.18	3.40
120	1.27	1.63	1.87	2.10	2.40	2.62	2.83	3.12	3.33
140	1.27	1.62	1.86	2.09	2.38	2.59	2.80	3.08	3.28
160	1.26	1.61	1.85	2.07	2.36	2.57	2.78	3.05	3.25
180	1.26	1.61	1.84	2.07	2.35	2.56	2.76	3.02	3.22
200	1.26	1.60	1.84	2.06	2.34	2.54	2.74	3.00	3.20
300	1.26	1.59	1.82	2.04	2.31	2.51	2.70	2.95	3.14
500	1.25	1.58	1.81	2.02	2.28	2.48	2.66	2.91	3.09
∞	1.25	1.57	1.79	1.99	2.25	2.43	2.61	2.84	3.01

Table 18: Critical Values of the F Distribution

Numerator DF = 12

ν	$\alpha(2)$: 0.50 $\alpha(1)$: 0.25	0.20 0.10	0.10 0.05	0.05 0.025	0.02 0.01	0.01 0.005	0.005 0.0025	0.002 0.001	0.001 0.0005
1	9.41	60.7	244.0	977.0	6110.0	24400.0	97700.0	611000.0	2440000.0
2	3.39	9.41	19.4	39.4	99.4	199.0	399.0	999.0	2000.0
3	2.45	5.22	8.74	14.3	27.1	43.4	69.3	128.0	204.0
4	2.08	3.90	5.91	8.75	14.4	20.7	29.7	47.4	67.4
5	1.89	3.27	4.68	6.52	9.89	13.4	18.0	26.4	35.2
6	1.77	2.90	4.00	5.37	7.72	10.0	12.9	18.0	23.0
7	1.68	2.67	3.57	4.67	6.47	8.18	10.3	13.7	17.0
8	1.62	2.50	3.28	4.20	5.67	7.01	8.61	11.2	13.6
9	1.58	2.38	3.07	3.87	5.11	6.23	7.52	9.57	11.4
10	1.54	2.28	2.91	3.62	4.71	5.66	6.75	8.45	9.94
11	1.51	2.21	2.79	3.43	4.40	5.24	6.18	7.63	8.88
12	1.49	2.15	2.69	3.28	4.16	4.91	5.74	7.00	8.09
13	1.47	2.10	2.60	3.15	3.96	4.64	5.40	6.52	7.48
14	1.45	2.05	2.53	3.05	3.80	4.43	5.12	6.13	6.99
15	1.44	2.02	2.48	2.96	3.67	4.25	4.88	5.81	6.59
16	1.43	1.99	2.42	2.89	3.55	4.10	4.69	5.55	6.26
17	1.41	1.96	2.38	2.82	3.46	3.97	4.52	5.32	5.98
18	1.40	1.93	2.34	2.77	3.37	3.86	4.38	5.13	5.75
19	1.40	1.91	2.31	2.72	3.30	3.76	4.26	4.97	5.55
20	1.39	1.89	2.28	2.68	3.23	3.68	4.15	4.82	5.37
21	1.38	1.87	2.25	2.64	3.17	3.60	4.06	4.70	5.21
22	1.37	1.86	2.23	2.60	3.12	3.54	3.97	4.58	5.08
23	1.37	1.84	2.20	2.57	3.07	3.47	3.89	4.48	4.96
24	1.36	1.83	2.18	2.54	3.03	3.42	3.83	4.39	4.85
25	1.36	1.82	2.16	2.51	2.99	3.37	3.76	4.31	4.75
26	1.35	1.81	2.15	2.49	2.96	3.33	3.71	4.24	4.66
27	1.35	1.80	2.13	2.47	2.93	3.28	3.66	4.17	4.58
28	1.34	1.79	2.12	2.45	2.90	3.25	3.61	4.11	4.51
29	1.34	1.78	2.10	2.43	2.87	3.21	3.56	4.05	4.44
30	1.34	1.77	2.09	2.41	2.84	3.18	3.52	4.00	4.38
35	1.32	1.74	2.04	2.34	2.74	3.05	3.36	3.79	4.13
40	1.31	1.71	2.00	2.29	2.66	2.95	3.25	3.64	3.95
45	1.30	1.70	1.97	2.25	2.61	2.88	3.16	3.53	3.82
50	1.30	1.68	1.95	2.22	2.56	2.82	3.09	3.44	3.71
60	1.29	1.66	1.92	2.17	2.50	2.74	2.99	3.32	3.57
70	1.28	1.64	1.89	2.14	2.45	2.68	2.92	3.23	3.46
80	1.27	1.63	1.88	2.11	2.42	2.64	2.87	3.16	3.39
90	1.27	1.62	1.86	2.09	2.39	2.61	2.83	3.11	3.33
100	1.27	1.61	1.85	2.08	2.37	2.58	2.80	3.07	3.28
120	1.26	1.60	1.83	2.05	2.34	2.54	2.75	3.02	3.22
140	1.26	1.59	1.82	2.04	2.31	2.52	2.72	2.98	3.17
160	1.26	1.59	1.81	2.03	2.30	2.50	2.69	2.95	3.14
180	1.25	1.58	1.81	2.02	2.28	2.48	2.67	2.92	3.11
200	1.25	1.58	1.80	2.01	2.27	2.47	2.66	2.90	3.09
300	1.25	1.57	1.78	1.99	2.24	2.43	2.61	2.85	3.02
500	1.24	1.56	1.77	1.97	2.22	2.40	2.58	2.81	2.97
∞	1.24	1.55	1.75	1.94	2.18	2.36	2.53	2.74	2.90

Table 19: Critical Values of the *F* Distribution

Numerator DF = 13

	α(2) : 0.50	0.20	0.10	0.05	0.02	0.01	0.005	0.002	0.001
ν	α(1) : 0.25	0.10	0.05	0.025	0.01	0.005	0.0025	0.001	0.0005
1	9.44	60.9	245.0	980.0	6130.0	24500.0	98000.0	613000.0	2450000.0
2	3.40	9.41	19.4	39.4	99.4	199.0	399.0	999.0	2000.0
3	2.45	5.21	8.73	14.3	27.0	43.3	69.1	128.0	204.0
4	2.08	3.89	5.89	8.71	14.3	20.6	29.5	47.2	67.1
5	1.89	3.26	4.66	6.49	9.82	13.3	17.9	26.2	34.9
6	1.77	2.89	3.98	5.33	7.66	9.95	12.8	17.8	22.7
7	1.68	2.65	3.55	4.63	6.41	8.10	10.1	13.6	16.8
8	1.62	2.49	3.26	4.16	5.61	6.94	8.51	11.1	13.4
9	1.58	2.36	3.05	3.83	5.05	6.15	7.43	9.44	11.3
10	1.54	2.27	2.89	3.58	4.65	5.59	6.66	8.32	9.80
11	1.51	2.19	2.76	3.39	4.34	5.16	6.09	7.51	8.74
12	1.49	2.13	2.66	3.24	4.10	4.84	5.66	6.89	7.96
13	1.47	2.08	2.58	3.12	3.91	4.57	5.31	6.41	7.35
14	1.45	2.04	2.51	3.01	3.75	4.36	5.03	6.02	6.86
15	1.43	2.00	2.45	2.92	3.61	4.18	4.80	5.71	6.47
16	1.42	1.97	2.40	2.85	3.50	4.03	4.61	5.44	6.14
17	1.41	1.94	2.35	2.79	3.40	3.90	4.44	5.22	5.86
18	1.40	1.92	2.31	2.73	3.32	3.79	4.30	5.03	5.63
19	1.39	1.89	2.28	2.68	3.24	3.70	4.18	4.87	5.43
20	1.38	1.87	2.25	2.64	3.18	3.61	4.07	4.72	5.25
21	1.37	1.86	2.22	2.60	3.12	3.54	3.98	4.60	5.10
22	1.37	1.84	2.20	2.56	3.07	3.47	3.89	4.49	4.97
23	1.36	1.83	2.18	2.53	3.02	3.41	3.82	4.39	4.84
24	1.36	1.81	2.15	2.50	2.98	3.35	3.75	4.30	4.74
25	1.35	1.80	2.14	2.48	2.94	3.30	3.69	4.22	4.64
26	1.35	1.79	2.12	2.45	2.90	3.26	3.63	4.14	4.55
27	1.34	1.78	2.10	2.43	2.87	3.22	3.58	4.08	4.47
28	1.34	1.77	2.09	2.41	2.84	3.18	3.53	4.01	4.40
29	1.33	1.76	2.08	2.39	2.81	3.15	3.49	3.96	4.33
30	1.33	1.75	2.06	2.37	2.79	3.11	3.45	3.91	4.27
35	1.32	1.72	2.01	2.30	2.69	2.98	3.29	3.70	4.02
40	1.31	1.70	1.97	2.25	2.61	2.89	3.17	3.55	3.85
45	1.30	1.68	1.94	2.21	2.55	2.82	3.08	3.44	3.71
50	1.29	1.66	1.92	2.18	2.51	2.76	3.01	3.35	3.61
60	1.28	1.64	1.89	2.13	2.44	2.68	2.91	3.23	3.46
70	1.27	1.62	1.86	2.10	2.40	2.62	2.84	3.14	3.36
80	1.27	1.61	1.84	2.07	2.36	2.58	2.79	3.07	3.29
90	1.26	1.60	1.83	2.05	2.33	2.54	2.75	3.02	3.23
100	1.26	1.59	1.82	2.04	2.31	2.52	2.72	2.99	3.19
120	1.26	1.58	1.80	2.01	2.28	2.48	2.67	2.93	3.12
140	1.25	1.57	1.79	2.00	2.26	2.45	2.64	2.89	3.07
160	1.25	1.57	1.78	1.99	2.24	2.43	2.62	2.86	3.04
180	1.25	1.56	1.77	1.98	2.23	2.42	2.60	2.83	3.01
200	1.24	1.56	1.77	1.97	2.22	2.40	2.58	2.82	2.99
300	1.24	1.55	1.75	1.95	2.19	2.37	2.54	2.76	2.93
500	1.24	1.54	1.74	1.93	2.17	2.34	2.50	2.72	2.88
∞	1.23	1.52	1.72	1.90	2.13	2.29	2.45	2.66	2.81

Table 20: Critical Values of the F Distribution

Numerator DF = 14

ν	$\alpha(2)$: 0.50	0.20	0.10	0.05	0.02	0.01	0.005	0.002	0.001
	$\alpha(1)$: 0.25	0.10	0.05	0.025	0.01	0.005	0.0025	0.001	0.0005
1	9.47	61.1	245.0	983.0	6140.0	24600.0	98300.0	614000.0	2460000.0
2	3.41	9.42	19.4	39.4	99.4	199.0	399.0	999.0	2000.0
3	2.45	5.20	8.71	14.3	26.9	43.2	69.0	128.0	203.0
4	2.08	3.88	5.87	8.68	14.2	20.5	29.4	46.9	66.8
5	1.89	3.25	4.64	6.46	9.77	13.2	17.8	26.1	34.7
6	1.76	2.88	3.96	5.30	7.60	9.88	12.7	17.7	22.0
7	1.68	2.64	3.53	4.60	6.36	8.03	10.1	13.4	16.6
8	1.62	2.48	3.24	4.13	5.56	6.87	8.43	10.9	13.3
9	1.57	2.35	3.03	3.80	5.01	6.09	7.35	9.33	11.1
10	1.54	2.26	2.86	3.55	4.60	5.53	6.58	8.22	9.67
11	1.51	2.18	2.74	3.36	4.29	5.10	6.02	7.41	8.62
12	1.48	2.12	2.64	3.21	4.05	4.77	5.58	6.79	7.84
13	1.46	2.07	2.55	3.08	3.86	4.51	5.24	6.31	7.23
14	1.44	2.02	2.48	2.98	3.70	4.30	4.96	5.93	6.75
15	1.43	1.99	2.42	2.89	3.56	4.12	4.73	5.62	6.36
16	1.42	1.95	2.37	2.82	3.45	3.97	4.54	5.35	6.03
17	1.41	1.93	2.33	2.75	3.35	3.84	4.37	5.13	5.76
18	1.40	1.90	2.29	2.70	3.27	3.73	4.23	4.94	5.53
19	1.39	1.88	2.26	2.65	3.19	3.64	4.11	4.78	5.33
20	1.38	1.86	2.22	2.60	3.13	3.55	4.00	4.64	5.15
21	1.37	1.84	2.20	2.56	3.07	3.48	3.91	4.51	5.00
22	1.36	1.83	2.17	2.53	3.02	3.41	3.82	4.40	4.87
23	1.36	1.81	2.15	2.50	2.97	3.35	3.75	4.30	4.75
24	1.35	1.80	2.13	2.47	2.93	3.30	3.68	4.21	4.64
25	1.35	1.79	2.11	2.44	2.89	3.25	3.62	4.13	4.54
26	1.34	1.77	2.09	2.42	2.86	3.20	3.56	4.06	4.46
27	1.34	1.76	2.08	2.39	2.82	3.16	3.51	3.99	4.38
28	1.33	1.75	2.06	2.37	2.79	3.12	3.46	3.93	4.30
29	1.33	1.75	2.05	2.36	2.77	3.09	3.42	3.88	4.24
30	1.33	1.74	2.04	2.34	2.74	3.06	3.38	3.82	4.18
35	1.31	1.70	1.99	2.27	2.64	2.93	3.22	3.62	3.93
40	1.30	1.68	1.95	2.21	2.56	2.83	3.10	3.47	3.76
45	1.29	1.66	1.92	2.17	2.51	2.76	3.02	3.36	3.63
50	1.28	1.64	1.89	2.14	2.46	2.70	2.95	3.27	3.52
60	1.27	1.62	1.86	2.09	2.39	2.62	2.85	3.15	3.38
70	1.27	1.60	1.84	2.06	2.35	2.56	2.78	3.06	3.28
80	1.26	1.59	1.82	2.03	2.31	2.52	2.73	3.00	3.20
90	1.26	1.58	1.80	2.02	2.29	2.49	2.69	2.95	3.14
100	1.25	1.57	1.79	2.00	2.27	2.46	2.65	2.91	3.10
120	1.25	1.56	1.78	1.98	2.23	2.42	2.61	2.85	3.03
140	1.24	1.55	1.76	1.96	2.21	2.40	2.58	2.81	2.99
160	1.24	1.55	1.75	1.95	2.20	2.38	2.55	2.78	2.95
180	1.24	1.54	1.75	1.94	2.18	2.36	2.53	2.76	2.93
200	1.24	1.54	1.74	1.93	2.17	2.35	2.52	2.74	2.91
300	1.23	1.53	1.72	1.91	2.14	2.31	2.47	2.69	2.84
500	1.23	1.52	1.71	1.89	2.12	2.28	2.44	2.64	2.79
∞	1.22	1.50	1.69	1.87	2.08	2.24	2.39	2.58	2.72

Table 21: Critical Values of the F Distribution

Numerator DF = 15

ν	$\alpha(2)$: 0.50	0.20	0.10	0.05	0.02	0.01	0.005	0.002	0.001
	$\alpha(1)$: 0.25	0.10	0.05	0.025	0.01	0.005	0.0025	0.001	0.0005
1	9.49	61.2	246.0	985.0	6160.0	24600.0	98500.0	616000.0	2460000.0
2	3.41	9.42	19.4	39.4	99.4	199.0	399.0	999.0	2000.0
3	2.46	5.20	8.70	14.3	26.9	43.1	68.8	127.0	203.0
4	2.08	3.87	5.86	8.66	14.2	20.4	29.3	46.8	66.5
5	1.89	3.24	4.62	6.43	9.72	13.1	17.7	25.9	34.5
6	1.76	2.87	3.94	5.27	7.56	9.81	12.7	17.6	22.4
7	1.68	2.63	3.51	4.57	6.31	7.97	9.98	13.3	16.5
8	1.62	2.46	3.22	4.10	5.52	6.81	8.35	10.8	13.1
9	1.57	2.34	3.01	3.77	4.96	6.03	7.28	9.24	11.0
10	1.53	2.24	2.85	3.52	4.56	5.47	6.51	8.13	9.56
11	1.50	2.17	2.72	3.33	4.25	5.05	5.95	7.32	8.52
12	1.48	2.10	2.62	3.18	4.01	4.72	5.52	6.71	7.74
13	1.46	2.05	2.53	3.05	3.82	4.46	5.17	6.23	7.13
14	1.44	2.01	2.46	2.95	3.66	4.25	4.89	5.85	6.65
15	1.43	1.97	2.40	2.86	3.52	4.07	4.67	5.54	6.26
16	1.41	1.94	2.35	2.79	3.41	3.92	4.47	5.27	5.94
17	1.40	1.91	2.31	2.72	3.31	3.79	4.31	5.05	5.67
18	1.39	1.89	2.27	2.67	3.23	3.68	4.17	4.87	5.44
19	1.38	1.86	2.23	2.62	3.15	3.59	4.05	4.70	5.24
20	1.37	1.84	2.20	2.57	3.09	3.50	3.94	4.56	5.07
21	1.37	1.83	2.18	2.53	3.03	3.43	3.85	4.44	4.92
22	1.36	1.81	2.15	2.50	2.98	3.36	3.76	4.33	4.78
23	1.35	1.80	2.13	2.47	2.93	3.30	3.69	4.23	4.66
24	1.35	1.78	2.11	2.44	2.89	3.25	3.62	4.14	4.56
25	1.34	1.77	2.09	2.41	2.85	3.20	3.56	4.06	4.46
26	1.34	1.76	2.07	2.39	2.81	3.15	3.50	3.99	4.37
27	1.33	1.75	2.06	2.36	2.78	3.11	3.45	3.92	4.29
28	1.33	1.74	2.04	2.34	2.75	3.07	3.40	3.86	4.22
29	1.32	1.73	2.03	2.32	2.73	3.04	3.36	3.80	4.15
30	1.32	1.72	2.01	2.31	2.70	3.01	3.32	3.75	4.09
35	1.31	1.69	1.96	2.23	2.60	2.88	3.16	3.55	3.85
40	1.30	1.66	1.92	2.18	2.52	2.78	3.04	3.40	3.68
45	1.29	1.64	1.89	2.14	2.46	2.71	2.96	3.29	3.55
50	1.28	1.63	1.87	2.11	2.42	2.65	2.89	3.20	3.45
60	1.27	1.60	1.84	2.06	2.35	2.57	2.79	3.08	3.30
70	1.26	1.59	1.81	2.03	2.31	2.51	2.72	2.99	3.20
80	1.26	1.57	1.79	2.00	2.27	2.47	2.67	2.93	3.12
90	1.25	1.56	1.78	1.98	2.24	2.44	2.63	2.88	3.07
100	1.25	1.56	1.77	1.97	2.22	2.41	2.60	2.84	3.02
120	1.24	1.55	1.75	1.94	2.19	2.37	2.55	2.78	2.96
140	1.24	1.54	1.74	1.93	2.17	2.35	2.52	2.74	2.91
160	1.24	1.53	1.73	1.92	2.15	2.33	2.49	2.71	2.88
180	1.23	1.53	1.72	1.91	2.14	2.31	2.48	2.69	2.85
200	1.23	1.52	1.72	1.90	2.13	2.30	2.46	2.67	2.83
300	1.23	1.51	1.70	1.88	2.10	2.26	2.42	2.62	2.77
500	1.22	1.50	1.69	1.86	2.07	2.23	2.38	2.58	2.72
∞	1.22	1.49	1.67	1.83	2.04	2.19	2.33	2.51	2.65

Table 22: Critical Values of the F Distribution

Numerator DF = 16

ν	$\alpha(2)$: 0.50 $\alpha(1)$: 0.25	0.20 0.10	0.10 0.05	0.05 0.025	0.02 0.01	0.01 0.005	0.005 0.0025	0.002 0.001	0.001 0.0005
1	9.52	61.3	246.0	987.0	6170.0	24700.0	98700.0	617000.0	2470000.0
2	3.41	9.43	19.4	39.4	99.4	199.0	399.0	999.0	2000.0
3	2.46	5.20	8.69	14.2	26.8	43.0	68.7	127.0	202.0
4	2.08	3.86	5.84	8.63	14.2	20.4	29.2	46.6	66.2
5	1.88	3.23	4.60	6.40	9.68	13.1	17.6	25.8	34.3
6	1.76	2.86	3.92	5.24	7.52	9.76	12.6	17.4	22.3
7	1.68	2.62	3.49	4.54	6.28	7.91	9.91	13.2	16.4
8	1.62	2.45	3.20	4.08	5.48	6.76	8.29	10.8	13.0
9	1.57	2.33	2.99	3.74	4.92	5.98	7.21	9.15	10.9
10	1.53	2.23	2.83	3.50	4.52	5.42	6.45	8.05	9.46
11	1.50	2.16	2.70	3.30	4.21	5.00	5.89	7.24	8.43
12	1.48	2.09	2.60	3.15	3.97	4.67	5.46	6.63	7.65
13	1.46	2.04	2.51	3.03	3.78	4.41	5.11	6.16	7.05
14	1.44	2.00	2.44	2.92	3.62	4.20	4.84	5.78	6.57
15	1.42	1.96	2.38	2.84	3.49	4.02	4.61	5.46	6.18
16	1.41	1.93	2.33	2.76	3.37	3.87	4.42	5.20	5.86
17	1.40	1.90	2.29	2.70	3.27	3.75	4.25	4.99	5.59
18	1.39	1.87	2.25	2.64	3.19	3.64	4.11	4.80	5.36
19	1.38	1.85	2.21	2.59	3.12	3.54	3.99	4.64	5.16
20	1.37	1.83	2.18	2.55	3.05	3.46	3.89	4.49	4.99
21	1.36	1.81	2.16	2.51	2.99	3.38	3.79	4.37	4.84
22	1.36	1.80	2.13	2.47	2.94	3.31	3.71	4.26	4.71
23	1.35	1.78	2.11	2.44	2.89	3.25	3.63	4.16	4.59
24	1.34	1.77	2.09	2.41	2.85	3.20	3.56	4.07	4.48
25	1.34	1.76	2.07	2.38	2.81	3.15	3.50	3.99	4.39
26	1.33	1.75	2.05	2.36	2.78	3.11	3.45	3.92	4.30
27	1.33	1.74	2.04	2.34	2.75	3.07	3.40	3.86	4.22
28	1.32	1.73	2.02	2.32	2.72	3.03	3.35	3.80	4.15
29	1.32	1.72	2.01	2.30	2.69	2.99	3.31	3.74	4.08
30	1.32	1.71	1.99	2.28	2.66	2.96	3.27	3.69	4.02
35	1.30	1.67	1.94	2.21	2.56	2.83	3.11	3.48	3.78
40	1.29	1.65	1.90	2.15	2.48	2.74	2.99	3.34	3.61
45	1.28	1.63	1.87	2.11	2.43	2.66	2.90	3.23	3.48
50	1.27	1.61	1.85	2.08	2.38	2.61	2.84	3.14	3.38
60	1.26	1.59	1.82	2.03	2.31	2.53	2.74	3.02	3.23
70	1.26	1.57	1.79	2.00	2.27	2.47	2.67	2.93	3.13
80	1.25	1.56	1.77	1.97	2.23	2.43	2.62	2.87	3.06
90	1.25	1.55	1.76	1.95	2.21	2.39	2.58	2.82	3.00
100	1.24	1.54	1.75	1.94	2.19	2.37	2.55	2.78	2.96
120	1.24	1.53	1.73	1.92	2.15	2.33	2.50	2.72	2.89
140	1.23	1.52	1.72	1.90	2.13	2.30	2.47	2.68	2.84
160	1.23	1.52	1.71	1.89	2.11	2.28	2.44	2.65	2.81
180	1.23	1.51	1.70	1.88	2.10	2.26	2.42	2.63	2.78
200	1.23	1.51	1.69	1.87	2.09	2.25	2.41	2.61	2.76
300	1.22	1.49	1.68	1.85	2.06	2.21	2.36	2.56	2.70
500	1.22	1.49	1.66	1.83	2.04	2.19	2.33	2.52	2.65
∞	1.21	1.47	1.64	1.80	2.00	2.14	2.28	2.45	2.58

Table 23: Critical Values of the F Distribution

Numerator DF = 17

ν	$\alpha(2)$: 0.50	0.20	0.10	0.05	0.02	0.01	0.005	0.002	0.001
	$\alpha(1)$: 0.25	0.10	0.05	0.025	0.01	0.005	0.0025	0.001	0.0005
1	9.53	61.5	247.0	989.0	6180.0	24700.0	98900.0	618000.0	2470000.0
2	3.42	9.43	19.4	39.4	99.4	199.0	399.0	999.0	2000.0
3	2.46	5.19	8.68	14.2	26.8	42.9	68.6	127.0	202.0
4	2.08	3.86	5.83	8.61	14.1	20.3	29.1	46.5	66.0
5	1.88	3.22	4.59	6.38	9.64	13.0	17.5	25.7	34.2
6	1.76	2.85	3.91	5.22	7.48	9.71	12.5	17.4	22.1
7	1.67	2.61	3.48	4.52	6.24	7.87	9.85	13.1	16.3
8	1.61	2.45	3.19	4.05	5.44	6.72	8.23	10.7	12.9
9	1.57	2.32	2.97	3.72	4.89	5.94	7.16	9.08	10.8
10	1.53	2.22	2.81	3.47	4.49	5.38	6.40	7.98	9.38
11	1.50	2.15	2.69	3.28	4.18	4.96	5.84	7.17	8.34
12	1.47	2.08	2.58	3.13	3.94	4.63	5.40	6.57	7.57
13	1.45	2.03	2.50	3.00	3.75	4.37	5.06	6.09	6.97
14	1.44	1.99	2.43	2.90	3.59	4.16	4.79	5.71	6.49
15	1.42	1.95	2.37	2.81	3.45	3.98	4.56	5.40	6.11
16	1.41	1.92	2.32	2.74	3.34	3.83	4.37	5.14	5.79
17	1.39	1.89	2.27	2.67	3.24	3.71	4.21	4.92	5.52
18	1.38	1.86	2.23	2.62	3.16	3.60	4.07	4.74	5.29
19	1.37	1.84	2.20	2.57	3.08	3.50	3.94	4.58	5.09
20	1.37	1.82	2.17	2.52	3.02	3.42	3.84	4.44	4.92
21	1.36	1.80	2.14	2.48	2.96	3.34	3.74	4.31	4.77
22	1.35	1.79	2.11	2.45	2.91	3.27	3.66	4.20	4.64
23	1.35	1.77	2.09	2.42	2.86	3.21	3.58	4.10	4.52
24	1.34	1.76	2.07	2.39	2.82	3.16	3.52	4.02	4.41
25	1.33	1.75	2.05	2.36	2.78	3.11	3.46	3.94	4.32
26	1.33	1.73	2.03	2.34	2.75	3.07	3.40	3.86	4.23
27	1.33	1.72	2.02	2.31	2.71	3.03	3.35	3.80	4.15
28	1.32	1.71	2.00	2.29	2.68	2.99	3.30	3.74	4.08
29	1.32	1.71	1.99	2.27	2.66	2.95	3.26	3.68	4.02
30	1.31	1.70	1.98	2.26	2.63	2.92	3.22	3.63	3.96
35	1.30	1.66	1.92	2.18	2.53	2.79	3.06	3.43	3.72
40	1.29	1.64	1.89	2.13	2.45	2.70	2.95	3.28	3.54
45	1.28	1.62	1.86	2.09	2.39	2.62	2.86	3.17	3.41
50	1.27	1.60	1.83	2.06	2.35	2.57	2.79	3.09	3.31
60	1.26	1.58	1.80	2.01	2.28	2.49	2.69	2.96	3.17
70	1.25	1.56	1.77	1.97	2.23	2.43	2.62	2.88	3.07
80	1.25	1.55	1.75	1.95	2.20	2.39	2.57	2.81	3.00
90	1.24	1.54	1.74	1.93	2.17	2.35	2.53	2.76	2.94
100	1.24	1.53	1.73	1.91	2.15	2.33	2.50	2.73	2.89
120	1.23	1.52	1.71	1.89	2.12	2.29	2.45	2.67	2.83
140	1.23	1.51	1.70	1.87	2.10	2.26	2.42	2.63	2.78
160	1.23	1.50	1.69	1.86	2.08	2.24	2.40	2.60	2.75
180	1.22	1.50	1.68	1.85	2.07	2.22	2.38	2.58	2.72
200	1.22	1.49	1.67	1.84	2.06	2.21	2.36	2.56	2.70
300	1.22	1.48	1.66	1.82	2.03	2.17	2.32	2.50	2.64
500	1.21	1.47	1.64	1.80	2.00	2.14	2.28	2.46	2.59
∞	1.21	1.46	1.62	1.78	1.97	2.10	2.23	2.40	2.52

Table 24: Critical Values of the F Distribution

Numerator DF = 18

ν	$\alpha(2)$: 0.50 $\alpha(1)$: 0.25	0.20 0.10	0.10 0.05	0.05 0.025	0.02 0.01	0.01 0.005	0.005 0.0025	0.002 0.001	0.001 0.0005
1	9.55	61.6	247.0	990.0	6190.0	24800.0	99100.0	619000.0	2480000.0
2	3.42	9.44	19.4	39.4	99.4	199.0	399.0	999.0	2000.0
3	2.46	5.19	8.67	14.2	26.8	42.9	68.5	127.	202.0
4	2.08	3.85	5.82	8.59	14.1	20.3	29.0	46.3	65.8
5	1.88	3.22	4.58	6.36	9.61	13.0	17.4	25.6	34.0
6	1.76	2.85	3.90	5.20	7.45	9.66	12.4	17.3	22.0
7	1.67	2.61	3.47	4.50	6.21	7.83	9.79	13.1	16.2
8	1.61	2.44	3.17	4.03	5.41	6.68	8.18	10.6	12.8
9	1.56	2.31	2.96	3.70	4.86	5.90	7.11	9.01	10.7
10	1.53	2.22	2.80	3.45	4.46	5.34	6.35	7.91	9.30
11	1.50	2.14	2.67	3.26	4.15	4.92	5.79	7.11	8.27
12	1.47	2.08	2.57	3.11	3.91	4.59	5.36	6.51	7.50
13	1.45	2.02	2.48	2.98	3.72	4.33	5.02	6.03	6.90
14	1.43	1.98	2.41	2.88	3.56	4.12	4.74	5.66	6.43
15	1.42	1.94	2.35	2.79	3.42	3.95	4.51	5.35	6.04
16	1.40	1.91	2.30	2.72	3.31	3.80	4.32	5.09	5.72
17	1.39	1.88	2.26	2.65	3.21	3.67	4.16	4.87	5.45
18	1.38	1.85	2.22	2.60	3.13	3.56	4.02	4.68	5.23
19	1.37	1.83	2.18	2.55	3.05	3.46	3.90	4.52	5.03
20	1.36	1.81	2.15	2.50	2.99	3.38	3.79	4.38	4.86
21	1.36	1.79	2.12	2.46	2.93	3.31	3.70	4.26	4.71
22	1.35	1.78	2.10	2.43	2.88	3.24	3.62	4.15	4.58
23	1.34	1.76	2.08	2.39	2.83	3.18	3.54	4.05	4.46
24	1.34	1.75	2.05	2.36	2.79	3.12	3.47	3.96	4.35
25	1.33	1.74	2.04	2.34	2.75	3.08	3.41	3.88	4.26
26	1.33	1.72	2.02	2.31	2.72	3.03	3.36	3.81	4.17
27	1.32	1.71	2.00	2.29	2.68	2.99	3.31	3.75	4.10
28	1.32	1.70	1.99	2.27	2.65	2.95	3.26	3.69	4.02
29	1.31	1.69	1.97	2.25	2.63	2.92	3.22	3.63	3.96
30	1.31	1.69	1.96	2.23	2.60	2.89	3.18	3.58	3.90
35	1.29	1.65	1.91	2.16	2.50	2.76	3.02	3.38	3.66
40	1.28	1.62	1.87	2.11	2.42	2.66	2.90	3.23	3.49
45	1.27	1.60	1.84	2.07	2.36	2.59	2.82	3.12	3.36
50	1.27	1.59	1.81	2.03	2.32	2.53	2.75	3.04	3.26
60	1.26	1.56	1.78	1.98	2.25	2.45	2.65	2.91	3.11
70	1.25	1.55	1.75	1.95	2.20	2.39	2.58	2.83	3.01
80	1.24	1.53	1.73	1.92	2.17	2.35	2.53	2.76	2.94
90	1.24	1.52	1.72	1.91	2.14	2.32	2.49	2.71	2.88
100	1.23	1.52	1.71	1.89	2.12	2.29	2.46	2.68	2.84
120	1.23	1.50	1.69	1.87	2.09	2.25	2.41	2.62	2.78
140	1.22	1.50	1.68	1.85	2.07	2.22	2.38	2.58	2.73
160	1.22	1.49	1.67	1.84	2.05	2.20	2.35	2.55	2.70
180	1.22	1.48	1.66	1.83	2.04	2.19	2.34	2.53	2.67
200	1.22	1.48	1.66	1.82	2.03	2.18	2.32	2.51	2.65
300	1.21	1.47	1.64	1.80	1.99	2.14	2.28	2.46	2.59
500	1.21	1.46	1.62	1.78	1.97	2.11	2.24	2.41	2.54
∞	1.20	1.44	1.60	1.75	1.93	2.06	2.19	2.35	2.47

Table 25: Critical Values of the F Distribution

Numerator DF = 19

ν	$\alpha(2)$: 0.50 $\alpha(1)$: 0.25	0.20 0.10	0.10 0.05	0.05 0.025	0.02 0.01	0.01 0.005	0.005 0.0025	0.002 0.001	0.001 0.0005
1	9.57	61.7	248.0	992.0	6200.0	24800.0	99200.0	620000.0	2480000.0
2	3.42	9.44	19.4	39.4	99.4	199.0	399.0	999.0	2000.0
3	2.46	5.19	8.67	14.2	26.7	42.8	68.4	127.0	201.0
4	2.08	3.85	5.81	8.58	14.0	20.2	28.9	46.2	65.7
5	1.88	3.21	4.57	6.34	9.58	12.9	17.4	25.5	33.9
6	1.76	2.84	3.88	5.18	7.42	9.62	12.4	17.2	21.9
7	1.67	2.60	3.46	4.48	6.18	7.79	9.74	13.0	16.1
8	1.61	2.43	3.16	4.02	5.38	6.64	8.13	10.5	12.8
9	1.56	2.30	2.95	3.68	4.83	5.86	7.06	8.95	10.7
10	1.53	2.21	2.79	3.44	4.43	5.31	6.31	7.86	9.23
11	1.49	2.13	2.66	3.24	4.12	4.89	5.75	7.06	8.20
12	1.47	2.07	2.56	3.09	3.88	4.56	5.32	6.45	7.43
13	1.45	2.01	2.47	2.96	3.69	4.30	4.98	5.98	6.84
14	1.43	1.97	2.40	2.86	3.53	4.09	4.70	5.60	6.37
15	1.41	1.93	2.34	2.77	3.40	3.91	4.47	5.29	5.98
16	1.40	1.90	2.29	2.70	3.28	3.76	4.28	5.04	5.66
17	1.39	1.87	2.24	2.63	3.19	3.64	4.12	4.82	5.40
18	1.38	1.84	2.20	2.58	3.10	3.53	3.98	4.63	5.17
19	1.37	1.82	2.17	2.53	3.03	3.43	3.86	4.47	4.97
20	1.36	1.80	2.14	2.48	2.96	3.35	3.76	4.33	4.80
21	1.35	1.78	2.11	2.44	2.90	3.27	3.66	4.21	4.65
22	1.35	1.77	2.08	2.41	2.85	3.21	3.58	4.10	4.52
23	1.34	1.75	2.06	2.37	2.80	3.15	3.50	4.00	4.41
24	1.33	1.74	2.04	2.35	2.76	3.09	3.44	3.92	4.30
25	1.33	1.73	2.02	2.32	2.72	3.04	3.38	3.84	4.21
26	1.32	1.71	2.00	2.29	2.69	3.00	3.32	3.77	4.12
27	1.32	1.70	1.99	2.27	2.66	2.96	3.27	3.70	4.04
28	1.31	1.69	1.97	2.25	2.63	2.92	3.22	3.64	3.97
29	1.31	1.68	1.96	2.23	2.60	2.88	3.18	3.59	3.91
30	1.31	1.68	1.95	2.21	2.57	2.85	3.14	3.53	3.85
35	1.29	1.64	1.89	2.14	2.47	2.72	2.98	3.33	3.61
40	1.28	1.61	1.85	2.09	2.39	2.63	2.87	3.19	3.44
45	1.27	1.59	1.82	2.04	2.34	2.56	2.78	3.08	3.31
50	1.26	1.58	1.80	2.01	2.29	2.50	2.71	2.99	3.21
60	1.25	1.55	1.76	1.96	2.22	2.42	2.61	2.87	3.06
70	1.24	1.54	1.74	1.93	2.18	2.36	2.54	2.78	2.96
80	1.24	1.52	1.72	1.90	2.14	2.32	2.49	2.72	2.89
90	1.23	1.51	1.70	1.88	2.11	2.28	2.45	2.67	2.83
100	1.23	1.50	1.69	1.87	2.09	2.26	2.42	2.63	2.79
120	1.22	1.49	1.67	1.84	2.06	2.22	2.37	2.58	2.73
140	1.22	1.48	1.66	1.83	2.04	2.19	2.34	2.54	2.68
160	1.22	1.48	1.65	1.82	2.02	2.17	2.32	2.51	2.65
180	1.21	1.47	1.64	1.81	2.01	2.15	2.30	2.48	2.62
200	1.21	1.47	1.64	1.80	2.00	2.14	2.28	2.46	2.60
300	1.21	1.46	1.62	1.77	1.97	2.10	2.24	2.41	2.54
500	1.20	1.45	1.61	1.76	1.94	2.07	2.02	2.37	2.49
∞	1.20	1.43	1.59	1.73	1.90	2.03	2.15	2.31	2.42

Table 26: Critical Values of the F Distribution

Numerator DF = 20

ν	$\alpha(2)$: 0.50 $\alpha(1)$: 0.25	0.20 0.10	0.10 0.05	0.05 0.025	0.02 0.01	0.01 0.005	0.005 0.0025	0.002 0.001	0.001 0.0005
1	9.58	61.7	248.0	993.0	6210.0	24800.0	99300.0	621000.0	2480000.0
2	3.43	9.44	19.4	39.4	99.4	199.0	399.0	999.0	2000.0
3	2.46	5.18	8.66	14.2	26.7	42.8	68.3	126.0	201.0
4	2.08	3.84	5.80	8.56	14.0	20.2	28.9	46.1	65.5
5	1.88	3.21	4.56	6.33	9.55	12.9	17.3	25.4	33.8
6	1.76	2.84	3.87	5.17	7.40	9.59	12.3	17.1	21.8
7	1.67	2.59	3.44	4.47	6.16	7.75	9.70	12.9	16.0
8	1.61	2.42	3.15	4.00	5.36	6.61	8.09	10.5	12.7
9	1.56	2.30	2.94	3.67	4.81	5.83	7.02	8.90	10.6
10	1.52	2.20	2.77	3.42	4.41	5.27	6.27	7.80	9.17
11	1.49	2.12	2.65	3.23	4.10	4.86	5.71	7.01	8.14
12	1.47	2.06	2.54	3.07	3.86	4.53	5.28	6.40	7.37
13	1.45	2.01	2.46	2.95	3.66	4.27	4.94	5.93	6.78
14	1.43	1.96	2.39	2.84	3.51	4.06	4.66	5.56	6.31
15	1.41	1.92	2.33	2.76	3.37	3.88	4.44	5.25	5.93
16	1.40	1.89	2.28	2.68	3.26	3.73	4.25	4.99	5.61
17	1.39	1.86	2.23	2.62	3.16	3.61	4.09	4.78	5.34
18	1.38	1.84	2.19	2.56	3.08	3.50	3.95	4.59	5.12
19	1.37	1.81	2.16	2.51	3.00	3.40	3.83	4.43	4.92
20	1.36	1.79	2.12	2.46	2.94	3.32	3.72	2.29	4.75
21	1.35	1.78	2.10	2.42	2.88	3.24	3.63	4.17	4.60
22	1.34	1.76	2.07	2.39	2.83	3.18	3.54	4.06	4.47
23	1.34	1.74	2.05	2.36	2.78	3.12	3.47	3.96	4.36
24	1.33	1.73	2.03	2.33	2.74	3.06	3.40	3.87	4.25
25	1.33	1.72	2.01	2.30	2.70	3.01	3.34	3.79	4.16
26	1.32	1.71	1.99	2.28	2.66	2.97	3.28	3.72	4.07
27	1.32	1.70	1.97	2.25	2.63	2.93	3.23	3.66	3.99
28	1.31	1.69	1.96	2.23	2.60	2.89	3.19	3.60	3.92
29	1.31	1.68	1.94	2.21	2.57	2.86	3.14	3.54	3.86
30	1.30	1.67	1.93	2.20	2.55	2.82	3.11	3.49	3.80
35	1.29	1.63	1.88	2.12	2.44	2.69	2.95	3.29	3.56
40	1.28	1.61	1.84	2.07	2.37	2.60	2.83	3.14	3.39
45	1.27	1.58	1.81	2.03	2.31	2.53	2.74	3.04	3.26
50	1.26	1.57	1.78	1.99	2.27	2.47	2.68	2.95	3.16
60	1.25	1.54	1.75	1.94	2.20	2.39	2.58	2.83	3.02
70	1.24	1.53	1.72	1.91	2.15	2.33	2.51	2.74	2.92
80	1.23	1.51	1.70	1.88	2.12	2.29	2.46	2.68	2.85
90	1.23	1.50	1.69	1.86	2.09	2.25	2.42	2.63	2.79
100	1.23	1.49	1.68	1.85	2.07	2.23	2.38	2.59	2.75
120	1.22	1.48	1.66	1.82	2.03	2.19	2.34	2.53	2.68
140	1.22	1.47	1.65	1.81	2.01	2.16	2.31	2.49	2.64
160	1.21	1.47	1.64	1.80	1.99	2.14	2.28	2.47	2.60
180	1.21	1.46	1.63	1.79	1.98	2.12	2.26	2.44	2.58
200	1.21	1.46	1.62	1.78	1.97	2.11	2.25	2.42	2.56
300	1.20	1.45	1.61	1.75	1.94	2.07	2.20	2.37	2.49
500	1.20	1.44	1.59	1.74	1.92	2.04	2.17	2.33	2.45
∞	1.19	1.42	1.57	1.71	1.88	2.00	2.12	2.27	2.37

Table 27: Critical Values of the F Distribution

Numerator DF = 22

ν	$\alpha(2)$: 0.50 $\alpha(1)$: 0.25	0.20 0.10	0.10 0.05	0.05 0.025	0.02 0.01	0.01 0.005	0.005 0.0025	0.002 0.001	0.001 0.0005
1	9.61	61.9	249.0	995.0	6220.0	24900.0	99600.0	622000.0	2490000.0
2	3.43	9.45	19.5	39.5	99.5	199.0	399.0	999.0	2000.0
3	2.46	5.18	8.65	14.1	26.6	42.7	68.2	126.0	201.0
4	2.08	3.84	5.79	8.53	14.0	20.1	28.7	45.9	65.3
5	1.88	3.20	4.54	6.30	9.51	12.8	17.2	25.3	33.6
6	1.76	2.83	3.86	5.14	7.35	9.53	12.3	17.0	21.7
7	1.67	2.58	3.43	4.44	6.11	7.69	9.62	12.8	15.9
8	1.61	2.41	3.13	3.97	5.32	6.55	8.02	10.4	12.6
9	1.56	2.29	2.92	3.64	4.77	5.78	6.95	8.80	10.5
10	1.52	2.19	2.75	3.39	4.36	5.22	6.20	7.71	9.06
11	1.49	2.11	2.63	3.20	4.06	4.80	5.64	6.92	8.04
12	1.46	2.05	2.52	3.04	3.82	4.48	5.21	6.32	7.27
13	1.44	1.99	2.44	2.92	3.62	4.22	4.87	5.85	6.68
14	1.42	1.95	2.37	2.81	3.46	4.01	4.60	5.48	6.21
15	1.41	1.91	2.31	2.73	3.33	3.83	4.37	5.17	5.83
16	1.39	1.88	2.25	2.65	3.22	3.68	4.18	4.91	5.52
17	1.38	1.85	2.21	2.59	3.12	3.56	4.02	4.70	5.25
18	1.37	1.82	2.17	2.53	3.03	3.45	3.88	4.51	5.03
19	1.36	1.80	2.13	2.48	2.96	3.35	3.76	4.35	4.83
20	1.35	1.78	2.10	2.43	2.90	3.27	3.66	4.21	4.67
21	1.35	1.76	2.07	2.39	2.84	3.19	3.56	4.09	4.52
22	1.34	1.74	2.05	2.36	2.78	3.12	3.48	3.98	4.39
23	1.33	1.73	2.02	2.33	2.74	3.06	3.41	3.89	4.27
24	1.33	1.71	2.00	2.30	2.70	3.01	3.34	3.80	4.17
25	1.32	1.70	1.98	2.27	2.66	2.96	3.28	3.72	4.07
26	1.32	1.69	1.97	2.24	2.62	2.92	3.22	3.65	3.99
27	1.31	1.68	1.95	2.22	2.59	2.88	3.17	3.58	3.91
28	1.31	1.67	1.93	2.20	2.56	2.84	3.13	3.52	3.84
29	1.30	1.66	1.92	2.18	2.53	2.80	3.08	3.47	3.77
30	1.30	1.65	1.91	2.16	2.51	2.77	3.04	3.42	3.71
35	1.28	1.62	1.85	2.09	2.40	2.64	2.89	3.22	3.48
40	1.27	1.59	1.81	2.03	2.33	2.55	2.77	3.07	3.31
45	1.26	1.57	1.78	1.99	2.27	2.47	2.68	2.96	3.18
50	1.25	1.55	1.76	1.96	2.22	2.42	2.62	2.88	3.08
60	1.24	1.53	1.72	1.91	2.15	2.33	2.52	2.75	2.94
70	1.23	1.51	1.70	1.88	2.11	2.28	2.45	2.67	2.84
80	1.23	1.49	1.68	1.85	2.07	2.23	2.39	2.61	2.77
90	1.22	1.48	1.66	1.83	2.04	2.20	2.35	2.56	2.71
100	1.22	1.48	1.65	1.81	2.02	2.17	2.32	2.52	2.67
120	1.21	1.46	1.63	1.79	1.99	2.13	2.28	2.46	2.60
140	1.21	1.45	1.62	1.77	1.97	2.11	2.24	2.42	2.56
160	1.21	1.45	1.61	1.76	1.95	2.09	2.22	2.39	2.52
180	1.20	1.44	1.60	1.75	1.94	2.07	2.20	2.37	2.50
200	1.20	1.44	1.60	1.74	1.93	2.06	2.19	2.35	2.48
300	1.20	1.43	1.58	1.72	1.89	2.02	2.14	2.30	2.41
500	1.19	1.42	1.56	1.70	1.87	1.99	2.11	2.26	2.37
∞	1.18	1.40	1.54	1.67	1.83	1.95	2.05	2.19	2.30

Table 28: Critical Values of the F Distribution

Numerator DF = 24

ν	$\alpha(2)$: 0.50 $\alpha(1)$: 0.25	0.20 0.10	0.10 0.05	0.05 0.025	0.02 0.01	0.01 0.005	0.005 0.0025	0.002 0.001	0.001 0.0005
1	9.63	62.0	249.0	997.0	6230.0	24900.0	99800.0	623000.0	2490000.0
2	3.43	9.45	19.5	39.5	99.5	199.0	399.0	999.0	2000.0
3	2.46	5.18	8.64	14.1	26.6	42.6	68.1	126.0	200.0
4	2.08	3.83	5.77	8.51	13.9	20.0	28.7	45.8	65.0
5	1.88	3.19	4.53	6.28	9.47	12.8	17.1	25.1	33.4
6	1.75	2.82	3.84	5.12	7.31	9.47	12.2	16.9	21.5
7	1.67	2.58	3.41	4.41	6.07	7.64	9.56	12.7	15.8
8	1.60	2.40	3.12	3.95	5.28	6.50	7.95	10.3	12.5
9	1.56	2.28	2.90	3.61	4.73	5.73	6.89	8.72	10.4
10	1.52	2.18	2.74	3.37	4.33	5.17	6.14	7.64	8.96
11	1.49	2.10	2.61	3.17	4.02	4.76	5.58	6.85	7.95
12	1.46	2.04	2.51	3.02	3.78	4.43	5.16	6.25	7.19
13	1.44	1.98	2.42	2.89	3.59	4.17	4.82	5.78	6.60
14	1.42	1.94	2.35	2.79	3.43	3.96	4.55	5.41	6.13
15	1.41	1.90	2.29	2.70	3.29	3.79	4.32	5.10	5.75
16	1.39	1.87	2.24	2.63	3.18	3.64	4.13	4.85	5.44
17	1.38	1.84	2.19	2.56	3.08	3.51	3.97	4.63	5.18
18	1.37	1.81	2.15	2.50	3.00	3.40	3.83	4.45	4.95
19	1.36	1.79	2.11	2.45	2.92	3.31	3.71	4.29	4.76
20	1.35	1.77	2.08	2.41	2.86	3.22	3.61	4.15	4.59
21	1.34	1.75	2.05	2.37	2.80	3.15	3.51	4.03	4.44
22	1.33	1.73	2.03	2.33	2.75	3.08	3.43	3.92	4.31
23	1.33	1.72	2.01	2.30	2.70	3.02	3.35	3.82	4.20
24	1.32	1.70	1.98	2.27	2.66	2.97	3.29	3.74	4.09
25	1.32	1.69	1.96	2.24	2.62	2.92	3.23	3.66	4.00
26	1.31	1.68	1.95	2.22	2.58	2.87	3.17	3.59	3.92
27	1.31	1.67	1.93	2.19	2.55	2.83	3.12	3.52	3.84
28	1.30	1.66	1.91	2.17	2.52	2.79	3.07	3.46	3.77
29	1.30	1.65	1.90	2.15	2.49	2.76	3.03	3.41	3.70
30	1.29	1.64	1.89	2.14	2.47	2.73	2.99	3.36	3.64
35	1.28	1.60	1.83	2.06	2.36	2.60	2.83	3.16	3.41
40	1.26	1.57	1.79	2.01	2.29	2.50	2.72	3.01	3.24
45	1.26	1.55	1.76	1.96	2.23	2.43	2.63	2.90	3.11
50	1.25	1.54	1.74	1.93	2.18	2.37	2.56	2.82	3.01
60	1.24	1.51	1.70	1.88	2.12	2.29	2.46	2.69	2.87
70	1.23	1.49	1.67	1.85	2.07	2.23	2.39	2.61	2.77
80	1.22	1.48	1.65	1.82	2.03	2.19	2.34	2.54	2.70
90	1.22	1.47	1.64	1.80	2.00	2.15	2.30	2.50	2.64
100	1.21	1.46	1.63	1.78	1.98	2.13	2.27	2.46	2.60
120	1.21	1.45	1.61	1.76	1.95	2.09	2.23	2.40	2.53
140	1.20	1.44	1.60	1.74	1.93	2.06	2.19	2.36	2.49
160	1.20	1.43	1.59	1.73	1.91	2.04	2.17	2.33	2.45
180	1.20	1.43	1.58	1.72	1.90	2.02	2.15	2.31	2.43
200	1.19	1.42	1.57	1.71	1.89	2.01	2.13	2.29	2.41
300	1.19	1.41	1.55	1.69	1.85	1.97	2.09	2.24	2.35
500	1.18	1.40	1.54	1.67	1.83	1.94	2.05	2.20	2.30
∞	1.18	1.38	1.52	1.64	1.79	1.90	2.00	2.13	2.23

Table 29: Critical Values of the F Distribution

Numerator DF = 26

ν	$\alpha(2)$: 0.50 $\alpha(1)$: 0.25	0.20 0.10	0.10 0.05	0.05 0.025	0.02 0.01	0.01 0.005	0.005 0.0025	0.002 0.001	0.001 0.0005
1	9.64	62.1	249.0	999.0	6240.0	25000.0	99900.0	624000.0	2500000.0
2	3.44	9.45	19.5	39.5	99.5	199.0	399.0	999.0	2000.0
3	2.46	5.17	8.63	14.1	26.6	42.6	68.0	126.0	200.0
4	2.08	3.83	5.76	8.49	13.9	20.0	28.6	45.6	64.9
5	1.88	3.18	4.52	6.26	9.43	12.7	17.1	25.0	33.3
6	1.75	2.81	3.83	5.10	7.28	9.43	12.1	16.8	21.4
7	1.67	2.57	3.40	4.39	6.04	7.60	9.50	12.7	15.7
8	1.60	2.40	3.10	3.93	5.25	6.46	7.90	10.2	12.4
9	1.55	2.27	2.89	3.59	4.70	5.69	6.84	8.66	10.3
10	1.52	2.17	2.72	3.34	4.30	5.13	6.09	7.57	8.89
11	1.48	2.09	2.59	3.15	3.99	4.72	5.54	6.78	7.87
12	1.46	2.03	2.49	3.00	3.75	4.39	5.11	6.19	7.12
13	1.44	1.97	2.41	2.87	3.56	4.13	4.77	5.72	6.53
14	1.42	1.93	2.33	2.77	3.40	3.92	4.50	5.35	6.07
15	1.40	1.89	2.27	2.68	3.26	3.75	4.27	5.04	5.69
16	1.39	1.86	2.22	2.60	3.15	3.60	4.09	4.79	5.37
17	1.38	1.83	2.17	2.54	3.05	3.47	3.92	4.57	5.11
18	1.36	1.80	2.13	2.48	2.97	3.36	3.79	4.39	4.89
19	1.35	1.78	2.10	2.43	2.89	3.27	3.67	4.23	4.70
20	1.35	1.76	2.07	2.89	2.83	3.18	3.56	4.09	4.53
21	1.34	1.74	2.04	2.34	2.77	3.11	3.47	3.97	4.38
22	1.33	1.72	2.01	2.31	2.72	3.04	3.38	3.86	4.25
23	1.32	1.70	1.99	2.28	2.67	2.98	3.31	3.77	4.14
24	1.32	1.69	1.97	2.25	2.63	2.93	3.24	3.68	4.03
25	1.31	1.68	1.95	2.22	2.59	2.88	3.18	3.60	3.94
26	1.31	1.67	1.93	2.19	2.55	2.84	3.13	3.53	3.85
27	1.30	1.65	1.91	2.17	2.52	2.79	3.08	3.47	3.78
28	1.30	1.64	1.90	2.15	2.49	2.76	3.03	3.41	3.71
29	1.29	1.63	1.88	2.13	2.46	2.72	2.99	3.35	3.64
30	1.29	1.63	1.87	2.11	2.44	2.69	2.95	3.30	3.58
35	1.27	1.59	1.82	2.04	2.33	2.56	2.79	3.10	3.35
40	1.26	1.56	1.77	1.98	2.26	2.46	2.67	2.96	3.18
45	1.25	1.54	1.74	1.94	2.20	2.39	2.59	2.85	3.05
50	1.24	1.52	1.72	1.91	2.15	2.33	2.52	2.76	2.95
60	1.23	1.50	1.68	1.86	2.08	2.25	2.42	2.64	2.81
70	1.22	1.48	1.65	1.82	2.03	2.19	2.35	2.56	2.71
80	1.22	1.47	1.63	1.79	2.00	2.15	2.30	2.49	2.64
90	1.21	1.45	1.62	1.77	1.97	2.12	2.26	2.44	2.58
100	1.21	1.45	1.61	1.76	1.95	2.09	2.23	2.41	2.54
120	1.20	1.43	1.59	1.73	1.92	2.05	2.18	2.35	2.48
140	1.20	1.42	1.57	1.72	1.89	2.02	2.15	2.31	2.43
160	1.19	1.42	1.57	1.70	1.88	2.00	2.12	2.28	2.40
180	1.19	1.41	1.56	1.69	1.86	1.98	2.10	2.26	2.37
200	1.19	1.41	1.55	1.68	1.85	1.97	2.09	2.24	2.35
300	1.18	1.39	1.53	1.66	1.82	1.93	2.04	2.18	2.29
500	1.18	1.38	1.52	1.64	1.79	1.90	2.01	2.14	2.24
∞	1.17	1.37	1.50	1.61	1.76	1.86	1.95	2.08	2.17

Table 30: Critical Values of the F Distribution

Numerator DF = 28

ν	$\alpha(2)$: 0.50	0.20	0.10	0.05	0.02	0.01	0.005	0.002	0.001
	$\alpha(1)$: 0.25	0.10	0.05	0.025	0.01	0.005	0.0025	0.001	0.0005
1	9.66	62.2	250.0	1000.0	6250.0	25000.0	100000.0	625000.0	2500000.0
2	3.44	9.46	19.5	39.5	99.5	199.0	399.0	999.0	2000.0
3	2.46	5.17	8.62	14.1	26.5	42.5	67.9	126.0	200.0
4	2.08	3.82	5.75	8.48	13.9	19.9	28.5	45.5	64.7
5	1.88	3.18	4.50	6.24	9.40	12.7	17.0	24.9	33.2
6	1.75	2.81	3.82	5.08	7.25	9.30	12.1	16.7	21.3
7	1.66	2.56	3.39	4.38	6.02	7.57	9.45	12.6	15.6
8	1.60	2.39	3.09	3.91	5.22	6.43	7.86	10.2	12.3
9	1.55	2.26	2.87	3.58	4.67	5.65	6.80	8.60	10.2
10	1.51	2.16	2.71	3.33	4.27	5.10	6.05	7.52	8.82
11	1.48	2.08	2.58	3.13	3.96	4.68	5.49	6.73	7.81
12	1.46	2.02	2.48	2.98	3.72	4.36	5.07	6.14	7.05
13	1.43	1.96	2.39	2.85	3.53	4.10	4.73	5.67	6.47
14	1.42	1.92	2.32	2.75	3.37	3.89	4.46	5.30	6.01
15	1.40	1.88	2.26	2.66	3.24	3.72	4.23	4.99	5.63
16	1.39	1.85	2.21	2.58	3.12	3.57	4.05	4.74	5.32
17	1.37	1.82	2.16	2.52	3.03	3.44	3.89	4.53	5.05
18	1.36	1.79	2.12	2.46	2.94	3.33	3.75	4.34	4.83
19	1.35	1.77	2.08	2.41	2.87	3.24	3.63	4.18	4.64
20	1.34	1.75	2.05	2.37	2.80	3.15	3.52	4.05	4.47
21	1.33	1.73	2.02	2.33	2.74	3.08	3.43	3.93	4.33
22	1.33	1.71	2.00	2.29	2.69	3.01	3.35	3.82	4.20
23	1.32	1.69	1.97	2.26	2.64	2.95	3.27	3.72	4.08
24	1.31	1.68	1.95	2.23	2.60	2.90	3.20	3.63	3.98
25	1.31	1.67	1.93	2.20	2.56	2.85	3.14	3.56	3.89
26	1.30	1.66	1.91	2.17	2.53	2.80	3.09	3.49	3.80
27	1.30	1.64	1.90	2.15	2.49	2.76	3.04	3.42	3.72
28	1.29	1.63	1.88	2.13	2.46	2.72	2.99	3.36	3.65
29	1.29	1.62	1.87	2.11	2.44	2.69	2.95	3.31	3.59
30	1.29	1.62	1.85	2.09	2.41	2.66	2.91	3.26	3.53
35	1.27	1.58	1.80	2.02	2.30	2.53	2.75	3.06	3.30
40	1.26	1.55	1.76	1.96	2.23	2.43	2.64	2.91	3.13
45	1.25	1.53	1.73	1.92	2.17	2.36	2.55	2.80	3.00
50	1.24	1.51	1.70	1.89	2.12	2.30	2.48	2.72	2.90
60	1.23	1.49	1.66	1.83	2.05	2.22	2.38	2.60	2.76
70	1.22	1.47	1.64	1.80	2.01	2.16	2.31	2.51	2.66
80	1.21	1.45	1.62	1.77	1.97	2.11	2.26	2.45	2.59
90	1.21	1.44	1.60	1.75	1.94	2.08	2.22	2.40	2.53
100	1.20	1.43	1.59	1.74	1.92	2.05	2.19	2.36	2.49
120	1.20	1.42	1.57	1.71	1.89	2.01	2.14	2.30	2.42
140	1.19	1.41	1.56	1.69	1.86	1.99	2.11	2.26	2.38
160	1.19	1.40	1.55	1.68	1.85	1.97	2.08	2.23	2.35
180	1.19	1.40	1.54	1.67	1.83	1.95	2.06	2.21	2.32
200	1.18	1.39	1.53	1.66	1.82	1.94	2.05	2.19	2.30
300	1.18	1.38	1.51	1.64	1.79	1.90	2.00	2.14	2.24
500	1.17	1.37	1.50	1.62	1.76	1.87	1.97	2.10	2.19
∞	1.17	1.35	1.48	1.59	1.72	1.82	1.91	2.03	2.12

Table 31: Critical Values of the F Distribution

Numerator DF = 30

	$\alpha(2)$: 0.50	0.20	0.10	0.05	0.02	0.01	0.005	0.002	0.001
ν	$\alpha(1)$: 0.25	0.10	0.05	0.025	0.01	0.005	0.0025	0.001	0.0005
1	9.67	62.3	250.0	1000.0	6260.0	25000.0	100000.0	626000.0	2500000.0
2	3.44	9.46	19.5	39.5	99.5	199.0	399.0	999.0	2000.0
3	2.47	5.17	8.62	14.1	26.5	42.5	67.8	125.0	200.0
4	2.08	3.82	5.75	8.46	13.8	19.9	28.5	45.4	64.6
5	1.88	3.17	4.50	6.23	9.38	12.7	17.0	24.9	33.1
6	1.75	2.80	3.81	5.07	7.23	9.36	12.0	16.7	21.2
7	1.66	2.56	3.38	4.36	5.99	7.53	9.41	12.5	15.5
8	1.60	2.38	3.08	3.89	5.20	6.40	7.82	10.1	12.2
9	1.55	2.25	2.86	3.56	4.65	5.62	6.76	8.55	10.2
10	1.51	2.16	2.70	3.31	4.25	5.07	6.01	7.47	8.76
11	1.48	2.08	2.57	3.12	3.94	4.65	5.46	6.68	7.75
12	1.45	2.01	2.47	2.96	3.70	4.33	5.03	6.09	7.00
13	1.43	1.96	2.38	2.84	3.51	4.07	4.70	5.63	6.42
14	1.41	1.91	2.31	2.73	3.35	3.86	4.42	5.25	5.95
15	1.40	1.87	2.25	2.64	3.21	3.69	4.20	4.95	5.58
16	1.38	1.84	2.19	2.57	3.10	3.54	4.01	4.70	5.27
17	1.37	1.81	2.15	2.50	3.00	3.41	3.85	4.48	5.01
18	1.36	1.78	2.11	2.44	2.92	3.30	3.71	4.30	4.78
19	1.35	1.76	2.07	2.39	2.84	3.21	3.59	4.14	4.59
20	1.34	1.74	2.04	2.35	2.78	3.12	3.49	4.00	4.42
21	1.33	1.72	2.01	2.31	2.72	3.05	3.40	3.88	4.28
22	1.32	1.70	1.98	2.27	2.67	2.98	3.31	3.78	4.15
23	1.32	1.69	1.96	2.24	2.62	2.92	3.24	3.68	4.03
24	1.31	1.67	1.94	2.21	2.58	2.87	3.17	3.59	3.93
25	1.31	1.66	1.92	2.18	2.54	2.82	3.11	3.52	3.84
26	1.30	1.65	1.90	2.16	2.50	2.77	3.06	3.44	3.75
27	1.30	1.64	1.88	2.13	2.47	2.73	3.00	3.38	3.68
28	1.29	1.63	1.87	2.11	2.44	2.69	2.96	3.32	3.61
29	1.29	1.62	1.85	2.09	2.41	2.66	2.92	3.27	3.54
30	1.28	1.61	1.84	2.07	2.39	2.63	2.88	3.22	3.49
35	1.27	1.57	1.79	2.00	2.28	2.50	2.72	3.02	3.25
40	1.25	1.54	1.74	1.94	2.20	2.40	2.60	2.87	3.08
45	1.24	1.52	1.71	1.90	2.14	2.33	2.51	2.76	2.96
50	1.23	1.50	1.69	1.87	2.10	2.27	2.45	2.68	2.86
60	1.22	1.48	1.65	1.82	2.03	2.29	2.35	2.55	2.71
70	1.21	1.46	1.62	1.78	1.98	2.13	2.28	2.47	2.62
80	1.21	1.44	1.60	1.75	1.94	2.08	2.22	2.41	2.54
90	1.20	1.43	1.59	1.73	1.92	2.05	2.18	2.36	2.49
100	1.20	1.42	1.57	1.71	1.89	2.02	2.15	2.32	2.44
120	1.19	1.41	1.55	1.69	1.86	1.98	2.11	2.26	2.38
140	1.19	1.40	1.54	1.67	1.84	1.96	2.07	2.22	2.33
160	1.18	1.39	1.53	1.66	1.82	1.93	2.05	2.19	2.30
180	1.18	1.39	1.52	1.65	1.81	1.92	2.03	2.17	2.27
200	1.18	1.38	1.52	1.64	1.79	1.91	2.01	2.15	2.25
300	1.17	1.37	1.50	1.62	1.76	1.87	1.97	2.10	2.19
500	1.17	1.36	1.48	1.60	1.74	1.84	1.93	2.05	2.14
∞	1.16	1.34	1.46	1.57	1.70	1.79	1.88	1.99	2.07

Table 32: Critical Values of the F Distribution

Numerator DF = 40

ν	$\alpha(2)$: 0.50 $\alpha(1)$: 0.25	0.20 0.10	0.10 0.05	0.05 0.025	0.02 0.01	0.01 0.005	0.005 0.0025	0.002 0.001	0.001 0.0005
1	9.71	62.5	251.0	1010.0	6290.0	25100.00	101000.0	629000.0	2510000.0
2	3.45	9.47	19.5	39.5	99.5	199.0	399.0	999.0	2000.0
3	2.47	5.16	8.59	14.0	26.4	42.3	67.5	125.0	199.0
4	2.08	3.80	5.72	8.41	13.7	19.8	28.2	45.1	64.1
5	1.88	3.16	4.46	6.18	9.29	12.5	16.8	24.6	32.7
6	1.75	2.78	3.77	5.01	7.14	9.24	11.9	16.4	21.0
7	1.66	2.54	3.34	4.31	5.91	7.42	9.26	12.3	15.2
8	1.59	2.36	3.04	3.84	5.12	6.29	7.68	9.92	12.0
9	1.54	2.23	2.83	3.51	4.57	5.52	6.62	8.37	9.94
10	1.51	2.13	2.66	3.26	4.17	4.97	5.88	7.30	8.55
11	1.47	2.05	2.53	3.06	3.86	4.55	5.33	6.52	7.55
12	1.45	1.99	2.43	2.91	3.62	4.23	4.91	5.93	6.81
13	1.42	1.93	2.34	2.78	3.43	3.97	4.57	5.47	6.23
14	1.41	1.89	2.27	2.67	3.27	3.76	4.30	5.10	5.77
15	1.39	1.85	2.20	2.59	3.13	3.58	4.08	4.80	5.40
16	1.37	1.81	2.15	2.51	3.02	3.44	3.89	4.54	5.09
17	1.36	1.78	2.10	2.44	2.92	3.31	3.73	4.33	4.83
18	1.35	1.75	2.06	2.38	2.84	3.20	3.59	4.15	4.61
19	1.34	1.73	2.03	2.33	2.76	3.11	3.47	3.99	4.42
20	1.33	1.71	1.99	2.29	2.69	3.02	3.37	3.86	4.25
21	1.32	1.69	1.96	2.25	2.64	2.95	3.27	3.74	4.11
22	1.31	1.67	1.94	2.21	2.58	2.88	3.19	3.63	3.98
23	1.31	1.66	1.91	2.18	2.54	2.82	3.12	3.53	3.87
24	1.30	1.64	1.89	2.15	2.49	2.77	3.05	3.45	3.76
25	1.29	1.63	1.87	2.12	2.45	2.72	2.99	3.37	3.67
26	1.29	1.61	1.85	2.09	2.42	2.67	2.93	3.30	3.59
27	1.28	1.60	1.84	2.07	2.38	2.63	2.88	3.23	3.51
28	1.28	1.59	1.82	2.05	2.35	2.59	2.84	3.18	3.44
29	1.27	1.58	1.81	2.03	2.33	2.56	2.79	3.12	3.38
30	1.27	1.57	1.79	2.01	2.30	2.52	2.76	3.07	3.32
35	1.25	1.53	1.74	1.93	2.19	2.39	2.60	2.87	3.09
40	1.24	1.51	1.69	1.88	2.11	2.30	2.48	2.73	2.92
45	1.23	1.48	1.66	1.83	2.05	2.22	2.39	2.62	2.79
50	1.22	1.46	1.63	1.80	2.01	2.16	2.32	2.53	2.69
60	1.21	1.44	1.59	1.74	1.94	2.08	2.22	2.41	2.55
70	1.20	1.42	1.57	1.71	1.89	2.02	2.15	2.32	2.45
80	1.19	1.40	1.54	1.68	1.85	1.97	2.10	2.26	2.38
90	1.19	1.39	1.53	1.66	1.82	1.94	2.06	2.21	2.32
100	1.18	1.38	1.52	1.64	1.80	1.91	2.02	2.17	2.28
120	1.18	1.37	1.50	1.61	1.76	1.87	1.98	2.11	2.21
140	1.17	1.36	1.48	1.60	1.74	1.84	1.94	2.07	2.17
160	1.17	1.35	1.47	1.58	1.72	1.82	1.92	2.04	2.13
180	1.16	1.34	1.46	1.57	1.71	1.80	1.90	2.02	2.11
200	1.16	1.34	1.46	1.56	1.69	1.79	1.88	2.00	2.09
300	1.15	1.32	1.43	1.54	1.66	1.75	1.84	1.94	2.02
500	1.15	1.31	1.42	1.52	1.63	1.72	1.80	1.90	1.98
∞	1.14	1.30	1.39	1.48	1.59	1.67	1.74	1.84	1.90

Table 33: Critical Values of the F Distribution

Numerator DF = 50

ν	$\alpha(2)$: 0.50 $\alpha(1)$: 0.25	0.20 0.10	0.10 0.05	0.05 0.025	0.02 0.01	0.01 0.005	0.005 0.0025	0.002 0.001	0.001 0.0005
1	9.74	62.7	252.0	1010.0	6300.0	25200.00	101000.0	630000.0	2520000.0
2	3.46	9.47	19.5	39.5	99.5	199.0	399.0	999.0	2000.0
3	2.47	5.15	8.58	14.0	26.4	42.2	67.4	125.0	198.0
4	2.08	3.80	5.70	8.38	13.7	19.7	28.1	44.9	63.8
5	1.88	3.15	4.44	6.14	9.24	12.5	16.7	24.4	32.5
6	1.75	2.77	3.75	4.98	7.09	9.17	11.8	16.3	20.8
7	1.66	2.52	3.32	4.28	5.86	7.35	9.17	12.2	15.1
8	1.59	2.35	3.02	3.81	5.07	6.22	7.59	9.80	11.8
9	1.54	2.22	2.80	3.47	4.52	5.45	6.54	8.26	9.81
10	1.50	2.12	2.64	3.22	4.12	4.90	5.80	7.19	8.43
11	1.47	2.04	2.51	3.03	3.81	4.49	5.25	6.42	7.43
12	1.44	1.97	2.40	2.87	3.57	4.17	4.83	5.83	6.69
13	1.42	1.92	2.31	2.74	3.38	3.91	4.50	5.37	6.11
14	1.40	1.87	2.24	2.64	3.22	3.70	4.23	5.00	5.66
15	1.38	1.83	2.18	2.55	3.08	3.52	4.00	4.70	5.29
16	1.37	1.79	2.12	2.47	2.97	3.37	3.81	4.45	4.98
17	1.36	1.76	2.08	2.41	2.87	3.25	3.65	4.24	4.72
18	1.34	1.74	2.04	2.35	2.78	3.14	3.52	4.06	4.50
19	1.33	1.71	2.00	2.30	2.71	3.04	3.40	3.90	4.31
20	1.32	1.69	1.97	2.25	2.64	2.96	3.29	3.77	4.15
21	1.32	1.67	1.94	2.21	2.58	2.88	3.20	3.64	4.00
22	1.31	1.65	1.91	2.17	2.53	2.82	3.12	3.54	3.88
23	1.30	1.64	1.88	2.14	2.48	2.76	3.04	3.44	3.76
24	1.29	1.62	1.86	2.11	2.44	2.70	2.98	3.36	3.66
25	1.29	1.61	1.84	2.08	2.40	2.65	2.91	3.28	3.57
26	1.28	1.59	1.82	2.05	2.36	2.61	2.86	3.21	3.49
27	1.28	1.58	1.81	2.03	2.33	2.57	2.81	3.14	3.41
28	1.27	1.57	1.79	2.01	2.30	2.53	2.76	3.09	3.34
29	1.27	1.56	1.77	1.99	2.27	2.49	2.72	3.03	3.28
30	1.26	1.55	1.76	1.97	2.25	2.46	2.68	2.98	3.22
35	1.24	1.51	1.70	1.89	2.14	2.33	2.52	2.78	2.98
40	1.23	1.48	1.66	1.83	2.06	2.23	2.40	2.64	2.82
45	1.22	1.46	1.63	1.79	2.00	2.16	2.31	2.53	2.69
50	1.21	1.44	1.60	1.75	1.95	2.10	2.24	2.44	2.59
60	1.20	1.41	1.56	1.70	1.88	2.01	2.14	2.32	2.45
70	1.19	1.39	1.53	1.66	1.83	1.95	2.07	2.23	2.35
80	1.18	1.38	1.51	1.63	1.79	1.90	2.02	2.16	2.28
90	1.18	1.36	1.49	1.61	1.76	1.87	1.98	2.11	2.22
100	1.17	1.35	1.48	1.59	1.74	1.84	1.94	2.08	2.18
120	1.16	1.34	1.46	1.56	1.70	1.80	1.89	2.02	2.11
140	1.16	1.33	1.44	1.55	1.67	1.77	1.86	1.98	2.06
160	1.15	1.32	1.43	1.53	1.66	1.75	1.83	1.95	2.03
180	1.15	1.32	1.42	1.52	1.64	1.73	1.81	1.92	2.00
200	1.15	1.31	1.41	1.51	1.63	1.71	1.80	1.90	1.98
300	1.14	1.29	1.39	1.48	1.59	1.67	1.75	1.85	1.92
500	1.14	1.28	1.38	1.46	1.57	1.64	1.71	1.80	1.87
∞	1.13	1.26	1.35	1.43	1.52	1.59	1.65	1.73	1.79

Table 34: Critical Values of the F Distribution

Numerator DF = 60

ν	$\alpha(2)$: 0.50	0.20	0.10	0.05	0.02	0.01	0.005	0.002	0.001
	$\alpha(1)$: 0.25	0.10	0.05	0.025	0.01	0.005	0.0025	0.001	0.0005
1	9.76	62.8	252.0	1010.0	6310.0	25300.00	101000.0	631000.0	2530000.0
2	3.46	9.47	19.5	39.5	99.5	199.0	399.0	999.0	2000.0
3	2.47	5.15	8.57	14.0	26.3	42.1	67.3	124.00	198.0
4	2.08	3.79	5.69	8.36	13.7	19.6	28.0	44.7	63.6
5	1.87	3.14	4.43	6.12	9.20	12.4	16.6	24.3	32.4
6	1.74	2.76	3.74	4.96	7.06	9.12	11.7	16.2	20.7
7	1.65	2.51	3.30	4.25	5.82	7.31	9.12	12.1	15.0
8	1.59	2.34	3.01	3.78	5.03	6.18	7.54	9.73	11.8
9	1.54	2.21	2.79	3.45	4.48	5.41	6.49	8.19	9.72
10	1.50	2.11	2.62	3.20	4.08	4.86	5.75	7.12	8.34
11	1.47	2.03	2.49	3.00	3.78	4.45	5.20	6.35	7.35
12	1.44	1.96	2.38	2.85	3.54	4.12	4.78	5.76	6.61
13	1.42	1.90	2.30	2.72	3.34	3.87	4.44	5.30	6.04
14	1.40	1.86	2.22	2.61	3.18	3.66	4.17	4.94	5.58
15	1.38	1.82	2.16	2.52	3.05	3.48	3.95	4.64	5.21
16	1.36	1.78	2.11	2.45	2.93	3.33	3.76	4.39	4.91
17	1.35	1.75	2.06	2.38	2.83	3.21	3.60	4.18	4.65
18	1.34	1.72	2.02	2.32	2.75	3.10	3.47	4.00	4.43
19	1.33	1.70	1.98	2.27	2.67	3.00	3.35	3.84	4.24
20	1.32	1.68	1.95	2.22	2.61	2.92	3.24	3.70	4.08
21	1.31	[illegible]	1.92	2.18	2.55	2.84	3.15	3.58	3.93
22	1.30	1.64	1.89	2.14	2.50	2.77	3.07	3.48	3.81
23	1.30	1.62	1.86	2.11	2.45	2.71	2.99	3.38	3.69
24	1.29	1.61	1.84	2.08	2.40	2.66	2.92	3.29	3.59
25	1.28	1.59	1.82	2.05	2.36	2.61	2.86	3.22	3.50
26	1.28	1.58	1.80	2.03	2.33	2.56	2.81	3.15	3.42
27	1.27	1.57	1.79	2.00	2.29	2.52	2.76	3.08	3.34
28	1.27	1.56	1.77	1.98	2.26	2.48	2.71	3.02	3.27
29	1.26	1.55	1.75	1.96	2.23	2.45	2.67	2.97	3.21
30	1.26	1.54	1.74	1.94	2.21	2.42	2.63	2.92	3.15
35	1.24	1.50	1.68	1.86	2.10	2.28	2.47	2.72	2.91
40	1.22	1.47	1.64	1.80	2.02	2.18	2.35	2.57	2.75
45	1.21	1.44	1.60	1.76	1.96	2.11	2.26	2.46	2.62
50	1.20	1.42	1.58	1.72	1.91	2.05	2.19	2.38	2.52
60	1.19	1.40	1.53	1.67	1.84	1.96	2.09	2.25	2.38
70	1.18	1.37	1.50	1.63	1.78	1.90	2.01	2.16	2.28
80	1.17	1.36	1.48	1.60	1.75	1.85	1.96	2.10	2.20
90	1.17	1.35	1.46	1.58	1.72	1.82	1.92	2.05	2.15
100	1.16	1.34	1.45	1.56	1.69	1.79	1.89	2.01	2.10
120	1.16	1.32	1.43	1.53	1.66	1.75	1.84	1.95	2.04
140	1.15	1.31	1.41	1.51	1.63	1.72	1.80	1.91	1.99
160	1.15	1.30	1.40	1.50	1.61	1.69	1.77	1.88	1.95
180	1.14	1.29	1.39	1.48	1.60	1.68	1.75	1.85	1.93
200	1.14	1.29	1.39	1.47	1.58	1.66	1.74	1.83	1.90
300	1.13	1.27	1.36	1.45	1.55	1.62	1.69	1.78	1.84
500	1.13	1.26	1.35	1.42	1.52	1.58	1.65	1.73	1.79
∞	1.12	1.24	1.32	1.39	1.47	1.53	1.59	1.66	1.71

Table 35: Critical Values of the F Distribution

Numerator DF = 70

ν	$\alpha(2)$: 0.50 $\alpha(1)$: 0.25	0.20 0.10	0.10 0.05	0.05 0.025	0.02 0.01	0.01 0.005	0.005 0.0025	0.002 0.001	0.001 0.0005
1	9.77	62.8	252.0	1010.0	6320.0	25300.00	101000.0	632000.0	2530000.0
2	3.46	9.48	19.5	39.5	99.5	199.0	399.0	999.0	2000.0
3	2.47	5.15	8.57	14.0	26.3	42.1	67.2	124.0	198.0
4	2.08	3.79	5.68	8.35	13.6	19.6	28.0	44.6	63.4
5	1.87	3.14	4.42	6.11	9.18	12.4	16.6	24.3	32.3
6	1.74	2.76	3.73	4.94	7.03	9.09	11.7	16.1	20.6
7	1.65	2.51	3.29	4.24	5.80	7.28	9.07	12.1	14.9
8	1.59	2.33	2.99	3.77	5.01	6.15	7.49	9.67	11.7
9	1.54	2.20	2.78	3.43	4.46	5.38	6.45	8.13	9.65
10	1.50	2.10	2.61	3.18	4.06	4.83	5.71	7.07	8.28
11	1.46	2.02	2.48	2.99	3.75	4.41	5.16	6.30	7.29
12	1.44	1.95	2.37	2.83	3.51	4.09	4.74	5.71	6.55
13	1.41	1.90	2.28	2.70	3.32	3.84	4.41	5.26	5.98
14	1.39	1.85	2.21	2.60	3.16	3.62	4.14	4.89	5.53
15	1.38	1.81	2.15	2.51	3.02	3.45	3.91	4.59	5.16
16	1.36	1.77	2.09	2.43	2.91	3.30	3.73	4.34	4.85
17	1.35	1.74	2.05	2.36	2.81	3.18	3.57	4.13	4.60
18	1.34	1.71	2.00	2.30	2.72	3.07	3.43	3.95	4.38
19	1.33	1.69	1.97	2.25	2.65	2.97	3.31	3.79	4.19
20	1.32	1.67	1.93	2.20	2.58	2.88	3.20	3.66	4.03
21	1.31	1.65	1.90	2.16	2.52	2.81	3.11	3.54	3.88
22	1.30	1.63	1.88	2.13	2.47	2.74	3.03	3.43	3.76
23	1.29	1.61	1.85	2.09	2.42	2.68	2.95	3.34	3.64
24	1.28	1.60	1.83	2.06	2.38	2.63	2.89	3.25	3.54
25	1.28	1.58	1.81	2.03	2.34	2.58	2.83	3.17	3.45
26	1.27	1.57	1.79	2.01	2.30	2.53	2.77	3.10	3.37
27	1.27	1.56	1.77	1.98	2.27	2.49	2.72	3.04	3.29
28	1.26	1.55	1.75	1.96	2.24	2.45	2.67	2.98	3.22
29	1.26	1.54	1.74	1.94	2.21	2.42	2.63	2.92	3.16
30	1.25	1.53	1.72	1.92	2.18	2.38	2.59	2.87	3.10
35	1.23	1.49	1.66	1.84	2.07	2.25	2.43	2.67	2.86
40	1.22	1.46	1.62	1.78	1.99	2.15	2.31	2.53	2.70
45	1.21	1.43	1.59	1.74	1.93	2.08	2.22	2.42	2.57
50	1.20	1.41	1.56	1.70	1.88	2.02	2.15	2.33	2.47
60	1.19	1.38	1.52	1.64	1.81	1.93	2.05	2.21	2.33
70	1.18	1.36	1.49	1.60	1.75	1.86	1.97	2.12	2.22
80	1.17	1.34	1.46	1.57	1.71	1.82	1.92	2.05	2.15
90	1.16	1.33	1.44	1.55	1.68	1.78	1.88	2.00	2.09
100	1.16	1.32	1.43	1.53	1.66	1.75	1.84	1.96	2.05
120	1.15	1.31	1.41	1.50	1.62	1.71	1.79	1.90	1.98
140	1.14	1.29	1.39	1.48	1.60	1.68	1.76	1.86	1.93
160	1.14	1.29	1.38	1.47	1.58	1.65	1.73	1.83	1.90
180	1.14	1.28	1.37	1.46	1.56	1.64	1.71	1.80	1.87
200	1.13	1.27	1.36	1.45	1.55	1.62	1.69	1.78	1.85
300	1.13	1.26	1.34	1.42	1.51	1.58	1.64	1.72	1.78
500	1.12	1.24	1.32	1.39	1.48	1.54	1.60	1.68	1.73
∞	1.11	1.22	1.29	1.36	1.43	1.49	1.54	1.60	1.65

Table 36: Critical Values of the F Distribution

Numerator DF = 80

ν	$\alpha(2)$: 0.50 $\alpha(1)$: 0.25	0.20 0.10	0.10 0.05	0.05 0.025	0.02 0.01	0.01 0.005	0.005 0.0025	0.002 0.001	0.001 0.0005
1	9.78	62.9	253.0	1010.0	6330.0	25300.0	101000.0	633000.0	2530000.0
2	3.46	9.48	19.5	39.5	99.5	199.0	399.0	999.0	2000.0
3	2.47	5.15	8.56	14.0	26.3	42.1	67.1	124.0	198.0
4	2.08	3.78	5.67	8.33	13.6	19.5	27.9	44.6	63.3
5	1.87	3.13	4.41	6.10	9.16	12.3	16.5	24.2	32.2
6	1.74	2.75	3.72	4.93	7.01	9.06	11.6	16.1	20.5
7	1.65	2.50	3.29	4.23	5.78	7.25	9.04	12.0	14.8
8	1.59	2.33	2.99	3.76	4.99	6.12	7.46	9.63	11.6
9	1.54	2.20	2.77	3.42	4.44	5.36	6.42	8.09	9.61
10	1.50	2.09	2.60	3.17	4.04	4.80	5.68	7.03	8.23
11	1.46	2.01	2.47	2.97	3.73	4.39	5.13	6.26	7.25
12	1.44	1.95	2.36	2.82	3.49	4.07	4.71	5.68	6.51
13	1.41	1.89	2.27	2.69	3.30	3.81	4.38	5.22	5.94
14	1.39	1.84	2.20	2.58	3.14	3.60	4.11	4.86	5.49
15	1.37	1.80	2.14	2.49	3.00	3.43	3.89	4.56	5.12
16	1.36	1.77	2.08	2.42	2.89	3.28	3.70	4.31	4.81
17	1.35	1.74	2.03	2.35	2.79	3.15	3.54	4.10	4.56
18	1.33	1.71	1.99	2.29	2.70	3.04	3.40	3.92	4.34
19	1.32	1.68	1.96	2.24	2.63	2.95	3.28	3.76	4.15
20	1.31	1.66	1.92	2.19	2.56	2.86	3.18	3.62	3.99
21	1.30	1.64	1.89	2.15	2.50	2.79	3.08	3.50	3.84
22	1.30	1.62	1.86	2.11	2.45	2.72	3.00	3.40	3.72
23	1.29	1.61	1.84	2.08	2.40	2.66	2.93	3.30	3.60
24	1.28	1.59	1.82	2.05	2.36	2.60	2.86	3.32	3.50
25	1.28	1.58	1.80	2.02	2.32	2.55	2.80	3.14	3.41
26	1.27	1.56	1.78	1.99	2.28	2.51	2.74	3.07	3.33
27	1.26	1.55	1.76	1.97	2.25	2.47	2.69	3.00	3.25
28	1.26	1.54	1.74	1.94	2.22	2.43	2.64	2.94	3.18
29	1.25	1.53	1.73	1.92	2.19	2.39	2.60	2.89	3.12
30	1.25	1.52	1.71	1.90	2.16	2.36	2.56	2.84	3.06
35	1.23	1.48	1.65	1.82	2.05	2.22	2.40	2.64	2.83
40	1.22	1.45	1.61	1.76	1.97	2.12	2.28	2.49	2.66
45	1.21	1.42	1.57	1.72	1.91	2.05	2.19	2.38	2.53
50	1.20	1.40	1.54	1.68	1.86	1.99	2.12	2.30	2.43
60	1.18	1.37	1.50	1.63	1.78	1.90	2.02	2.17	2.29
70	1.17	1.35	1.47	1.59	1.73	1.84	1.94	2.08	2.18
80	1.16	1.33	1.45	1.55	1.69	1.79	1.89	2.01	2.11
90	1.16	1.32	1.43	1.53	1.66	1.75	1.84	1.96	2.05
100	1.15	1.31	1.41	1.51	1.63	1.72	1.81	1.92	2.01
120	1.14	1.29	1.39	1.48	1.60	1.68	1.76	1.86	1.94
140	1.14	1.28	1.38	1.46	1.57	1.65	1.72	1.82	1.89
160	1.13	1.27	1.36	1.45	1.55	1.62	1.69	1.79	1.85
180	1.13	1.27	1.35	1.43	1.53	1.61	1.67	1.76	1.83
200	1.13	1.26	1.35	1.42	1.52	1.59	1.66	1.74	1.80
300	1.12	1.24	1.32	1.39	1.48	1.55	1.61	1.68	1.74
500	1.11	1.23	1.30	1.37	1.45	1.51	1.56	1.63	1.68
∞	1.10	1.21	1.27	1.33	1.40	1.45	1.50	1.56	1.60

Table 37: Critical Values of the F Distribution

Numerator DF = 90

ν	$\alpha(2)$: 0.50	0.20	0.10	0.05	0.02	0.01	0.005	0.002	0.001
	$\alpha(1)$: 0.25	0.10	0.05	0.025	0.01	0.005	0.0025	0.001	0.0005
1	9.79	63.0	253.0	1010.0	6330.0	25300.0	101000.0	633000.0	2530000.0
2	3.46	9.48	19.5	39.5	99.5	199.0	399.0	999.0	2000.0
3	2.47	5.15	8.56	14.0	26.3	42.0	67.1	124.0	197.0
4	2.08	3.78	5.67	8.33	13.6	19.5	27.9	44.5	63.2
5	1.87	3.13	4.41	6.09	9.14	12.3	16.5	24.2	32.1
6	1.74	2.75	3.72	4.92	7.00	9.04	11.6	16.1	20.4
7	1.65	2.50	3.28	4.22	5.77	7.23	9.01	12.0	14.8
8	1.59	2.32	2.98	3.75	4.97	6.10	7.44	9.60	11.6
9	1.53	2.19	2.76	3.41	4.43	5.34	6.40	8.06	9.57
10	1.49	2.09	2.59	3.16	4.03	4.79	5.66	7.00	8.20
11	1.46	2.01	2.46	2.96	3.72	4.37	5.11	6.23	7.21
12	1.43	1.94	2.36	2.81	3.48	4.05	4.69	5.65	6.48
13	1.41	1.89	2.27	2.68	3.28	3.79	4.36	5.19	5.91
14	1.39	1.84	2.19	2.57	3.12	3.58	4.09	4.83	5.45
15	1.37	1.80	2.13	2.48	2.99	3.41	3.86	4.53	5.09
16	1.36	1.76	2.07	2.40	2.87	3.26	3.68	4.28	4.78
17	1.34	1.73	2.03	2.34	2.78	3.13	3.52	4.07	4.53
18	1.33	1.70	1.98	2.28	2.69	3.02	3.38	3.89	4.31
19	1.32	1.68	1.95	2.23	2.61	2.93	3.26	3.73	4.12
20	1.31	1.65	1.91	2.18	2.55	2.84	3.16	3.60	3.96
21	1.30	1.63	1.88	2.14	2.49	2.77	3.06	3.48	3.81
22	1.29	1.62	1.86	2.10	2.43	2.70	2.98	3.37	3.69
23	1.29	1.60	1.83	2.07	2.39	2.64	2.90	3.28	3.57
24	1.28	1.58	1.81	2.03	2.34	2.58	2.84	3.19	3.47
25	1.27	1.57	1.79	2.01	2.30	2.53	2.78	3.11	3.38
26	1.27	1.56	1.77	1.98	2.26	2.49	2.72	3.04	3.30
27	1.26	1.54	1.75	1.95	2.23	2.45	2.67	2.98	3.22
28	1.26	1.53	1.73	1.93	2.20	2.41	2.62	2.92	3.15
29	1.25	1.52	1.72	1.91	2.17	2.37	2.58	2.86	3.09
30	1.25	1.51	1.70	1.89	2.14	2.34	2.54	2.81	3.03
35	1.23	1.47	1.64	1.81	2.03	2.20	2.38	2.61	2.80
40	1.21	1.44	1.60	1.75	1.95	2.10	2.26	2.47	2.63
45	1.20	1.41	1.56	1.70	1.89	2.03	2.17	2.36	2.50
50	1.19	1.39	1.53	1.67	1.84	1.97	2.10	2.27	2.40
60	1.18	1.36	1.49	1.61	1.76	1.88	1.99	2.14	2.25
70	1.17	1.34	1.46	1.57	1.71	1.81	1.92	2.05	2.15
80	1.16	1.33	1.44	1.54	1.67	1.77	1.86	1.98	2.08
90	1.15	1.31	1.42	1.52	1.64	1.73	1.82	1.93	2.02
100	1.15	1.30	1.40	1.50	1.61	1.70	1.78	1.89	1.97
120	1.14	1.28	1.38	1.47	1.58	1.66	1.73	1.83	1.90
140	1.13	1.27	1.36	1.45	1.55	1.62	1.70	1.79	1.86
160	1.13	1.26	1.35	1.43	1.53	1.60	1.67	1.75	1.82
180	1.13	1.26	1.34	1.42	1.51	1.58	1.65	1.73	1.79
200	1.12	1.25	1.33	1.41	1.50	1.56	1.63	1.71	1.77
300	1.11	1.23	1.31	1.38	1.46	1.52	1.58	1.65	1.70
500	1.11	1.22	1.29	1.35	1.43	1.48	1.53	1.60	1.65
∞	1.10	1.20	1.26	1.31	1.38	1.43	1.47	1.52	1.56

Table 38: Critical Values of the F Distribution

Numerator DF = 100

ν	α(2) : 0.50	0.20	0.10	0.05	0.02	0.01	0.005	0.002	0.001
	α(1) : 0.25	0.10	0.05	0.025	0.01	0.005	0.0025	0.001	0.0005
1	9.80	63.0	253.0	1010.0	6330.0	25300.0	101000.0	633000.0	2530000.0
2	3.47	9.48	19.5	39.5	99.5	199.0	399.0	999.0	2000.0
3	2.47	5.14	8.55	14.0	26.2	42.0	67.1	124.0	197.0
4	2.08	3.78	5.66	8.32	13.6	19.5	27.9	44.5	63.2
5	1.87	3.13	4.41	6.08	9.13	12.3	16.5	24.1	32.1
6	1.74	2.75	3.71	4.92	6.99	9.03	11.6	16.0	20.4
7	1.65	2.50	3.27	4.21	5.75	7.22	8.99	12.0	14.8
8	1.58	2.32	2.97	3.74	4.96	6.09	7.42	9.57	11.6
9	1.53	2.19	2.76	3.40	4.41	5.32	6.38	8.04	9.54
10	1.49	2.09	2.59	3.15	4.01	4.77	5.64	6.98	8.17
11	1.46	2.01	2.46	2.96	3.71	4.36	5.09	6.21	7.19
12	1.43	1.94	2.35	2.80	3.47	4.04	4.67	5.63	6.45
13	1.41	1.88	2.26	2.67	3.27	3.78	4.34	5.17	5.88
14	1.39	1.83	2.19	2.56	3.11	3.57	4.07	4.81	5.43
15	1.37	1.79	2.12	2.47	2.98	3.39	3.85	4.51	5.06
16	1.36	1.76	2.07	2.40	2.86	3.25	3.66	4.26	4.76
17	1.34	1.73	2.02	2.33	2.76	3.12	3.50	4.05	4.50
18	1.33	1.70	1.98	2.27	2.68	3.01	3.36	3.87	4.28
19	1.32	1.67	1.94	2.22	2.60	2.91	3.24	3.71	4.10
20	1.31	1.65	1.91	2.17	2.54	2.83	3.14	3.58	3.93
21	1.30	1.63	1.88	2.13	2.48	2.75	3.04	3.46	3.79
22	1.29	1.61	1.85	2.09	2.42	2.69	2.96	3.35	3.66
23	1.29	1.59	1.82	2.06	2.37	2.62	2.89	3.25	3.55
24	1.28	1.58	1.80	2.02	2.33	2.57	2.82	3.17	3.45
25	1.27	1.56	1.78	2.00	2.29	2.52	2.76	3.09	3.36
26	1.27	1.55	1.76	1.97	2.25	2.47	2.70	3.02	3.27
27	1.26	1.54	1.74	1.94	2.22	2.43	2.65	2.96	3.20
28	1.25	1.53	1.73	1.92	2.19	2.39	2.60	2.90	3.13
29	1.25	1.52	1.71	1.90	2.16	2.36	2.56	2.84	3.06
30	1.25	1.51	1.70	1.88	2.13	2.32	2.52	2.79	3.01
35	1.23	1.47	1.63	1.80	2.02	2.19	2.36	2.59	2.77
40	1.21	1.43	1.59	1.74	1.94	2.09	2.24	2.44	2.60
45	1.20	1.41	1.55	1.69	1.88	2.01	2.15	2.33	2.47
50	1.19	1.39	1.52	1.66	1.82	1.95	2.08	2.25	2.37
60	1.18	1.36	1.48	1.60	1.75	1.86	1.97	2.12	2.23
70	1.16	1.34	1.45	1.56	1.70	1.80	1.90	2.03	2.13
80	1.16	1.32	1.43	1.53	1.65	1.75	1.84	1.96	2.05
90	1.15	1.30	1.41	1.50	1.62	1.71	1.80	1.91	1.99
100	1.14	1.29	1.39	1.48	1.60	1.68	1.76	1.87	1.95
120	1.14	1.28	1.37	1.45	1.56	1.64	1.71	1.81	1.88
140	1.13	1.26	1.35	1.43	1.53	1.60	1.67	1.76	1.83
160	1.13	1.26	1.34	1.42	1.51	1.58	1.64	1.73	1.79
180	1.12	1.25	1.33	1.40	1.49	1.56	1.62	1.70	1.76
200	1.12	1.24	1.32	1.39	1.48	1.54	1.60	1.68	1.74
300	1.11	1.22	1.30	1.36	1.44	1.50	1.55	1.62	1.67
500	1.10	1.21	1.28	1.34	1.41	1.46	1.51	1.57	1.62
∞	1.09	1.18	1.24	1.30	1.36	1.40	1.44	1.49	1.53

Table 39: Critical Values of the F Distribution

Numerator DF = 120

ν	$\alpha(2)$: 0.50 $\alpha(1)$: 0.25	0.20 0.10	0.10 0.05	0.05 0.025	0.02 0.01	0.01 0.005	0.005 0.0025	0.002 0.001	0.001 0.0005
1	9.80	63.1	253.0	1010.0	6340.0	25400.0	101000.0	634000.0	2540000.0
2	3.47	9.48	19.5	39.5	99.5	199.0	399.0	999.0	2000.0
3	2.47	5.14	8.55	13.9	26.2	42.0	67.0	124.0	197.0
4	2.08	3.78	5.66	8.31	13.6	19.5	27.8	44.4	63.1
5	1.87	3.12	4.40	6.07	9.11	12.3	16.4	24.1	32.0
6	1.74	2.74	3.70	4.90	6.97	9.00	11.6	16.0	20.3
7	1.65	2.49	3.27	4.20	5.74	7.19	8.96	11.9	14.7
8	1.58	2.32	2.97	3.73	4.95	6.06	7.39	9.53	11.5
9	1.53	2.18	2.75	3.39	4.40	5.30	6.35	8.00	9.49
10	1.49	2.08	2.58	3.14	4.00	4.75	5.61	6.94	8.12
11	1.46	2.00	2.45	2.94	3.69	4.34	5.07	6.18	7.14
12	1.43	1.93	2.34	2.79	3.45	4.01	4.65	5.59	6.41
13	1.41	1.88	2.25	2.66	3.25	3.76	4.31	5.14	5.84
14	1.39	1.83	2.18	2.55	3.09	3.55	4.04	4.77	5.39
15	1.37	1.79	2.11	2.46	2.96	3.37	3.82	4.47	5.02
16	1.35	1.75	2.06	2.38	2.84	3.22	3.63	4.23	4.72
17	1.34	1.72	2.01	2.32	2.75	3.10	3.47	4.02	4.46
18	1.33	1.69	1.97	2.26	2.66	2.99	3.34	3.84	4.25
19	1.32	1.67	1.93	2.20	2.58	2.89	3.22	3.68	4.06
20	1.31	1.64	1.90	2.16	2.52	2.81	3.11	3.54	3.90
21	1.30	1.62	1.87	2.11	2.46	2.73	3.02	3.42	3.75
22	1.29	1.60	1.84	2.08	2.40	2.66	2.93	3.32	3.62
23	1.28	1.59	1.81	2.04	2.35	2.60	2.86	3.22	3.51
24	1.28	1.57	1.79	2.01	2.31	2.55	2.79	3.14	3.41
25	1.27	1.56	1.77	1.98	2.27	2.50	2.73	3.06	3.32
26	1.26	1.54	1.75	1.95	2.23	2.45	2.68	2.99	3.24
27	1.26	1.53	1.73	1.93	2.20	2.41	2.62	2.92	3.16
28	1.25	1.52	1.71	1.91	2.17	2.37	2.58	2.86	3.09
29	1.25	1.51	1.70	1.89	2.14	2.33	2.53	2.81	3.03
30	1.24	1.50	1.68	1.87	2.11	2.30	2.49	2.76	2.97
35	1.22	1.46	1.62	1.79	2.00	2.16	2.33	2.56	2.73
40	1.21	1.42	1.58	1.72	1.92	2.06	2.21	2.41	2.56
45	1.20	1.40	1.54	4.68	1.85	1.99	2.12	2.30	2.44
50	1.19	1.38	1.51	1.64	1.80	1.93	2.05	2.21	2.34
60	1.17	1.35	1.47	1.58	1.73	1.83	1.94	2.08	2.19
70	1.16	1.32	1.44	1.54	1.67	1.77	1.87	1.99	2.09
80	1.15	1.31	1.41	1.51	1.63	1.72	1.81	1.92	2.01
90	1.15	1.29	1.39	1.48	1.60	1.68	1.76	1.87	1.95
100	1.14	1.28	1.38	1.46	1.57	1.65	1.73	1.83	1.90
120	1.13	1.26	1.35	1.43	1.53	1.61	1.68	1.77	1.83
140	1.13	1.25	1.33	1.41	1.50	1.57	1.64	1.72	1.78
160	1.12	1.24	1.32	1.39	1.48	1.55	1.61	1.69	1.75
180	1.12	1.23	1.31	1.38	1.47	1.53	1.59	1.66	1.72
200	1.11	1.23	1.30	1.37	1.45	1.51	1.57	1.64	1.69
300	1.10	1.21	1.28	1.34	1.41	1.46	1.51	1.58	1.62
500	1.10	1.19	1.26	1.31	1.38	1.42	1.47	1.53	1.57
∞	1.08	1.17	1.22	1.27	1.32	1.36	1.40	1.45	1.48

Table 40: Critical Values of the F Distribution

Numerator DF = 140

ν	$\alpha(2)$: 0.50	0.20	0.10	0.05	0.02	0.01	0.005	0.002	0.001
	$\alpha(1)$: 0.25	0.10	0.05	0.025	0.01	0.005	0.0025	0.001	0.0005
1	9.81	63.1	253.0	1010.0	6340.0	25400.0	101000.0	634000.0	2540000.0
2	3.47	9.48	19.5	39.5	99.5	199.0	399.0	999.0	2000.0
3	2.47	5.14	8.55	13.9	26.2	42.0	67.0	124.0	197.0
4	2.08	3.77	5.65	8.30	13.50	19.4	27.8	44.4	63.0
5	1.87	3.12	4.39	6.06	9.10	12.3	16.4	24.0	31.9
6	1.74	2.74	3.70	4.90	6.96	8.98	11.5	15.9	20.3
7	1.65	2.49	3.26	4.19	5.72	7.18	8.94	11.9	14.7
8	1.58	2.31	2.96	3.72	4.93	6.05	7.37	9.50	11.5
9	1.53	2.18	2.74	3.38	4.39	5.28	6.33	7.97	9.46
10	1.49	2.08	2.57	3.13	3.98	4.73	5.59	6.92	8.09
11	1.46	2.00	2.44	2.94	3.68	4.32	5.05	6.15	7.11
12	1.43	1.93	2.33	2.78	3.44	4.00	4.63	5.57	6.38
13	1.41	1.87	2.25	2.65	3.24	3.74	4.29	5.11	5.81
14	1.39	1.82	2.17	2.54	3.08	3.53	4.02	4.75	5.36
15	1.37	1.78	2.11	2.45	2.95	3.36	3.80	4.45	5.00
16	1.35	1.75	2.05	2.37	2.83	3.21	3.61	4.20	4.69
17	1.34	1.71	2.00	2.31	2.73	3.08	3.45	3.99	4.44
18	1.33	1.69	1.96	2.25	2.65	2.97	3.32	3.81	4.22
19	1.32	1.66	1.92	2.19	2.57	2.87	3.20	3.66	4.03
20	1.31	1.64	1.89	2.15	2.50	2.79	3.09	3.52	3.87
21	1.30	1.62	1.86	2.10	2.44	2.71	3.00	3.40	3.73
22	1.29	1.60	1.83	2.07	2.39	2.65	2.92	3.29	3.60
23	1.28	1.58	1.81	2.03	2.34	2.59	2.84	3.20	3.49
24	1.27	1.57	1.78	2.00	2.30	2.53	2.77	3.11	3.38
25	1.27	1.55	1.76	1.97	2.26	2.48	2.71	3.03	3.29
26	1.26	1.54	1.74	1.94	2.22	2.43	2.66	2.96	3.21
27	1.26	1.53	1.72	1.92	2.18	2.39	2.60	2.90	3.13
28	1.25	1.51	1.71	1.90	2.15	2.35	2.56	2.84	3.06
29	1.24	1.50	1.69	1.88	2.12	2.32	2.51	2.79	3.00
30	1.24	1.49	1.68	1.86	2.10	2.28	2.47	2.74	2.94
35	1.22	1.45	1.61	1.77	1.98	2.15	2.31	2.53	2.71
40	1.21	1.42	1.57	1.71	1.90	2.05	2.19	2.39	2.54
45	1.19	1.39	1.53	1.66	1.84	1.97	2.10	2.27	2.41
50	1.18	1.37	1.50	1.63	1.79	1.91	2.03	2.19	2.31
60	1.17	1.34	1.46	1.57	1.71	1.81	1.92	2.06	2.16
70	1.16	1.32	1.42	1.53	1.65	1.75	1.84	1.96	2.06
80	1.15	1.30	1.40	1.49	1.61	1.70	1.79	1.90	1.98
90	1.14	1.28	1.38	1.47	1.58	1.66	1.74	1.84	1.92
100	1.14	1.27	1.36	1.45	1.55	1.63	1.70	1.80	1.87
120	1.13	1.26	1.34	1.42	1.51	1.58	1.65	1.74	1.80
140	1.12	1.24	1.32	1.39	1.48	1.55	1.61	1.69	1.75
160	1.12	1.23	1.31	1.38	1.46	1.52	1.58	1.66	1.71
180	1.11	1.22	1.30	1.36	1.45	1.50	1.56	1.63	1.68
200	1.11	1.22	1.29	1.35	1.43	1.49	1.54	1.61	1.66
300	1.10	1.20	1.26	1.32	1.39	1.44	1.49	1.55	1.59
500	1.09	1.18	1.24	1.29	1.35	1.40	1.44	1.49	1.53
∞	1.08	1.16	1.20	1.25	1.30	1.33	1.37	1.41	1.44

Table 41: Critical Values of the F Distribution

Numerator DF = 200

ν	$\alpha(2)$: 0.50	0.20	0.10	0.05	0.02	0.01	0.005	0.002	0.001
	$\alpha(1)$: 0.25	0.10	0.05	0.025	0.01	0.005	0.0025	0.001	0.0005
1	9.82	63.2	254.0	1020.0	6350.0	25400.0	102000.0	635000.0	2540000.0
2	3.47	9.49	19.5	39.5	99.5	199.0	399.0	999.0	2000.0
3	2.47	5.14	8.54	13.9	26.2	41.9	66.9	124.0	197.0
4	2.08	3.77	5.65	8.29	13.5	19.4	27.7	44.3	62.3
5	1.87	3.12	4.39	6.05	9.08	12.2	16.4	24.0	31.8
6	1.74	2.73	3.69	4.88	6.93	8.95	11.5	15.9	20.2
7	1.65	2.48	3.25	4.18	5.70	7.15	8.90	11.8	14.6
8	1.58	2.31	2.95	3.70	4.91	6.02	7.33	9.45	11.4
9	1.53	2.17	2.73	3.37	4.36	5.26	6.29	7.93	9.40
10	1.49	2.07	2.56	3.12	3.96	4.71	5.56	6.87	8.04
11	1.46	1.99	2.43	2.92	3.66	4.29	5.01	6.10	7.06
12	1.43	1.92	2.32	2.76	3.41	3.97	4.59	5.52	6.33
13	1.40	1.86	2.23	2.63	3.22	3.71	4.26	5.07	5.76
14	1.38	1.82	2.16	2.53	3.06	3.50	3.99	4.71	5.31
15	1.37	1.77	2.10	2.44	2.92	3.33	3.77	4.41	4.95
16	1.35	1.74	2.04	2.36	2.81	3.18	3.58	4.16	4.64
17	1.34	1.71	1.99	2.29	2.71	3.05	3.42	3.95	4.39
18	1.32	1.68	1.95	2.23	2.62	2.94	3.28	3.77	4.17
19	1.31	1.65	1.91	2.18	2.55	2.85	3.16	3.61	3.98
20	1.30	1.63	1.88	2.13	2.48	2.76	3.06	3.48	3.82
21	1.29	1.61	1.84	2.09	2.42	2.68	2.96	3.36	3.68
22	1.28	1.59	1.82	2.05	2.36	2.62	2.88	3.25	3.55
23	1.28	1.57	1.79	2.01	2.32	2.56	2.81	3.16	3.44
24	1.27	1.56	1.77	1.98	2.27	2.50	2.74	3.07	3.34
25	1.26	1.54	1.75	1.95	2.23	2.45	2.68	2.99	3.24
26	1.26	1.53	1.73	1.92	2.19	2.40	2.62	2.92	3.16
27	1.25	1.52	1.71	1.90	2.16	2.36	2.57	2.86	3.09
28	1.25	1.50	1.69	1.88	2.13	2.32	2.52	2.80	3.02
29	1.24	1.49	1.67	1.86	2.10	2.29	2.48	2.74	2.95
30	1.24	1.48	1.66	1.84	2.07	2.25	2.44	2.69	2.89
35	1.22	1.44	1.60	1.75	1.96	2.11	2.27	2.49	2.66
40	1.20	1.41	1.55	1.69	1.87	2.01	2.15	2.34	2.49
45	1.19	1.38	1.51	1.64	1.81	1.93	2.06	2.23	2.36
50	1.18	1.36	1.48	1.60	1.76	1.87	1.99	2.14	2.26
60	1.16	1.33	1.44	1.54	1.68	1.78	1.88	2.01	2.11
70	1.15	1.30	1.40	1.50	1.62	1.71	1.80	1.92	2.00
80	1.14	1.28	1.38	1.47	1.58	1.66	1.74	1.85	1.93
90	1.13	1.27	1.36	1.44	1.55	1.62	1.70	1.79	1.86
100	1.13	1.26	1.34	1.42	1.52	1.59	1.66	1.75	1.82
120	1.12	1.24	1.32	1.39	1.48	1.54	1.60	1.68	1.74
140	1.11	1.22	1.30	1.36	1.45	1.51	1.56	1.64	1.69
160	1.11	1.21	1.28	1.35	1.42	1.48	1.53	1.60	1.65
180	1.10	1.21	1.27	1.33	1.41	1.46	1.51	1.57	1.62
200	1.10	1.20	1.26	1.32	1.39	1.44	1.49	1.55	1.60
300	1.09	1.18	1.23	1.28	1.35	1.39	1.43	1.48	1.52
500	1.08	1.16	1.21	1.25	1.31	1.35	1.38	1.43	1.46
∞	1.07	1.13	1.17	1.21	1.25	1.28	1.30	1.34	1.36

Table 42: Critical Values of the F Distribution

Numerator DF $= \infty$

ν	$\alpha(2)$: 0.50 $\alpha(1)$: 0.25	0.20 0.10	0.10 0.05	0.05 0.025	0.02 0.01	0.01 0.005	0.005 0.0025	0.002 0.001	0.001 0.0005
1	9.85	63.3	254.0	1020.0	6370.0	25500.0	102000.0	637000.0	2550000.0
2	3.48	9.49	19.5	39.5	99.5	199.0	399.0	999.0	2000.0
3	2.47	5.13	8.53	13.9	26.1	41.8	66.8	123.0	196.0
4	2.08	3.76	5.63	8.26	13.5	19.3	27.6	44.0	62.6
5	1.87	3.11	4.37	6.02	9.02	12.1	16.3	23.8	31.6
6	1.74	2.72	3.67	4.85	6.88	8.88	11.4	15.7	20.0
7	1.65	2.47	3.23	4.14	5.65	7.08	8.81	11.7	14.4
8	1.58	2.29	2.93	3.67	4.86	5.95	7.25	9.33	11.3
9	1.53	2.16	2.71	3.33	4.31	5.19	6.21	7.81	9.26
10	1.48	2.06	2.54	3.08	3.91	4.64	5.47	6.76	7.91
11	1.45	1.97	2.40	2.88	3.60	4.23	4.93	6.00	6.93
12	1.42	1.90	2.30	2.72	3.36	3.90	4.51	5.42	6.20
13	1.40	1.85	2.21	2.60	3.17	3.65	4.18	4.97	5.64
14	1.38	1.80	2.13	2.49	3.00	3.44	3.91	4.60	5.19
15	1.36	1.76	2.07	2.40	2.87	3.26	3.69	4.31	4.83
16	1.34	1.72	2.01	2.32	2.75	3.11	3.50	4.06	4.52
17	1.33	1.69	1.96	2.25	2.65	2.98	3.34	3.85	4.27
18	1.32	1.66	1.92	2.19	2.57	2.87	3.20	3.67	4.05
19	1.30	1.63	1.88	2.13	2.49	2.78	3.08	3.51	3.87
20	1.29	1.61	1.84	2.09	2.42	2.69	2.97	3.38	3.71
21	1.28	1.59	1.81	2.04	2.36	2.61	2.88	3.26	3.56
22	1.28	1.57	1.78	2.00	2.31	2.55	2.80	3.15	3.43
23	1.27	1.55	1.76	1.97	2.26	2.48	2.72	3.05	3.32
24	1.26	1.53	1..73	1.94	2.21	2.43	2.65	2.97	3.22
25	1.25	1.52	1.71	1.91	2.17	2.38	2.59	2.89	3.13
26	1.25	1.50	1.69	1.88	2.13	2.33	2.54	2.82	3.05
27	1.24	1.49	1.67	1.85	2.10	2.29	2.48	2.75	2.97
28	1.24	1.48	1.65	1.83	2.06	2.25	2.44	2.69	2.90
29	1.23	1.47	1.64	1.81	2.03	2.21	2.39	2.64	2.84
30	1.23	1.46	1.62	1.79	2.01	2.18	2.35	2.59	2.78
35	1.20	1.41	1.56	1.70	1.89	2.04	2.18	2.38	2.54
40	1.19	1.38	1.51	1.64	1.80	1.93	2.06	2.23	2.37
45	1.18	1.35	1.47	1.59	1.74	1.85	1.97	2.12	2.23
50	1.16	1.33	1.44	1.55	1.68	1.79	1.89	2.03	2.13
60	1.15	1.29	1.39	1.48	1.60	1.69	1.78	1.89	1.98
70	1.13	1.27	1.35	1.44	1.54	1.62	1.69	1.79	1.87
80	1.12	1.24	1.32	1.40	1.49	1.56	1.63	1.72	1.79
90	1.12	1.23	1.30	1.37	1.46	1.52	1.58	1.66	1.72
100	1.11	1.21	1.28	1.35	1.43	1.49	1.54	1.62	1.67
120	1.10	1.19	1.25	1.31	1.38	1.43	1.48	1.54	1.59
140	1.09	1.18	1.23	1.28	1.35	1.39	1.43	1.49	1.53
160	1.08	1.16	1.21	1.26	1.32	1.36	1.40	1.45	1.49
180	1.08	1.15	1.20	1.24	1.30	1.33	1.37	1.42	1.45
200	1.07	1.14	1.19	1.23	1.28	1.31	1.35	1.39	1.42
300	1.06	1.11	1.15	1.18	1.22	1.25	1.27	1.30	1.33
500	1.05	1.09	1.11	1.14	1.16	1.18	1.20	1.23	1.24
∞	1.00	1.00	1.00	1.00	1.00	1.00	1.00	1.00	1.00

Table 43: Critical Values of the Q' nonparametric Multiple Comparison Testing with a Control

k	$\alpha(2)$: 0.50	0.20	0.10	0.05	0.02	0.01	0.005	0.002	0.001
	$\alpha(1)$: 0.25	0.10	0.05	0.025	0.01	0.005	0.0025	0.001	0.0005
1	0.707	0.951	0.988	0.997	1.000	1.000	1.000	1.000	1.000
2	0.500	0.800	0.900	0.950	0.980	0.990	0.995	0.998	0.999
3	0.404	0.687	0.805	0.878	0.934	0.959	0.974	0.986	0.991
4	0.347	0.608	0.729	0.811	0.882	0.917	0.942	0.963	0.974
5	0.309	0.551	0.669	0.755	0.833	0.875	0.906	0.935	0.951
6	0.281	0.507	0.621	0.707	0.789	0.834	0.870	0.905	0.925
7	0.260	0.472	0.582	0.666	0.750	0.798	0.836	0.875	0.898
8	0.242	0.443	0.549	0.632	0.715	0.765	0.805	0.847	0.872
9	0.228	0.419	0.521	0.602	0.685	0.735	0.776	0.820	0.847
10	0.216	0.398	0.497	0.576	0.658	0.708	0.750	0.795	0.823
11	0.206	0.380	0.476	0.553	0.634	0.684	0.726	0.772	0.801
12	0.197	0.365	0.457	0.532	0.612	0.661	0.703	0.750	0.780
13	0.189	0.351	0.441	0.514	0.592	0.641	0.683	0.730	0.760
14	0.182	0.338	0.426	0.497	0.574	0.623	0.664	0.711	0.742
15	0.176	0.327	0.412	0.482	0.558	0.606	0.647	0.694	0.725
16	0.170	0.317	0.400	0.468	0.542	0.590	0.631	0.678	0.708
17	0.165	0.308	0.389	0.456	0.529	0.575	0.616	0.662	0.693
18	0.160	0.299	0.378	0.444	0.515	0.561	0.602	0.648	0.679
19	0.156	0.291	0.369	0.433	0.503	0.549	0.589	0.635	0.665
20	0.152	0.284	0.360	0.423	0.492	0.537	0.576	0.622	0.652
21	0.148	0.277	0.352	0.413	0.482	0.526	0.565	0.610	0.640
22	0.145	0.271	0.344	0.404	0.472	0.515	0.554	0.599	0.629
23	0.141	0.265	0.337	0.396	0.462	0.505	0.543	0.588	0.618
24	0.138	0.260	0.230	0.388	0.453	0.496	0.534	0.578	0.607
25	0.136	0.255	0.323	0.381	0.445	0.487	0.524	0.568	0.597
26	0.133	0.250	0.317	0.374	0.437	0.479	0.515	0.559	0.588
27	0.131	0.245	0.311	0.367	0.430	0.471	0.507	0.550	0.579
28	0.128	0.241	0.306	0.361	0.423	0.463	0.499	0.541	0.570
29	0.126	0.237	0.301	0.355	0.416	0.456	0.491	0.533	0.562
30	0.124	0.233	0.296	0.349	0.409	0.449	0.484	0.526	0.554
31	0.122	0.229	0.291	0.344	0.403	0.442	0.477	0.518	0.546
32	0.120	0.225	0.287	0.339	0.397	0.436	0.470	0.511	0.539
33	0.118	0.222	0.283	0.334	0.392	0.430	0.464	0.504	0.532
34	0.116	0.219	0.279	0.329	0.386	0.424	0.458	0.498	0.525
35	0.115	0.216	0.275	0.325	0.381	0.418	0.452	0.492	0.519
36	0.113	0.213	0.271	0.320	0.376	0.413	0.446	0.486	0.513
37	0.111	0.210	0.267	0.316	0.371	0.408	0.441	0.480	0.507
38	0.110	0.207	0.264	0.312	0.367	0.403	0.435	0.474	0.501
39	0.108	0.204	0.261	0.308	0.362	0.398	0.430	0.469	0.495
40	0.107	0.202	0.257	0.304	0.358	0.393	0.425	0.463	0.490
41	0.106	0.199	0.254	0.301	0.354	0.389	0.420	0.458	0.484
42	0.104	0.197	0.251	0.297	0.350	0.384	0.416	0.453	0.479
43	0.103	0.195	0.248	0.294	0.346	0.380	0.411	0.449	0.474
44	0.012	0.192	0.246	0.291	0.342	0.376	0.407	0.444	0.469
45	0.101	0.190	0.243	0.288	0.338	0.372	0.403	0.439	0.465
46	0.100	0.188	0.240	0.285	0.335	0.368	0.399	0.435	0.460
47	0.099	0.186	0.238	0.282	0.331	0.365	0.395	0.431	0.456
48	0.098	0.184	0.235	0.279	0.328	0.361	0.391	0.427	0.451
49	0.097	0.182	0.233	0.276	0.325	0.358	0.387	0.423	0.447
50	0.096	0.181	0.231	0.273	0.322	0.354	0.384	0.419	0.443
52	0.094	0.177	0.226	0.268	0.316	0.348	0.377	0.411	0.435
54	0.092	0.174	0.222	0.263	0.310	0.341	0.370	0.404	0.428
56	0.090	0.171	0.218	0.259	0.305	0.336	0.364	0.398	0.421
58	0.089	0.168	0.214	0.254	0.300	0.330	0.358	0.391	0.414
60	0.087	0.165	0.211	0.250	0.295	0.325	0.352	0.385	0.408

Table 44: Critical Values of the Q' nonparametric Multiple Comparison Testing with a Control

k	α(2) : 0.50	0.20	0.10	0.05	0.02	0.01	0.005	0.002	0.001
	α(1) : 0.25	0.10	0.05	0.025	0.01	0.005	0.0025	0.001	0.0005
62	0.086	0.162	0.207	0.246	0.290	0.320	0.347	0.379	0.402
64	0.084	0.160	0.204	0.242	0.286	0.315	0.342	0.374	0.396
66	0.083	0.157	0.201	0.239	0.282	0.310	0.337	0.368	0.390
68	0.082	0.155	0.198	0.235	0.278	0.306	0.332	0.363	0.385
70	0.081	0.153	0.195	0.232	0.274	0.302	0.327	0.358	0.380
72	0.080	0.151	0.193	0.229	0.270	0.298	0.323	0.354	0.375
74	0.079	0.149	0.190	0.226	0.266	0.294	0.319	0.349	0.370
76	0.078	0.147	0.188	0.223	0.263	0.290	0.315	0.345	0.365
78	0.077	0.145	0.185	0.220	0.260	0.286	0.311	0.340	0.361
80	0.076	0.143	0.183	0.217	0.257	0.283	0.307	0.336	0.357
82	0.075	0.141	0.181	0.215	0.253	0.280	0.304	0.333	0.353
84	0.074	0.140	0.179	0.212	0.251	0.276	0.300	0.329	0.349
86	0.073	0.138	0.177	0.210	0.248	0.273	0.297	0.325	0.345
88	0.072	0.136	0.174	0.207	0.245	0.270	0.293	0.321	0.341
90	0.071	0.135	0.173	0.205	0.242	0.267	0.290	0.318	0.338
92	0.070	0.133	0.171	0.203	0.240	0.264	0.287	0.315	0.334
94	0.070	0.132	0.169	0.201	0.237	0.262	0.284	0.312	0.331
96	0.069	0.131	0.167	0.199	0.235	0.259	0.281	0.308	0.327
98	0.068	0.129	0.165	0.197	0.232	0.256	0.279	0.305	0.324
100	0.068	0.128	0.164	0.195	0.230	0.254	0.276	0.303	0.321
105	0.066	0.125	0.160	0.190	0.225	0.248	0.270	0.296	0.314
110	0.064	0.122	0.156	0.186	0.220	0.242	0.264	0.289	0.307
115	0.063	0.119	0.153	0.182	0.215	0.237	0.258	0.283	0.300
120	0.062	0.117	0.150	0.178	0.210	0.232	0.253	0.277	0.294
125	0.060	0.114	0.147	0.174	0.206	0.228	0.248	0.272	0.289
130	0.059	0.112	0.144	0.171	0.202	0.223	0.243	0.267	0.283
135	0.058	0.110	0.141	0.168	0.199	0.219	0.239	0.262	0.278
140	0.057	0.108	0.139	0.165	0.195	0.215	0.234	0.257	0.273
145	0.056	0.106	0.136	0.162	0.192	0.212	0.230	0.253	0.269
150	0.055	0.105	0.134	0.159	0.189	0.208	0.227	0.249	0.264
160	0.053	0.101	0.130	0.154	0.183	0.202	0.220	0.241	0.256
170	0.052	0.098	0.126	0.150	0.177	0.196	0.213	0.234	0.249
180	0.050	0.095	0.122	0.145	0.172	0.190	0.207	0.228	0.242
190	0.049	0.093	0.119	0.142	0.168	0.185	0.202	0.222	0.236
200	0.048	0.091	0.116	0.138	0.164	0.181	0.197	0.216	0.230
250	0.043	0.081	0.104	1.124	0.146	0.162	0.176	0.194	0.206
300	0.039	0.074	0.095	0.113	0.134	0.148	0.161	0.177	0.188
350	0.036	0.068	0.088	0.105	0.124	0.137	0.149	0.164	0.175
400	0.034	0.064	0.082	0.098	0.116	0.128	0.140	0.154	0.164
450	0.032	0.060	0.077	0.092	0.109	0.121	0.132	0.145	0.154
500	0.030	0.057	0.074	0.088	0.104	0.115	0.125	0.138	0.146
600	0.028	0.052	0.067	0.080	0.095	0.105	0.114	0.126	0.134
700	0.026	0.048	0.062	0.074	0.088	0.097	0.106	0.116	0.124
800	0.024	0.045	0.058	0.069	0.082	0.091	0.099	0.109	0.116
900	0.022	0.043	0.055	0.065	0.077	0.086	0.093	0.103	0.109
1000	0.021	0.041	0.052	0.062	0.073	0.081	0.089	0.098	0.104

Table 45: Fisher's Tranformation for Correlation coefficients

r	0	1	2	3	4	5	6	7	8	9	r
0.000	0.0000	0.0010	0.0020	0.0030	0.0040	0.0050	0.0060	0.0070	0.0080	0.0090	0.000
0.010	0.0100	0.0110	0.0120	0.0130	0.0140	0.0150	0.0160	0.0170	0.0180	0.0190	0.010
0.020	0.0200	0.0210	0.0220	0.0230	0.0240	0.0250	0.0260	0.0270	0.0280	0.0290	0.020
0.030	0.0300	0.0310	0.0320	0.0330	0.0340	0.0350	0.0360	0.0370	0.0380	0.0390	0.030
0.040	0.0400	0.0410	0.0420	0.0430	0.0440	0.0450	0.0460	0.0470	0.0480	0.0490	0.040
0.050	0.0500	0.0510	0.0520	0.0530	0.0541	0.0551	0.0561	0.0571	0.0581	0.0591	0.050
0.060	0.0601	0.0611	0.0621	0.0631	0.0641	0.0651	0.0661	0.0671	0.0681	0.0691	0.060
0.070	0.0701	0.0711	0.0721	0.0731	0.0741	0.0751	0.0761	0.0772	0.0782	0.0792	0.070
0.080	0.0802	0.0812	0.0822	0.0832	0.0842	0.0852	0.0862	0.0872	0.0882	0.0892	0.080
0.090	0.0902	0.0913	0.0923	0.0933	0.0943	0.0953	0.0963	0.0973	0.0983	0.0993	0.090
0.100	0.1003	0.1013	0.1024	0.1034	0.1044	0.1054	0.1064	0.1074	0.1084	0.1094	0.100
0.110	0.1104	0.1115	0.1125	0.1135	0.1145	0.1155	0.1165	0.1175	0.1186	0.1196	0.110
0.120	0.1206	0.1216	0.1226	0.1236	0.1246	0.1257	0.1267	0.1277	0.1287	0.1297	0.120
0.130	0.1307	0.1318	0.1328	0.1338	0.1348	0.1358	0.1368	0.1379	0.1389	0.1399	0.130
0.140	0.1409	0.1419	0.1430	0.1440	0.1450	0.1460	0.1470	0.1481	0.1491	0.1501	0.140
0.150	0.1511	0.1522	0.1532	0.1542	0.1552	0.1563	0.1573	0.1583	0.1593	0.1604	0.150
0.160	0.1614	0.1624	0.1634	0.1645	0.1655	0.1665	0.1675	0.1686	0.1696	0.1706	0.160
0.170	0.1717	0.1727	0.1737	0.1748	0.1758	0.1768	0.1779	0.1789	0.1799	0.1809	0.170
0.180	0.1820	0.1830	0.1840	0.1851	0.1861	0.1872	0.1882	0.1892	0.1903	0.1913	0.180
0.190	0.1923	0.1934	0.1944	0.1955	0.1965	0.1975	0.1986	0.1996	0.2006	0.2017	0.190
0.200	0.2027	0.2038	0.2048	0.2059	0.2069	0.2079	0.2090	0.2100	0.2111	0.2121	0.200
0.210	02132	0.2142	0.2153	0.2163	0.2174	0.2184	0.2195	0.2205	0.2216	0.2226	0.210
0.220	02237	0.2247	0.2258	0.2268	0.2279	0.2289	0.2300	0.2310	0.2321	0.2331	0.220
0.230	0.2342	0.2352	0.2363	0.2374	0.2384	0.2395	0.2405	0.2416	0.2427	0.2437	0.230
0.240	02448	0.2458	0.2469	0.2480	0.2490	0.2501	0.2512	0.2522	0.2533	0.2543	0.240
0.250	0.2554	02565	0.2575	0.2586	0.2597	0.2608	0.2618	0.2629	0.2640	0.2650	0.250
0.260	0.2661	0.2672	0.2683	0.2693	0.2704	0.2715	0.2726	0.2736	0.2747	0.2758	0.260
0.270	0.2769	0.2779	0.2790	0.2801	0.2812	0.2823	0.2833	0.2844	0.2855	0.2866	0.270
0.280	0.2877	0.2888	0.2899	02909	0.2920	0.2931	0.2942	0.2953	0.2964	0.2975	0.280
0.290	0.2986	0.2997	0.3008	0.3018	0.3029	0.3040	0.3051	0.3062	0.3073	0.3084	0.290
0.300	0.3095	0.3106	0.3117	0.3128	0.3139	0.3150	0.3161	0.3172	0.3183	0.3194	0.300
0.310	0.3205	0.3217	0.3228	0.3239	0.3250	0.3261	0.3272	0.3283	0.3294	0.3305	0.310
0.320	0.3316	0.3328	0.3339	0.3350	0.3361	0.3372	0.3383	0.3395	0.3406	0.3417	0.320
0.330	0.3428	0.3440	0.3451	0.3462	0.3473	0.3484	0.3496	0.3507	0.3518	0.3530	0.330
0.340	0.3541	0.3552	0.3564	0.3575	0.3586	0.3598	0.3609	0.3620	0.3632	0.3643	0.340
0.350	0.3654	0.3666	0.3677	0.3689	0.3700	0.3712	0.3723	0.3734	0.3746	0.3757	0.350
0.360	0.3769	0.3780	0.3792	0.3803	0.3815	0.3826	0.3838	0.3850	0.3861	0.3873	0.360
0.370	0.3884	0.3896	0.3907	0.3919	0.3931	0.3942	0.3954	0.3966	0.3977	0.3989	0.376
0.380	0.4001	0.4012	0.4024	0.4036	0.4047	0.4059	0.4071	0.4083	0.4094	0.4106	0.380
0.390	0.4118	0.4130	0.4142	0.4153	0.4165	0.4177	0.4189	0.4201	0.4213	0.4225	0.390
0.400	0.4236	0.4248	0.4260	0.4272	0.4284	0.4296	0.4308	0.4320	0.4332	0.4344	0.400
0.410	0.4356	0.4368	0.4380	0.4392	0.4404	0.4416	0.4428	0.4441	0.4453	0.4465	0.410
0.420	0.4477	0.4489	0.4501	0.4513	0,4526	0.4538	0.4550	0.4562	0.4574	0.4587	0.420
0.430	0.4599	0.4611	0.4624	0.4636	0.4648	0.4660	0.4673	0.4685	0.4698	0.4710	0.430
0.440	0.4722	0.4735	0.4747	0.4760	0.4772	0.4784	0.4797	0.4809	0.4822	0.4834	0.440
0.450	0.4847	0.4860	0.4872	0.4885	0.4897	0.4910	0.4922	0.4935	0.4948	0.4960	0.450
0.460	0.4973	0.4986	0.4999	0.5011	0.5024	0.5037	0.5049	0.5062	0.5075	0.5088	0.460
0.470	0.5101	0.5114	0.5126	0.5139	0.5152	0.5165	0.5178	0.5191	0.5204	0.5217	0.470
0.480	0.5230	0.5243	0.5256	0.5269	0.5282	0.5295	0.5308	0.5321	0.5334	0.5347	0.480
0.490	0.5361	0.5374	0.5387	0.5400	0.5413	0.5427	0.5440	0.5453	0.5466	0.5480	0.490
0.500	0.5493	0.5506	0.5520	0.5533	0.5547	0.5560	0.5573	0.5587	0.5600	0.5614	0.500
0.510	0.5627	0.5641	0.5654	0.5668	0.5681	0.5695	0.5709	0.5722	0.5736	0.5750	0.510
0:520	0.5763	0.5777	0.5791	0.5805	0.5818	0.5832	0.5846	0.5860	0.5874	0.5888	0.520
0.530	0.5901	0.5915	0.5929	0.5943	0.5957	0.5971	0.5985	0.5999	0.6013	0.6027	0.530
0.540	0.6042	0.6056	0.6070	0.6084	0.6098	0.6112	0.6127	0.6141	0.6155	0.6169	0.540

Table 46: Fisher's Tranformation for Correlation coefficients

r	0	1	2	3	4	5	6	7	8	9	r
0.550	0.6184	0.6198	0.6213	0.6227	0.6241	0.6256	0.6270	0.6285	0.6299	0.6314	0.550
0.560	0.6328	0.6343	0.6358	0.6372	0.6387	0.6401	0.6416	0.6431	0.6446	0.6460	0.560
0.570	0.6475	0.6490	0.6505	0.6520	0.6535	0.6550	0.6565	0.6580	0.6595	0.6610	0.570
0.580	0.6625	0.6640	0.6655	0.6670	0.6685	0.6700	0.6716	0.6731	0.6746	0.6761	0.580
0.590	0.6777	0.6792	0.6807	0.6823	0.6838	0.6854	0.6869	0.6885	0.6900	0.6916	0.590
0.600	0.6931	0.6947	0.6963	0.6978	0.6994	0.7010	0.7026	0.7042	0.7057	0.7073	0.600
0.610	0.7089	0.7105	0.7121	0.7137	0.7153	0.7169	0.7185	0.7201	0.7218	0.7234	0.610
0.620	0.7250	0.7266	0.7283	0.7299	0.7315	0.7332	0.7348	0.7365	0.7381	0.7398	0.620
0.630	0.7414	0.7431	0.7447	0.7464	0.7481	0.7497	0.7514	0.7531	0.7548	0.7565	0.630
0.640	0.7582	0.7599	0.7616	0.7633	0.7650	0.7667	0.7684	0.7701	0.7718	0.7736	0.640
0.650	0.7753	0.7770	0.7788	0.7805	0.7823	0.7840	0.7858	0.7875	0.7893	0.7910	0.650
0.660	0.7928	0.7946	0.7964	0.7981	0.7999	0.8017	0.8035	0.8053	0.8071	0.8089	0.660
0.670	0.8107	0.8126	0.8144	0.8162	0.8180	0.8199	0.8217	0.8236	0.8254	0.8273	0.670
0.680	0.8291	0.8310	0.8328	0.8347	0.8366	0.8385	0.8404	0.8422	0.8441	0.8460	0.680
0.690	0.8480	0.8499	0.8518	0.8537	0.8556	0.8576	0.8595	0.8614	0.8634	0.8653	0.690
0.700	0.8673	0.8693	0.8712	0.8732	0.8752	0.8772	0.8792	0.8812	0.8832	0.8852	0.700
0.710.	0.8872	0.8892	0.8912	0.8933	0.8953	0.8973	0.8994	0.9014	0.9035	0.9056	0.710
0.720	0.9076	0.9097	0.9118	0.9139	0.9160	0.9181	0.9202	0.9223	0.9245	0.9266	0.720
0.730	0.9287	0.9309	0.9330	0.9352	0.9373	0.9395	0.9417	0.9439	0.9461	0.9483	0.730
0.740	0.9505	0.9527	0.9549	0.9571	0.9594	0.9616	0.9639	0.9661	0.9684	0.9707	0.740
0.750	0.9730	0.9752	0.9775	0.9798	0.9822	0.9845	0.9868	0.9891	0.9915	0.9938	0.750
0.760	0.9962	0.9986	1.0010	1.0034	1.0058	1.0082	1.0106	1.0130	1.0154	1.0179	0.760
0.770	1.0203	1.0228	1.0253	1.0277	1.0302	1.0327	1.0352	1.0378	1.0403	1.0428	0.770
0.780	1.0454	1.0479	1.0505	1.0531	1.0557	1.0583	1.0609	1.0635	1.0661	1.0688	0.780
0.790	1.0714	1.0741	1.0768	1.0795	1.0822	1.0849	1.0876	1.0903	1.0931	1.0958	0.790
0.800	1.0986	1.1014	1.1042	1.1070	1.1098	1.1127	1.1155	1.1184	1.1212	1.1241	0.800
0.810	1.1270	1.1299	1.1329	1.1358	1.1388	1.1417	1.1447	1.1477	1.1507	1.1538	0.810
0.820	1.1568	1.1599	1.1630	1.1660	1.1691	1.1723	1.1754	1.1786	1.1817	1.1849	0.820
0.830	1.1881	1.1914	1.1946	1.1979	1.2011	1.2044	1.2077	1.2111	1.2144	1.2178	0.830
0.840	1.2212	1.2246	1.2280	1.2314	1.2349	1.2384	1.2419	1.2454	1.2490	1.2526	0.840
0.850	1.2561	1.2598	1.2634	1.2671	1.2707	1.2744	1.2782	1.2819	1.2857	1.2895	0.850
0.860	1.2933	1.2972	1.3011	1.3050	1.3089	1.3129	1.3169	1.3209	1.3249	1.3290	0.860
0.870	1.3331	1.3372	1.3414	1.3456	1.3498	1.3540	1.3583	1.3626	1.3670	1.3713	0.870
0.880	1.3758	1.3802	1.3847	1.3892	1.3938	1.3984	1.4030	1.4077	1.4124	1.4171	0.880
0.890	1.4219	1.4268	1.4316	1.4365	1.4415	1.4465	1.4516	1.4566	1.4618	1.4670	0.890
0.900	1.4722	1.4775	1.4828	1.4882	1.4937	1.4992	1.5047	1.5103	1.5160	1.5217	0.900
0.910	1.5275	1.5334	1.5393	1.5453	1.5513	1.5574	1.5636	1.5698	1.5762	1.5825	0.910
0.920	1.5890	1.5956	1.6022	1.6089	1.6157	1.6226	1.6296	1.6366	1.6438	1.6510	0.920
0.930	1.6584	1.6658	1.6734	1.6811	1.6888	1.6967	1.7047	1.7129	1.7211	1.7295	0.930
0.940	1.7380	1.7467	1.7555	1.7645	1.7736	1.7828	1.7923	1.8019	1.8116	1.8216	0.940
0.950	1.8318	1.8421	1.8527	1.8635	1.8745	1.8857	1.8972	1.9090	1.9210	1.9333	0.950
0.960	1.9459	1.9588	1.9721	1.9856	1.9996	2.0139	2.0287	2.0439	2.0595	2.0756	0.960
0.970	2.0923	2.1095	2.1273	2.1457	2.1648	2.1847	2.2054	2.2269	2.2494	2.2729	0.970
0.980	2.2975	2.3234	2.3507	2.3795	2.4101	2.4426	2.4774	2.5147	2.5549	2.5987	0.980
0.990	2.6466	2.6995	2.7587	2.8257	2.9030	2.9944	3.1062	3.2502	3.4531	3.7997	0.990

Table 47: Correlation coefficients r Corresponding to Fisher's z Transformation

r	0	1	2	3	4	5	6	7	8	9	r
0.00	0.0000	0.0010	0.0020	0.0030	0.0040	0.0050	0.0060	0.0070	0.0080	0.0090	0.00
0.01	0.0100	0.0110	0.0120	0.0130	0.0140	0.0150	0.0160	0.0170	0.0180	0.0190	0.01
0.02	0.0200	0.0210	0.0220	0.0230	0.0240	0.0250	0.0260	0.0270	0.0280	0.0290	0.02
0.03	0.0300	0.0310	0.0320	0.0330	0.0340	0.0350	0.0360	0.0370	0.0380	0.0390	0.03
0.04	0.0400	0.0410	0.0420	0.0430	0.0440	0.0450	0.0460	0.0470	0.0480	0.0490	0.04
0.05	0.0500	0.0510	0.0520	0.0530	0.0539	0.0549	0.0559	0.0569	0.0579	0.0589	0.05
0.06	0.0599	0.0609	0.0619	0.0629	0.0639	0.0649	0.0659	0.0669	0.0679	0.0689	0.06
0.07	0.0699	0.0709	0.0719	0.0729	0.0739	0.0749	0.0759	0.0768	0.0778	0.0788	0.07
0.08	0.0798	0.0808	0.0818	0.0828	0.0838	0.0848	0.0858	0.0868	0.0878	0.0888	0,08
0.09	0.0898	0.0907	0.0917	0.0927	0.0937	0.0947	0.0957	0.0967	0.0977	0.0987	0.09
0.10	0.0997	0.1007	0.1016	0.1026	0.1036	0.1046	0.1056	0.1066	0.1076	0.1086	0.10
0.11	0.1096	0.1105	0.1115	0.1125	0.1135	0.1145	0.1155	0.1165	0.1175	0.1184	0.11
0.12	0.1194	0.1204	0.1214	0.1224	0.1234	0.1244	0.1253	0.1263	0.1273	0.1283	0.12
0.13	0.1293	0.1303	0.1312	0.1322	0.1332	0.1342	0.1352	0.1361	0.1371	0.1381	0.13
0.14	0.1391	0.1401	0.1411	0.1420	.0.1430	0.1440	0.1450	0.1460	0.1469	0.1479	0.14
0.15	0.1489	0.1499	0.1508	0.1518	0.1528	0.1538	0.1547	0.1557	0.1567	0.1577	0.15
0.16	0.1586	0.1596	0.1606	0.1616	0.1625	0.1635	0.1645	0.1655	0.1664	0.1674	0.16
0.17	O.1684	0.1694	0.1703	0.1713	0.1723	0.1732	0.1742	0.1752	0.1761	0.1771	0.17
0.18	0.1781	0.1790	0.1800	0.1810	0.1820	0.1829	0.1839	0.1849	0.1858	0.1868	0.18
0.19	0.1877	0.1887	0.1897	0.1906	0.1916	0.1926	0.1935	0.1945	0.1955	0.1964	0.19
0.20	0.1974	0.1983	0.1993	0.2003	0.2012	0.2022	0.2031	0.2041	0.2051	0.2060	0.20
0.21	0.2070	0.2079	0.2089	0.2098	0.2108	0.2117	0.2127	0.2137	0.2146	0.2156	0.21
0.22	0.2165	0.2175	0.2184	0.2194	0.2203	0.2213	0.2222	0.2232	0.2241	0.2251	0.22
0.23	0.2260	0.2270	0.2279	0.2289	0.2298	0.2308	0.2317	0.2327	0.2336	0.2346	0.23
0.24	0.2355	0.2364	0.2374	0.2383	0.2393	0.2402	0.2412	0.2421	0.2430	0.2440	0.24
0.25	0.2449	0.2459	0.2468	0.2477	0.2487	0.2496	0.2506	0.2515	0.2524	0.2534	0.25
0.26	0.2543	0.2552	0.2562	0.2571	0.2580	0.2590	0.2599	0.2608	0.2618	0.2627	0.26
0.27	0.2636	0.2646	0.2655	0.2664	0.2673	0.2683	0.2692	0.2701	0.2711	0.2720	0.27
0.28	0.2729	0.2738	0.2748	0.2757	0.2766	0.2775	0.2784	0.2794	0.2803	0.2812	0.28
0.29	0.2821	0.2831	0.2840	0.2849	0.2858	0.2867	0.2876	0.2886	0.2895	0.2904	0.29
0.30	0.2913	0.2922	0.2931	0.2941	0.2950	0.2959	0.2968	0.2977	0.2986	0.2995	0.30
0.31	0.3004	0.3013	0.3023	0.3032	0.3041	0.3050	0.3059	0.3068	0.3077	0.3086	0.31
0.32	0.3095	0.3104	0.3113	0.3122	0.3131	0.3140	0.3149	0.3158	0.3167	0.3176	0.32
0.33	0.3185	0.3194	0.3203	0.3212	0.3221	0.3230	0.3239	0.3248	0.3257	0.3266	0.33
0.34	0.3275	0.3284	0.3293	0.3302	0.3310	0.3319	0.3328	0.3337	0.3346	0.3355	0.34
0.35	0.3364	0.3373	0.3381	0.3390	0.3399	0.3408	0.3417	0.3426	0.3435	0.3443	0.35
0.36	0.3452	0.3461	0.3470	0.3479	0.3487	0.3496	0.3505	0.3514	0.3522	0.3531	0.36
0.37	0.3540	0.3549	0.3557	0.3566	0.3575	0.3584	0.3592	0.3601	0.3610	0.3618	0.37
0.38	0.3627	0.3636	0.3644	0.3653	0.3662	0.3670	0.3679	0.3688	0.3696	0.3705	0.38
0.39	0.3714	0.3722	0.3731	0.3739	0.3748	0.3757	0.3765	0.3774	0.3782	0.3791	0.39
0.40	0.3799	0.3808	0.3817	0.3825	0.3834	0.3842	0.3851	0.3859	0.3868	0.3876	0.40
0.41	0.3885	0.3893	0.3902	0.3910	0.3919	0.3927	0.3936	0.3944	0.3952	0.3961	0.41
0.42	0.3969	0.3978	0.3986	0.3995	0.4003	0.4011	0.4020	0.4028	0.4036	0.4045	0.42
0.43	0.4053	0.4062	0.4070	0.4078	0.4087	0.4095	0.4103	0.4112	0.4120	0.4128	0.43
0.44	0.4136	0.4145	0.4153	0.4161	0.4170	0.4178	0.4186	0.4194	0.4203	0.4211	0.44
0.45	0.4219	0.4227	0.4235	0.4244	0.4252	0.4260	0.4268	0.4276	0.4285	0.4293	0.45
0.46	0.4301	0.4309	0.4317	0.4325	0.4333	0.4342	0.4350	0.4358	0.4366	0.4374	0.46
0.47	0.4382	0.4390	0.4398	0.4406	0.4414	0.4422	0.4430	0.4438	0.4446	0.4454	0.47
0.48	0.4462	0.4470	0.4478	0.4486	0.4494	0.4502	0.4510	0.4518	0.4526	0.4534	0.48
0.49	0.4542	0.4550	0.4558	0.4566	0.4574	0.4582	0.4590	0.4598	0.4605	0.4613	0.49
0.50	0.4621	0.4629	0.4637	0.4645	0.4653	0.4660	0.4668	0.4676	0.4684	0.4692	0.50
0.51	0.4699	0.4707	0.4715	0.4723	0.4731	0.4738	0.4746	0.4754	0.4762	0.4769	0.51
0.52	0.4777	0.4785	0.4792	0.4800	0.4808	0.4815	0.4823	0.4831	0.4838	0.4846	0.52
0.53	0.4854	0.4861	0.4869	0.4877	0.4884	0.4892	0.4900	0.4907	0.4915	0.4922	0.53
0.54	0.4930	0.4937	0.4945	0.4953	0.4960	0.4968	0.4975	0.4983	0.4990	0.4998	0.54

Table 48: Correlation coefficients r Corresponding to Fisher's z Transformation

r	0	1	2	3	4	5	6	7	8	9	r
0.55	0.5005	0.5013	0.5020	0.5028	0.5035	0.5043	0.5050	0.5057	0.5065	0.5072	0.55
0.56	0.5080	0.5087	0.5095	0.5102	0.5109	0.5117	0.5124	0.5132	0.5139	0.5146	0.56
0.57	0.5154	0.5161	0.5168	0.5176	0.5183	0.5190	0.5198	0.5205	0.5212	0.5219	0.57
0.58	0.5227	0.5234	0.5241	0.5248	0.5256	0.5263	0.5270	0.5277	0.5285	0.5292	0.58
0.59	0.5299	0.5306	0.5313	0.5320	0.5328	0.5335	0.5342	0.5349	0.5356	0.5363	0.59
0.60	0.5370	0.5378	0.5385	0.5392	0.5399	0.5406	0.5413	0.5420	0.5427	0.5434	0.60
0.61	0.5441	0.5448	0.5455	0.5462	0.5469	0.5476	0.5483	0.5490	0.5497	0.5504	0.61
0.62	0.5511	0.5518	0.5525	0.5532	0.5539	0.5546	0.5553	0.5560	0.5567	0.5574	0.62
0.63	0.5581	0.5587	0.5594	0.5601	0.5608	0.5615	0.5622	0.5629	0.5635	0.5642	0.63
0.64	0.5649	0.5656	0.5663	0.5669	0.5676	0.5683	0.5690	0.5696	0.5703	0.5710	0.64
0.65	0.5717	0.5723	0.5730	0.5737	0.5744	0.5750	0.5757	0.5764	0.5770	0.5777	0.65
0.66	0.5784	0.5790	0.5797	0.5804	0.5810	0.5817	0.5823	0.5830	0.5837	0.5843	0.66
0.67	0.5850	0.5856	0.5863	0.5869	0.5876	0.5883	0.5889	0.5896	0.5902	0.5909	0.67
0.68	0.5915	0.5922	0.5928	0.5935	0.5941	0.5948	0.5954	0.5961	0.5967	0.5973	0.68
0.69	0.5980	0.5986	0.5993	0.5999	0.6005	0.6012	0.6018	0.6025	0.6031	0.6037	0.69
0.70	0.6044	0.6050	0.6056	0.6063	0.6069	0.6075	0.6082	0.6088	0.6094	0.6100	0.70
0.71	0.6107	0.6113	0.6119	0.6126	0.6132	0.6138	0.6144	0.6150	0.6157	0.6163	0.71
0.72	0.6169	0.6175	0.6181	0.6188	0.6194	0.6200	0.6206	0.6212	0.6218	0.6225	0.72
0.73	0.6231	0.6237	0.6243	0.6249	0.6255	0.6261	0.6267	0.6273	0.6279	0.6285	0.73
0.74	0.6291	0.6297	0.6304	0.6310	0.6316	0.6322	0.6328	0.6334	0.6340	0.6346	0.74
0.75	0.6351	0.6357	0.6363	0.6369	0.6375	0.6381	0.6387	0.6393	0.6399	0.6405	0.75
0.76	0.6411	0.6417	0.6423	0.6428	0.6434	0.6440	0.6446	0.6452	0.6458	0.6463	0.76
0.77	0.6469	0.6475	0.6481	0.6487	0.6492	0.6498	0.6504	0.6510	0.6516	0.6521	0.77
0.78	0.6527	0.6533	0.6539	0.6544	0.6550	0.6556	0.6561	0.6567	0.6573	0.6578	0.78
0.79	0.6584	0.6590	0.6595	0.6601	0.6607	0.6612	0.6618	0.6624	0.6629	0.6635	0.79
0.80	0.6640	0.6646	0.6652	0.6657	0.6663	0.6668	0.6674	0.6679	0.6685	0.6690	0.80
0.81	0.6696	0.6701	0.6707	0.6712	0.6718	0.6723	0.6729	0.6734	0.6740	0.6745	0.81
0.82	0.6751	0.6756	0.6762	0.6767	0.6772	0.6778	0.6783	0.6789	0.6794	0.6799	0.82
0.83	0.6805	0.6810	0.6815	0.6821	0.6826	0.6832	0.6837	0.6842	0.6847	0.6853	0.83
0.84	0.6858	0.6863	0.6869	0.6874	0.6879	0.6884	0.6890	0.6895	0.6900	0.6905	0.84
0.85	0.6911	0.6916	0.6921	0.6926	0.6932	0.6937	0.6942	0.6947	0.6952	0.6957	0.85
0.86	0.6963	0.6968	0.6973	0.6978	0.6983	0.6988	0.6993	0.6998	0.7004	0.7009	0.86
0.87	0.7014	0.7019	0.7024	0.7029	0.7034	0.7039	0.7044	0.7049	0.7054	0.7059	0.87
0.88	0.7064	0.7069	0.7074	0.7079	0.7084	0.7089	0.7094	0.7099	0.7104	0.7109	0.88
0.89	0.7114	0.7119	0.7124	0.7129	0.7134	0.7139	0.7143	0.7148	0.7153	0.7158	0.89
0.90	0.7163	0.7168	0.7173	0.7178	0.7182	0.7187	0.7192	0.7197	0.7202	0.7207	0.90
0.91	0.7211	0.7216	0.7221	0.7226	0.7230	0.7235	0.7240	0.7245	0.7249	0.7254	0.91
0.92	0.7259	0.7264	0.7268	0.7273	0.7278	0.7283	0.7287	0.7292	0.7297	0.7301	0.92
0.93	0.7306	0.7311	0.7315	0.7320	0.7325	0.7329	0.7334	0.7338	0.7343	0.7348	0.93
0.94	0.7352	0.7357	0.7361	0.7366	0.7371	0.7375	0.7380	0.7384	0.7389	0.7393	0.94
0.95	0.7398	0.7402	0.7407	0.7411	0.7416	0.7420	0.7425	0.7429	0.7434	0.7438	0.95
0.96	0.7443	0.7447	0.7452	0.7456	0.7461	0.7465	0.7469	0.7474	0.7478	0.7483	0.96
0.97	0.7487	0.7491	0.7496	0.7500	0.7505	0.7509	0.7513	0.7518	0.7522	0.7526	0.97
0.98	0.7531	0.7535	0.7539	0.7544	0.7548	0.7552	0.7557	0.7561	0.7565	0.7569	0.98
0.99	0.7574	0.7578	0.7582	0.7586	0.7591	0.7595	0.7599	0.7603	0.7608	0.7612	0.99
1.0	0.7616	0.7658	0.7699	0.7739	0.7779	0.7818	0.7857	0.7895	0.7932	0.7969	1.0
1.1	0.8005	0.8041	0.8076	0.8110	0.8144	0.8178	0.8210	0.8243	0.8274	0.8306	1.1
1.2	0.8337	0.8367	0.8397	0.8426	0.8455	0.8483	0.8511	0.8538	0.8565	0.8591	1.2
1.3	0.8617	0.8643	0.8668	0.8692	0.8717	0.8741	0.8764	0.8787	0.8809	0.8832	1.3
1.4	0.8854	0.8875	0.8896	0.8917	0.8937	0.8957	0.8977	0.8996	0.9015	0.9033	1.4
1.5	0.9051	0.9069	0.9087	0.9104	0.9121	0.9138	0.9154	0.9170	0.9186	0.9201	1.5
1.6	0.9217	0.9232	0.9246	0.9261	0.9275	0.9289	0.9302	0.9316	0.9329	0.9341	1.6
1.7	0.9354	0.9366	0.9379	0.9391	0.9402	0.9414	0.9425	0.9436	0.9447	0.9458	1.7
1.8	0.9468	0.9478	0.9488	0.9498	0.9508	0.9517	0.9527	0.9536	0.9545	0.9554	1.8
1.9	0.9562	0.9571	0.9579	0.9587	0.9595	0.9603	0.9611	0.9618	0.9626	0.9633	1.9

Table 49: Correlation coefficients r Corresponding to Fisher's z Transformation

z	0	1	2	3	4	5	6	7	8	9	z
2.0	0.9640	0.9647	0.9654	0.9661	0.9667	0.9674	0.9680	0.9687	0.9693	0.9699	2.0
2.1	0.9705	0.9710	0.9716	0.9721	0.9727	0.9732	0.9737	0.9743	0.9748	0.9753	2.1
2.2	0.9757	0.9762	0.9767	0.9771	0.9776	0.9780	0.9785	0.9789	0.9793	0.9797	2.2
2.3	0.9801	0.9805	0.9809	0.9812	0.9816	0.9810	0.9823	0.9827	0.9830	0.9833	2.3
2.4	0.9837	0.9840	0.9843	0.9846	0.9849	0.9852	0.9855	0.9858	0.9861	0.9863	2.4
2.5	0.9866	0.9869	0.9871	0.9874	0.9876	0.9879	0.9881	0.9884	0.9886	0.9888	2.5
2.6	0.9890	0.9892	0.9895	0.9897	0.9899	0.9901	0.9903	0.9905	0.9906	0.9908	2.6
2.7	0.9910	0.9912	0.9914	0.9915	0.9917	0.9919	0.9920	0.9922	0.9923	0.9925	2.7
2.8	0.9926	0.9928	0.9929	0.9931	0.9932	0.9933	0.9935	0.9936	0.9937	0.9938	2.8
2.9	0.9940	0.9941	0.9942	0.9943	0.9944	0.9945	0.9946	0.9947	0.9949	0.9950	2.9
3.0	0.9951	0.9952	0.9952	0.9953	0.9954	0.9955	0.9956	0.9957	0.9958	0.9959	3.0
3.1	0.9959	0.9960	0.9961	0.9962	0.9963	0.9963	0.9964	0.9965	0.9965	0.9966	3.1
3.2	0.9967	0.9967	0.9968	0.9969	0.9969	0.9970	0.9971	0.9971	0.9972	0.9972	3.2
3.3	0.9973	0.9973	0.9974	0.9974	0.9975	0.9975	0.9976	0.9976	0.9977	0.9977	3.3
3.4	0.9978	0.9978	0.9979	0.9979	0.9979	0.9980	0.9980	0.9981	0.9981	0.9981	3.4
3.5	0.9982	0.9982	0.9982	0.9983	0.9983	0.9984	0.9984	0.9984	0.9984	0.9985	3.5
3.6	0.9985	0.9985	0.9986	0.9986	0.9986	0.9986	0.9987	0.9987	0.9987	0.9988	3.6
3.7	0.9988	0.9988	0.9988	0.9988	0.9989	0.9989	0.9989	0.9989	0.9990	0.9990	3.7
3.8	0.9990	0.9990	0.9990	0.9991	0.9991	0.9991	0.9991	0.9991	0.9991	0.9992	3.8
3.9	0.9992	0.9992	0.9992	0.9992	0.9992	0.9993	0.9993	0.9993	0.9993	0.9993	3.9
4.0	0.9993	0.9993	0.9994	0.9994	0.9994	0.9994	0.9994	0.9994	0.9994	0.9994	4.0
4.1	0.9995	0.9995	0.9995	0.9995	0.9995	0.9995	0.9995	0.9995	0.9995	0.9995	4.1
4.2	0.9996	0.9996	0.9996	0.9996	0.9996	0.9996	0.9996	0.9996	0.9996	0.9996	4.2
4.3	0.9996	0.9996	0.9996	0.9997	0.9997	0.9997	0.9997	0.9997	0.9997	0.9997	4.3
4.4	0.9997	0.9997	0.9997	0.9997	0.9997	0.9997	0.9997	0.9997	0.9997	0.9997	4.4
4.5	0.9998	0.9998	0.9998	0.9998	0.9998	0.9998	0.9998	0.9998	0.9998	0.9998	4.5
4.6	0.9998	0.9998	0.9998	0.9998	0.9998	0.9998	0.9998	0.9998	0.9998	0.9998	4.6
4.7	0.9998	0.9998	0.9998	0.9998	0.9998	0.9999	0.9999	0.9999	0.9999	0.9999	4.7
4.8	0.9999	0.9999	0.9999	0.9999	0.9999	0.9999	0.9999	0.9999	0.9999	0.9999	4.8
4.9	0.9999	0.9999	0.9999	0.9999	0.9999	0.9999	0.9999	0.9999	0.9999	0.9999	4.9

Table 50: Ten Thousand Random Digits

	00-04	05-09	10-14	15-19	20-24	25-29	30-34	35-39	40-44	45-49
00	22808	04391	45529	53968	57136	98228	85485	13801	68194	56382
01	49305	36965	44849	64987	59501	35141	50159	57369	76913	75739
02	81934	19920	73316	69243	69605	17022	53264	83417	55193	92929
03	10840	13508	48120	22467	54505	70536	91206	81038	22418	34800
04	99555	73289	59605	37105	24621	44100	72832	12268	97089	68112
05	32677	45709	62337	35132	45128	96761	08745	53388	98353	46724
06	09401	75407	27704	11569	52842	83543	44750	03177	50511	15301
07	73424	31711	65519	74869	56744	10864	75315	89866	96563	75142
08	37075	81378	59472	71858	86903	66860	03757	32723	54273	45477
09	02060	37158	55244	44812	45369	78939	08048	28036	40946	03898
10	94719	43565	40028	79866	43137	28063	52513	66405	71511	66135
11	70234	48272	59621	88778	16536	36505	41724	24776	63971	01685
12	07972	71752	92745	86465	01845	27416	50519	48458	68460	63113
13	58521	64882	26993	48104	61307	73933	17214	44827	88306	78177
14	32580	45202	21148	09684	39411	04892	02055	75276	51831	85686
15	88796	30829	35009	22695	23694	11220	71006	26720	39476	60538
16	31525	82746	78935	82980	61236	28940	96341	13790	66247	33839
17	02747	35989	70387	89571	34570	17002	79223	96817	31681	15207
18	46651	28987	20625	61347	63981	41085	67412	29053	00724	14841
19	43598	14436	33521	55637	39789	26560	66404	71802	18763	80560
20	30596	92319	11474	64546	60030	73795	60809	24016	29166	36059
21	56198	64370	85771	62633	78240	05766	32419	35769	14057	80674
22	68266	67544	06464	84956	18431	04015	89049	15098	12018	89338
23	31107	28597	65102	75599	17496	87590	68848	33021	69855	54015
24	37555	05069	38680	87274	55152	21792	77219	48732	03377	01160
25	90463	27249	43845	94391	12145	36882	48906	52336	00780	74407
26	99189	88731	93531	52638	54989	04237	32978	59902	05463	09245
27	37631	74016	89072	59598	55356	27346	80856	80875	52850	36548
28	73829	21651	50141	76142	72303	06694	61697	76662	23745	96282
29	15634	89428	47090	12094	42134	62381	87236	90118	53463	46969
30	00571	45172	78532	63863	98597	15742	41967	11821	91389	07476
31	83374	10184	56384	27050	77700	13875	96607	76479	80535	17454
32	78666	85645	13181	08700	08289	62956	64439	39150	95690	18555
33	47890	88197	21368	65254	35917	54035	83028	84636	38186	50581
34	56238	13559	79344	83198	94642	35165	40188	21456	67024	62771
35	36369	32234	38129	59963	99237	72648	66504	99065	61161	16186
36	42934	34578	28968	74028	42164	56647	76806	61023	33099	48293
37	09010	15226	43474	30174	26727	39317	48508	55438	85336	40762
38	83897	90073	72941	85613	85569	24183	08247	15946	02957	68504
39	82206	01230	93252	89045	25141	91943	75531	87420	99012	80751
40	14115	32992	49046	41272	94040	44929	98531	27712	05106	35242
41	58968	88367	70927	74765	18635	85122	27722	95388	61523	91745
42	62601	04595	76926	11007	67631	64641	07994	04639	39314	83126
43	97030	71165	47032	85021	65554	66774	21560	04121	57297	85415
44	89074	31587	21360	41673	71192	85795	82757	52928	62586	02179
45	07806	81312	81215	99858	26762	28993	74951	64680	50934	32011
46	91540	86466	13229	76624	44092	96604	08590	89705	03424	48033
47	99279	27334	33804	77988	93592	90708	56780	70097	39907	51006
48	63224	05074	83941	25034	43516	22840	35230	66048	80754	46302
49	98361	97513	27529	66419	35328	19738	82366	38573	50967	12754

Table 51 (cont.): Ten Thousand Random Digits

	00-04	05-09	10-14	15-19	20-24	25-29	30-34	35-39	40-44	45-49
50	27791	82504	33523	27623	16597	32089	81596	78429	14111	68245
51	33147	46058	92388	10150	63224	26003	56427	29945	44546	50233
52	67243	10454	40269	44324	46013	00061	21622	68213	47749	76398
53	78176	70368	95523	09134	31178	33857	26171	07063	41984	99310
54	70199	70547	94431	45423	48695	01370	68065	61982	20200	27066
55	19840	01143	18606	07622	77282	68422	70767	33026	15135	91212
56	32970	28267	17695	20571	50227	69447	45535	16845	68283	15919
57	43233	53872	68520	70013	31395	60361	39034	59444	17066	07418
58	08514	23921	16685	89184	71512	82239	72947	69523	75618	79826
59	28595	51196	96108	84384	80359	02346	60581	01488	63177	47496
60	83334	81552	88223	29934	68663	23726	18429	84855	26897	94782
61	66112	95787	84997	91207	67576	27496	01603	22395	41546	68178
62	25245	14749	30653	42355	88625	37412	87384	09392	11273	28116
63	21861	22185	41576	15238	92294	50643	69848	48020	19785	41518
64	74506	40569	90770	40812	57730	84150	91500	53850	52104	37988
65	23271	39549	33042	10661	37312	50914	73027	21010	76788	64037
66	08548	16021	64715	08275	50987	67327	11431	31492	86970	47335
67	14236	80869	90798	85659	10079	28535	35938	10710	67046	74021
68	55270	49583	86467	40633	27952	27187	35058	66628	94372	75665
69	02301	05524	91801	23647	51330	35677	05972	90729	26650	81684
70	72843	03767	62590	92077	91552	76853	45812	15503	93138	87788
71	49248	43346	29503	22494	08051	09035	75802	63967	74257	00046
72	62598	99092	87806	42727	30659	10118	83000	96198	47155	00361
73	27510	69457	98616	62172	07056	61015	22159	65590	51082	34912
74	84167	66640	69100	22944	19833	23961	80834	37418	42284	12951
75	14722	88488	54999	55244	03301	37344	01053	79305	94771	95215
76	46696	05477	32442	18738	43021	72933	14995	30408	64043	67834
77	13938	09867	28949	94761	38149	38695	90165	82841	75399	09932
78	48778	56434	42495	07050	35250	09660	56192	34793	36146	96806
79	00571	71281	01563	66448	94560	55920	31580	26640	91262	30863
80	96050	57641	21798	14917	21836	15053	33566	51177	91786	12610
81	30870	81575	14019	07831	81840	25506	29358	88668	42742	62048
82	59153	29135	00712	73025	14263	17253	95662	75535	26170	95240
83	78283	70379	54969	05821	26485	28990	40207	00434	38863	61892
84	12175	95800	41106	93962	06245	00883	65337	75506	66294	62241
85	14192	39242	17961	29448	84078	14545	39417	83649	26495	41672
86	69060	38669	00849	24991	84252	41611	62773	63024	57079	59283
87	46154	11705	29355	71523	21377	36745	00766	21549	51796	81340
88	93419	54353	41269	07014	28352	77594	57293	59219	26098	63041
89	13201	04017	68889	81388	60829	46231	46161	01360	25839	52380
90	62264	99963	98226	29972	95169	07546	01574	94986	06123	52804
91	58030	30054	27479	70354	12351	33761	94357	81081	74418	74297
92	81242	26739	92304	81425	29052	37708	49370	46749	59613	50749
93	16372	70531	92036	54496	50521	83872	30064	67555	40354	23671
94	54191	04574	58634	91370	40041	77649	42030	42547	47593	07435
95	15933	92602	19496	18703	63380	58017	14665	88867	84807	44672
96	21518	77770	53826	97114	82062	34592	87400	64938	75540	54751
97	34524	64627	92997	21198	14976	07071	91566	44335	83237	24335
98	46557	67780	59432	23250	63352	43890	07109	07911	85956	62699
99	31929	13996	05126	83561	03244	33635	26952	01638	22788	26393

Table 52 (cont.): Ten Thousand Random Digits

	50-54	55-59	60-64	65-69	70-74	75-79	80-84	85-89	90-94	95-99
00	53330	26487	85005	06384	13822	83736	95876	71355	31226	56063
01	96990	62825	97110	73006	32661	63408	03893	10333	41902	69175
02	30385	16588	63609	09132	53081	14478	50813	22887	03746	10289
03	75252	66905	60536	13408	25158	35825	10447	47375	89249	91238
04	52615	66504	78496	90443	84414	31981	88768	49629	15174	99795
05	39992	51082	74547	31022	71980	40900	84729	34286	96944	49502
06	51788	87155	13272	92461	06466	25392	22330	17336	42528	78628
07	88569	35645	50602	94043	35316	66344	78004	89051	89025	12722
08	14513	34794	44976	71244	60548	03041	03300	46389	25340	23804
09	50257	53477	24546	01377	20292	85097	00660	39561	62367	61424
10	35170	69025	46214	27085	83416	48597	19494	49380	28469	77549
11	22225	83437	43912	30337	75784	77689	60425	85588	93438	61343
12	90103	12542	97828	85859	85859	64101	00924	89012	17889	01154
13	68240	89649	85705	18937	30114	89827	89460	01998	81745	31281
14	01589	18335	24024	39498	82052	07868	49486	25155	61730	08946
15	36375	61694	90654	16475	92703	59561	45517	90922	93357	00207
16.	11237	60921	51162	74153	94774	84150	39274	10089	45020	09624
17	48667	68353	40567	79819	48551	26789	07281	14669	00576	17435
18	99286	42806	02956	73762	40419	21676	67533	50553	21115	26742
19	44651	48349	13003	39656	99757	74964	00141	21387	66777	68533
20	83251	70164	05732	66842	77717	25305	36218	85600	23736	06629
21	41551	54630	88759	10085	48806	08724	50685	95638	20829	37264
22	68990	51280	51368	73661	21764	71552	69654	17776	51935	53169
23	63393	76820	33106	23322	16783	35630	50938	90047	97577	27699
24	93317	87564	32371	04190	27608	40658	11517	19646	82335	60088
25	48546	41090	69890	58014	04093	39286	12253	55859	83853	15023
26	31435	57566	99741	77250	43165	31150	20735	57406	85891	04806
27	56405	29392	76998	66849	29175	11641	85284	89978	73169	62140
28	70102	50882	85960	85955	03828	69417	55854	63173	60485	00327
29	92746	32004	52242	94763	32955	39848	09724	30029	45196	67606
30	67737	34389	57920	47081	60714	04935	48278	90687	99290	18554
31	35606	76646	14813	51114	52492	46778	08156	22372	59999	43938
32	64836	28649	45759	45788	43183	25275	25300	21548	33941	66314
33	86319	92367	37873	48993	71443	22768	69124	65611	79267	49709
34	90632	32314	24446	60301	31376	13575	99663	81929	39343	17648
35	83752	51966	43895	03129	37539	72989	52393	45542	70344	96712
36	56755	21142	86355	33569	63096	66780	97539	75150	25718	33724
37	14100	28857	60648	86304	97397	97210	74842	87483	51558	52883
38	69227	24872	48057	29318	74385	02097	63266	26950	73173	53025
39	77718	56967	36560	87155	26021	70903	32086	11722	32053	63723
40	09550	38799	88929	80877	87779	99905	17122	25985	16866	76005
41	12404	42453	88609	89148	85892	96045	10310	45021	62023	70061
42	07985	27418	92734	80000	58969	99011	73815	49705	68076	69605
43	58124	53830	08705	20916	46048	30342	86530	72608	93074	80937
44	46173	77223	75661	57691	24055	27568	41227	58542	73196	44886
45	13476	72301	85793	80516	59479	66985	24801	84009	71317	87321
46	82472	98647	17053	94591	36790	42275	51154	77765	01115	09331
47	55370	63433	80653	30739	68821	46854	41939	38962	20703	69424
48	89274	74795	82231	69384	53605	67860	01309	27273	76316	54253
49	55242	74511	62992	17981	17323	79325	35238	21393	13114	70084

Table 53 (cont.): Ten Thousand Random Digits

	50-54	55-59	60-64	65-69	70-74	75-79	80-84	85-89	90-94	95-99
50	03674	36059	46810	58367	82676	15051	57977	49410	02971	05797
51	26136	80623	96505	91089	02309	54743	15831	45538	96456	87272
52	61716	80405	84735	12997	86386	61606	75091	84996	76070	54923
53	67051	63246	99547	81223	52485	90333	24697	06266	07388	70389
54	17284	60347	87314	30218	87983	45426	84153	10569	64042	95618
55	12543	23999	95777	28105	66073	35174	67706	05181	35176	85558
56	45494	93037	29209	70724	86438	65354	71209	27909	85321	10216
57	39262	15415	93940	41615	43605	95675	53916	29580	07048	95838
58	29094	58703	92144	14287	50165	85661	95749	61118	36668	96852
59	77988	03222	57805	00725	91543	80021	16442	63360	33620	39324
60	02758	86823	52423	32355	96707	47448	06153	59430	43952	16775
61	46702	37467	66803	49344	59519	92717	97110	82087	36785	00880
62	61759	95153	80090	60626	55917	92812	63544	82295	50729	20116
63	82316	11402	28078	75325	43963	63105	99294	30285	64473	53613
64	92754	74241	14315	49697	61979	66711	61707	81589	53936	82115
65	37907	24080	31741	86653	81460	32304	99590	56644	41521	91172
66	16619	75264	12279	18996	16716	81959	65722	10058	91522	65410
67	66640	06195	84416	32836	53178	93810	36766	59778	26612	69017
68	45208	58525	07714	77120	67986	73140	12026	75550	84912	64691
69	00910	40237	91035	29125	03534	47246	64698	00608	39537	71755
70	19965	46945	59357	15551	20335	03145	21519	37882	99146	70161
71	37538	05747	54982	00494	51866	86172	82679	04152	56369	20356
72	38571	69663	03287	28101	46753	55715	93527	30508	19722	02072
73	76711	02864	00880	85518	25834	52317	48070	51582	03374	19540
74	07128	44400	48015	41449	21109	38948	21816	52089	64529	21510
75	00882	89357	80906	76476	58420	95793	34043	00991	38937	39859
76	96160	18580	40549	46562	45106	53768	76097	60504	85273	63076
77	13443	22235	46210	47755	05802	00311	15171	23818	89870	47578
78	99494	35395	71411	48281	92151	84465	63651	15969	61345	13324
79	90647	11809	96365	52409	17977	05971	35835	03889	43733	66100
80	33050	48785	92200	59319	36977	41111	28002	51580	10573	21763
81	21257	15066	72630	23206	03106	53140	50292	64012	83184	81304
82	45362	94324	81800	83980	97244	09691	08435	66723	06150	54972
83	93322	58684	95695	19090	98108	47678	98061	87193	99992	82870
84	20374	61803	62508	83696	54449	53649	86447	66115	90857	69114
85	00715	13209	17080	06890	38022	76469	27696	30778	31836	96676
86	85519	93677	90186	09579	98760	50320	98077	46048	79700	81431
87	71948	15871	84502	41330	46675	51342	93431	55566	90819	68923
88	43427	95500	02004	51802	59668	17806	87605	33010	20991	76269
89	64854	28815	74959	03531	77051	51807	89005	18898	23716	45862
90	62195	29095	23982	75883	41561	25897	43595	92703	86676	32038
91	61186	54041	60984	61602	18482	57941	59657	35924	21738	30646
92	88585	40218	69965	74354	62274	38948	44813	31558	40625	22477
93	15598	21389	79016	92151	21926	49901	16835	88055	30545	60306
94	27097	89653	21558	72731	66694	36703	92172	46129	32660	91356
95	40537	85697	78182	39711	59270	21934	78647	94801	78832	37287
96	74828	06544	13078	59528	31100	11132	91256	85899	72492	18200
97	43297	83195	66218	65838	63255	72093	38976	44892	96861	97848
98	32663	58127	73258	09220	49701	92357	43700	37214	56844	02048
99	45551	31330	08152	23712	23963	58274	94583	03761	73429	47328

INDEX

BIOCHEMICAL CALCULATIONS

BIOSTATISTICS

G

H

K

M

N

P

Q

R

S

T

V

Y